Practical Pharmaceutical Chemistry

Practical Pharmaceutical Chemistry

Fourth edition,
in two parts

A. H. BECKETT

O.B. E., Ph.D., D.Sc., F.P.S.,
C. Chem., F.R.S.C.

Emeritus Professor of Pharmaceutical Chemistry
King's College London, University of London

and

J. B. STENLAKE

C.B.E., Ph.D., D.Sc., F.P.S.,
C.Chem., F.R.S.C., F.R.S.E.

Honorary Professor of Pharmacy
University of Strathclyde, Glasgow

Part One

CBS

CBS Publishers & Distributors Pvt. Ltd.

New Delhi • Bengaluru • Chennai • Kochi • Kolkata • Mumbai
Hyderabad • Nagpur • Patna • Pune • Vijayawada

CBS Pubs ISBN: 81-239-0513-0
Athlone ISBN: 0-485-11322-8

First Indian Edition: 1997
Reprint: 1999, 2000, 2002, 2003, 2004, 2005

This edition has been published in India by arrangement with The Athlone Press, UK.

Published by **Satish Kumar Jain** and produced by **Varun Jain** for
CBS Publishers & Distributors Pvt. Ltd.,
4819/XI Prahlad Street, 24 Ansari Road, Daryaganj, New Delhi - 110002
delhi@cbspd.com, cbspubs@airtelmail.in • www.cbspd.com
Ph.: 23289259, 23266861, 23266867 • Fax: 011-23243014

Corporate Office: 204 FIE, Industrial Area, Patparganj, Delhi - 110 092
Ph: 49344934 • Fax: 011-49344935
E-mail: publishing@cbspd.com • publicity@cbspd.com

Branches:

- ***Bengaluru:*** 2975, 17th Cross, K.R. Road, Bansankari 2nd Stage, Bengaluru - 70 • Ph: +91-80-26771678/79 • Fax: +91-80-26771680 E-mail: cbsbng@gmail.com, bangalore@cbspd.com
- ***Chennai:*** No. 7, Subbaraya Street, Shenoy Nagar, Chennai - 600030 Ph: +91-44-26681266, 26680620 • Fax: +91-44-42032115 E-mail: chennai@cbspd.com
- ***Kochi:*** Ashana House, 39/1904, A.M. Thomas Road, Valanjambalam, Ernakulum, Kochi • Ph: +91-484-4059061-65 Fax: +91-484-4059065 • E-mail: cochin@cbspd.com
- ***Kolkata:*** 6-B, Ground Floor, Rameshwar Shaw Road, Kolkata - 700014 Ph: +91-33-22891126/7/8 • E-mail: kolkata@cbspd.com
- ***Mumbai:*** 83-C, Dr. E. Moses Road, Worli, Mumbai - 400018 Ph: +91-9833017933, 022-24902340/41 • E-mail: mumbai@cbspd.com

Representatives:

- Hyderabad: 0-9885175004
- Patna: 0-9334159340
- Jharkhand: 0-9811541605
- Nagpur: 0-9021734563
- Pune: 0-9623451994
- Uttarakhand: 0-9716462459

Printed at:
Neekunj Print Process, Delhi

Contents

Preface to the Fourth Edition

A spate of official publications, including the *British Pharmacopoeia* 1980, and its Addenda (1981, 1982, 1983 and 1986), the *British Pharmacopoeia (Veterinary)* 1985, the combined *United States Pharmacopoeia* XXI and *National Formulary* XVI and the second edition of the *European Pharmacopoeia*, call for yet another revision of *Practical Pharmaceutical Chemistry*. In this, the fourth edition, we have endeavoured to reflect the growing international convergence of policy and practice both in the change of subtitle to Pharmaceutical Analysis and Quality Control, and in the breadth of its content.

The objectives of this revision have been achieved in part by reducing the heavy dependence of earlier editions on the methods of the *British Pharmacopoeia*. Wherever possible, examples are based on drugs and dosage forms that are in widespread and common use in Britain, continental Europe and North America. Additionally, some reference to veterinary pharmaceuticals is made where they provide appropriate examples. As in previous editions, substances that are the subject of monographs in the British Pharmacopoeia are denoted by their British Approved Names (BAN), and are distinguished from United Adopted Names (Usan) where these are given by setting BANs in italics.

The discussion of drug registration has also been broadened to include reference to FDA and EEC procedures, the control of veterinary as well as human medicines, and the need in the United Kingdom for biologically based products to be manufactured in conformity with the requirements of the *Biological Compendium* (1977).

The detailed chapter-by-chapter revision in Part 1 encompasses the changeover in European analytical practice from NORMALITY to MOLARITY, which despite the reservations of some analysts, particularly in relation to oxidation-reduction titrations, is now virtually complete. The expansion of rapid complexometric titration methods, with the consequent almost complete demise of the older much slower gravimetric methods, is reflected in the regrouping of complexometric, argentometric and gravimetric methods into a single chapter. A brief treatment of variables in quantitative analysis appropriate to the application of chemical methods has been included for the first time, and the chapter on the analysis of dosage forms has been updated to better reflect both the range of products and the methods used in their control. In this respect the need for control analysts to embrace an appreciation of biological methods is reflected in the inclusion of short sections on sterility testing, on microbiological contamination and challenge tests for antimicrobial preservatives, on microbiological assays and on enzymes in pharmaceutical analysis.

There have been few, if any, major innovations in physical methods applicable to pharmaceutical analysis since the publication of the third edition in 1976. However, there have been a number of improvements and changes of emphasis. These are reflected in Part 2 of the present edition, which has been extensively revised in an endeavour to give a broader coverage of the most widely used techniques and a better balance of material that fairly reflects modern practice. In particular, the steady growth in importance of quantitative chromatographic techniques is recognised in the broader coverage and depth accorded to gas chromatography and high performance liquid chromatography, the inclusion of a short section on capillary column gas chromatography, and the transfer from Part 1 of important sections on ion exchange and size exclusion chromatography. The treatment of NMR spectroscopy has also been extended to include a brief introduction to ^{13}C NMR, and the coverage of radiopharmaceuticals increased to include radionuclide generators and quality control of radiopharmaceuticals – a subject of special interest to those engaged in hospital pharmacy. To balance these expansions, coverage of electrochemistry and polarography has been compressed into a single chapter – a degree of emphasis that is more in keeping with the rather modest role that these techniques continue to play in the practice of pharmaceutical analysis.

Two new chapters have been added to improve cohesion in Part 2. The first, by way of introduction, sets out the contribution and role of physical methods of analysis in the various phases of drug development, in quality control within the factory and in independent control laboratories, and in the clinic. The second, by way of conclusion, consists of a series of 'workshop style' exercises, with separate solutions, to illustrate and give practice in the application of spectroscopic techniques in structural elucidation and verification of identity.

Many analytical methods serve a common purpose in their suitability both for the quality control of pharmaceutical products and the study of their absorption, distribution, metabolism and excretion whether in laboratory animals or human subjects. The importance of developing sensitive, cold methods, as opposed to those based on radiolabelled compounds, is widely recognised in clinical pharmacology. Hospital pharmaceutical departments well equipped for essential quality control work are in a unique position with staff, equipment and laboratory facilities to undertake pharmacokinetic and metabolic studies, and even to provide a routine pharmacokinetic monitoring service if this is required. With such developments in mind, we have continued to feature applications of the various separation and spectroscopic methods to drug metabolism and pharmacokinetics together with relevant practical exercises.

J.B.S.

Part 1

Pharmaceutical Analyses and Quality Control

Revised by

J.B. STENLAKE
C.B.E., Ph.D., D.Sc., F.P.S., C.Chem., F.R.S.C., F.R.S.E.

with contribution by

DAVID WATT, M.Sc., Ph.D., F.P.S.
University of Strathclyde, Glasgow

Acknowledgements

To the National Physical Laboratory for permission to use Fig. 4.4 and Table 4.1, which are taken from their publication *Notes on Applied Sciences*, No. 7, HMSO, and to the *Journal of Pharmacy and Pharmacology* for permission to reproduce Figs. 11.3 and 11.4. We are indebted to Mr T. Moody for the preparation of drawings for Figs. 1.1, 1.2, 9.1, 10.2 and 11.5 and in particular, once again, to Sylvia Cohen for her invaluable help in the preparation and correction of the manuscript, without which a fourth edition would not have been possible.

1
Chemical purity and its control

Introduction

Modern medicines for human use are required to meet exacting standards which relate to their quality, safety and efficacy. The evaluation of safety and efficacy and their maintenance in practice is dependent upon the existence of adequate methods for quality control of the product. The standard of purity must, therefore, be strictly defined in such a way as to ensure that successive batches are consistent in composition, irrespective of whether they come from the same or different manufacturers. This is not to imply, however, that only the purest substances available are sufficiently safe to be used as medicines and, in practice, purity standards range widely, reflecting the many constraints on the extent of purification which it is reasonable to expect.

Chemical purity may be described as complete freedom from foreign matter. A state of absolute purity is virtually unattainable, but may be approached as closely as desired, provided sufficient care is taken in the manufacturing process. In practice, however, the level of purity which can be attained in the manufacture of pharmaceutical products depends partly on the cost-effectiveness of the process and the purification methods available and partly on the stability of the product. It would be impracticable, therefore, to lay down standards for drugs and medicines which would provide for the complete absence of even any one impurity and, in setting specifications, it is often necessary to strike a balance in order to obtain a product at reasonable cost yet sufficiently pure to ensure reasonable levels of safety and efficacy.

Similar considerations apply in setting standards for veterinary medicines. Despite views often expressed to the contrary, informed opinion now accepts that veterinary medicines should be of the same or similar quality to products for human use. Any departures from this criterion requiring either lowering or raising of standards in materials for veterinary use, whether on economic or scientific grounds, should not merely be justified by the proponents of such action but also be shown to have no deleterious effect on either the animal or the food product ultimately derived from that source.

The standards demanded of chemicals for pharmaceutical use are determined by a number of factors, which take account of the impurities likely to arise as a result of all known methods of manufacture. The essential criterion is safety in use; particular attention is directed towards impurities which might be toxic or which are capable of causing chemical interference when the substance is formulated (i.e. compounded to give a medicinal preparation in a form suitable for administration). The general stability of the substance is also important; thus, if it is chemically unstable, hygroscopic or, alternatively, should it effloresce, then standards must be so adjusted that materials stored with reasonable care and for reasonable periods of time will still comply with the imposed requirements. Standards must also guard against the possibility of either accidental contamination or intentional adulteration of pharmaceutical chemicals.

The source of impurities in pharmaceutical chemicals

A knowledge of those impurities which occur in pharmaceutical substances in general use is readily available from actual batch analyses and stability studies. Experience in the manufacture of any one particular substance often shows that not all the expected impurities are present in practice. But for a substance newly available it is important that one should be able to deduce the impurities with which it is likely to be contaminated. A list of the possible impurities can be readily compiled from a knowledge of the raw materials used, the method of manufacture and the stability of the product. To these must be added impurities which may arise from physical contamination or inadequate storage conditions.

Raw materials

Pharmaceutical chemicals may be either isolated from biological sources or prepared by synthesis from chemical starting materials. Biological sources include plant and animal tissues and microbiological fermentation. Whatever the process, the identity of the source material should be verified and its quality established. Additionally, the presence of any related or non-related substance present in the starting material which might, conceivably, be carried through the process to contaminate the final product should be identified. Thus, to take a simple inorganic example, rock salt contains small amounts of calcium and magnesium chlorides, so that *Sodium Chloride* prepared from this source will almost certainly contain traces of calcium and magnesium compounds. Likewise, digitoxin and gitoxin, which occur with the naturally occurring plant glycoside digoxin, are not completely separated from it by the extraction process and remain as contaminants of pharmaceutical grade *Digoxin*.

The manufacturing process

The majority of modern pharmaceutical chemicals are prepared by organic synthesis from starting materials which are, themselves, either synthetic organic chemicals or natural products isolated from biological sources. The process itself may, therefore, introduce impurities into the final product. These may arise from the following:

(a) the starting material and its impurities
(b) intermediates
(c) reagents, solvents and catalysts used in the process

(d) reaction vessels.

Starting materials and intermediates

The extent to which starting materials and intermediates may contaminate the product depends, to some extent, on whether the process is linear or convergent. In linear processes, the molecule is built in a direct linear sequence of steps, each of which leads to the modification or insertion of a functional group. A simple example of such a process is the synthesis of *Amitriptyline Hydrochloride* and *Embonate* from dibenzo[*a*,*d*]cycloheptan-5-one.

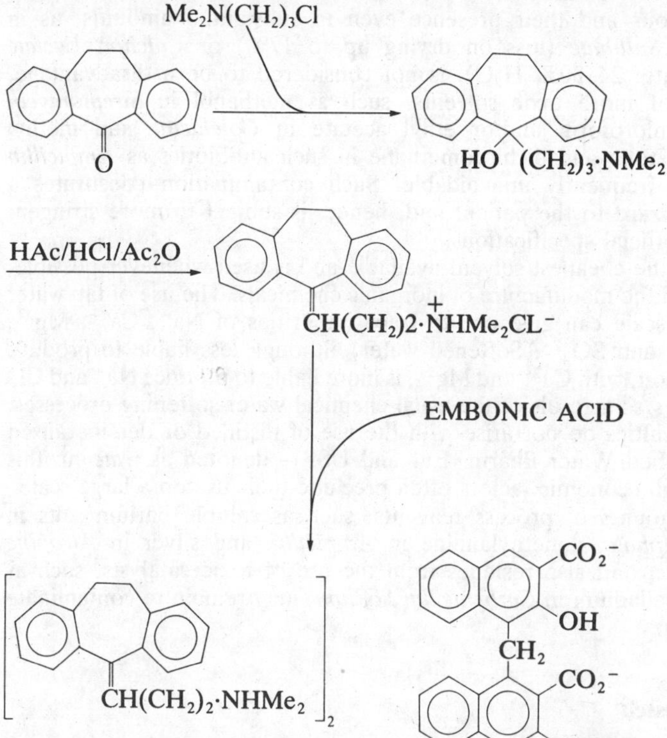

In this case the starting ketone, which is also a potential oxidative breakdown product, may be present in both the *Hydrochloride* and the *Embonate*, though only at an upper limit of *ca*. 0.5%. Additionally, specifications for the *Embonate* place a limit on the amount of chloride which may be present, arising from its immediate precursor, the *Hydrochloride*.

In general, linear syntheses consisting of more than three or four steps are unlikely to carry through significant amounts of starting materials or intermediates from the early stages into the final product, unless they are very closely similar in properties. In contrast, convergent processes in which the molecule is built up by condensation of two or more major fragments, each of which is the product of a short linear synthesis, provide fewer steps for elimination of contamination from starting materials and early stage intermediates.

Reagents, solvents and catalysts

Pharmaceutical chemicals are frequently produced in solvated crystalline form. Additionally, small amounts of solvents used in synthesis or for crystallisation of intermediates and in the final purification may contaminate the product. Common solvents, such as water and ethanol, are cheap and innocuous and their presence even in substantial amounts, as in *Vinblastine Sulphate* (loss on drying up to 17%) or *Calcium Lactate* (pentahydrate; 24–30% H_2O), is not considered to be a disadvantage. Retention of more toxic solvents, such as methanol in *Streptomycin Sulphate*, chloroform and/or ethyl acetate in *Colchicine* and methyl isopropyl ketone and dichloromethane in such antibiotics as *Ampicillin Sodium*, is frequently unavoidable. Such contamination constitutes a potential hazard to the patient and, hence, is subject to more stringent control in official specifications.

Water is the cheapest solvent available and is used whenever possible, especially in the manufacture of inorganic chemicals. The use of tap water on a large scale can give rise to trace impurities of Na^+, Ca^{2+}, Mg^{2+}, CO_3^{2-}, Cl^- and SO_4^{2-}. Softened water, although less liable to produce contamination, with Ca^{2+} and Mg^{2+}, is more liable to produce Na^+ and Cl^- as impurities as a result of the usual chemical water-softening processes. These difficulties do not arise with the use of distilled or demineralised water (Purified Water Pharm. Eur and USP – denoted as *water* in this text), though economic factors often preclude their use on a large scale.

Trace amounts of process reagents, such as soluble barium salts in *Barium Sulphate*, dimethylaniline in *Ampicillin* and silver in *Atropine Methonitrate*, and also residues from the use of toxic catalysts, such as nickel, or palladium in *Carbenicillin Sodium*, may remain to contaminate the product.

Reaction vessels

Modern reaction vessels are usually made from either stainless steel or hard borosilicate glass and, hence, present little, if any, hazard so far as product contamination is concerned. Vessels of copper, iron, galvanized iron (which gives rise to Zn^{2+} impurity) and lead may still be used in some older processes in aqueous solution, particularly for the production of inorganic compounds. Traces of metallic ions from such vessels may pass into solution and contaminate the product. Similarly, glass of an unsatisfactory standard (soft glass) may yield traces of alkali to the solvent. Likewise, plastic containers used for handling liquid and semisolid products in the course of manufacture may yield trace metal additives and antioxidants to the product (p.7).

Chemical and physical instability

Chemical instability

Impurity can also arise during storage as a result of chemical instability and a number of pharmaceutically important substances are known to undergo chemical decomposition when stored under non-ideal conditions. The nature of the decomposition, which is often catalysed by light, traces of acid or alkali, air oxidation, water vapour, carbon dioxide and traces of metallic ions, can frequently be predicted from a knowledge of the chemical properties of the substance. Thus, esters such as *Suxamethonium Chloride* (Succinylcholine Chloride) are prone to hydrolysis, especially when formulated in aqueous solution. Despite the fact that this product is theoretically neutral, strict limits of acidity are required, expressed either as **pH of solution** or as **titratable acidity**, to ensure adequate retention of potency. The acid-catalysed degradation of β-lactam antibiotics such as *Benzylpenicillin Potassium* (Penicillin G Potassium) is accelerated by heavy metal ions. Contact with metallic surfaces should, therefore, be avoided during preparation and storage, particularly of injection solutions. Similarly, alkali metal salts of weak acids such as *Pentabarbitone Sodium* (Pentabarbital Sodium) are subject to hydrolysis and decomposition by carbon dioxide. This leads to the formation of the parent barbiturate, pentabarbitone, which is insoluble in water, rendering the product unfit for the preparation of injection solutions. Oxidation of chloropromazine to the corresponding sulphoxide in *Chlorpromazine Elixir* and *Syrup* provides a typical example of oxidative decomposition with loss of potency which can occur under inadequate storage conditions.

The accent is on limitation or, if possible, total avoidance of all such decomposition by adopting suitable storage procedures and conditions. Thus, light-sensitive materials should be stored in darkened glass vessels or metal containers to inhibit photochemical decomposition. An interesting example of special precautions to exclude light is the use of opaque capsule shells to protect *Chlordiazepoxide* from decomposition by light. Material liable to oxidation or to attack by moisture or carbon dioxide may, if it is especially sensitive, require displacement of air from the container by nitrogen; for less sensitive products, it is usually sufficient to prescribe storage in a sealed container.

Oxidation is also preventable by the inclusion of appropriate antioxidants, including such phenols as *Butylated Hydroxyanisole* (BHA), *Butylated Hydroxytoluene* (BHT) and *Thymol*, which are capable of undergoing free radical oxidation at the expense of the material, e.g. *Liquid Paraffin*, they are being used to protect. *Sodium Metabisuphite*, another antioxidant, is suitable for use in aqueous solution, e.g. *Adrenaline Injection* (Epinephrine Injection), depending for its action on the oxidation which, as sodium bisulphite, it undergoes to sodium bisulphate. Exceptionally, suitable additives may be used to neutralise the toxic products of oxidation as in the stabilisation of *Chloroform*. When pure it readily forms traces of phosgene by a reaction which is light catalysed. In practice, therefore, a stable and non-toxic product is obtained by the addition of 1 or 2% of ethanol, which suppresses the decomposition reaction and, at the same time, converts any traces of phosgene to the harmless ethyl carbonate.

Physical changes

Change in physical form of a drug during storage is not unknown. Changes in crystal size and form, agglomeration and even caking of suspended particles, which are not always preventable, may lead to marked changes in the therapeutic efficiency of the product. Thus, particle size, and consequently surface area, may be a critical factor in determining the rate of absorption, and hence blood levels, of a drug of low solubility, such as *Griseofulvin*. Multidose suspensions, which are inefficient through rapid settling, or claying, likewise constitute a safety hazard, giving rise first to the possibility of underdosage and later to overdosage, as successive doses are withdrawn from the container. Injectable emulsions in which the globule size has increased during storage may similarly be dangerous in that they could be the cause of fat embolism.

Reaction with container materials

The possibility of reaction between container and contents constitutes a hazard which cannot be ignored. Creams and ointments liable to react with metal surfaces, e.g. *Salicylic Acid Ointment*, must not be packed in metal tubes unless they have been lacquered internally to inhibit reaction. Solutions of alkali-sensitive materials, particularly if subjected to any form of heat treatment during preparation, e.g. *Atropine Sulphate Injection*, which is sterilised by autoclaving, must be packed in glass ampoules which comply with the test for **hydrolytic resistance** of the European Pharmacopoeia (chemical resistance USP). This is the resistance offered by the glass to release of soluble mineral substances into freshly distilled water in the container and is determined by titration. Glass containers are recognised in three grades, I, II and III. Type I is neutral glass, with high hydrolytic resistance arising from the composition of the glass. Type II glass also has high hydrolytic resistance, but this is due to the surface treatment of the glass. It can be distinguished from Type I in a crushed-glass test. Type III glass has somewhat more limited hydrolytic resistance than glass of Types I and II.

All glass containers for injectable preparations must comply with the test for hydrolytic resistance. Aqueous injectable solutions must be packed in containers consisting of glass of either Type I or Type II. Type III glass may be used only for non-aqueous solutions and injection solids which are stable in this standard of glass.

Plastic containers and closures require careful evaluation because of the tendency to yield undesirable additives, such as plasticisers, a tendency which increases markedly in the presence of non-aqueous solvents. The plastics materials, mainly polyethylene, polypropylene, polystyrene and polyvinyl chloride, are used in conjunction with appropriate additives, which may consist of antioxidants, colours, antistatics, plasticisers, lubricants, impact modifiers and mould release agents. Not all these additives are used in the manufacture of any one type of container, but plastic containers for pharmaceutical use, irrespective of their composition, must be free from undesirable properties which affect the safety and efficacy of the medicament.

Plastic containers for injections should be sufficiently translucent to permit visual inspection of the contents and, if they are greater than 500 ml capacity, then products such as *Dextrose Intravenous Infusion* stored therein should also comply with tests which limit animal **toxicity** in the cat, **ether-soluble extractive**, and **metal additives** with special reference to barium and to the heavy metals tin, cadmium and lead.

Rubber closures, widely used in the packing of multidose injections, are, on the other hand, more prone to absorb medicaments, and also antioxidants and bactericides from solution, unless suitably pretreated by immersion in solutions of the compounds concerned. They should comply in general with the British Standard 3263: 1960 for Rubber Closures for Injectable Products, as modified by the requirements of the British Pharmacopoeia.

Temperature effects

The rate at which chemical and also physical change occurs in stored products is conditioned by temperature and labile products may have temperature storage requirements assigned to them to guard against unnecessary decomposition. The use of subjective instructions, such as **store in a cool place,** are capable of ambiguous interpretation on the part of the operator and more precise interpretations are generally placed on these phrases (Table 1.1).

Table 1.1 Recommended storage conditions (°C)

Storage conditions	European Pharmacopoeia	United States Pharmacopoeia
Deep-freeze (freezer)	−15 to 0	−20 to −10
Refrigerator	2 to 8	2 to 8
Cold		>8
Cool	8 to 15	8 to 15
Room temperature	15 to 25	Ambient
Controlled room temperature		15 to 30

Products required for tropical use, must, if prepared under temperate conditions, be capable of withstanding the effects of change to tropical temperatures. Thus, pastes and ointments must be so formulated as to retain their consistency at higher temperatures, and appropriate standards laid down for them. The high humidity often encountered in tropical countries also constitutes yet another hazard, in that it is conducive to mould contamination of the product, and suitable precautions may be necessary in formulation.

Manufacturing hazards

Even in a well run manufacturing house, certain hazards exist which can lead to product contamination. A well established system of checks on manufacture and insistence on adequate analytical control of the product may be sufficient to ensure that standards are well maintained. Considerable attention to the design of buildings and manufacturing areas, cleanliness and production procedures is also essential if the highest standards are to be achieved. Recommended practice in these important matters is set out in the so-called **Orange Guide** (*Guide to Good Pharmaceutical Manufacturing Practice*, published on behalf of the Department of Health and Social Security, Medicines Division, by Her Majesty's Stationery Office).

Control analysts will be aware of all the more likely hazards in achieving good quality control of pharmaceutical products. Specifications for drugs and formulated products should be designed to guard against common manufacturing hazards and, in particular, to exclude contamination from accidental inclusion of particulate matter, microbial contamination, or even labelling errors.

Particulate contamination

The presence of unwanted particulate matter can arise in a number of ways. These include accidental inclusion of atmospheric pollutants, such as aluminium oxide, silica, sulphur and soot, or from glass, porcelain, metallic and plastic fragments from sieves, granulating, tabletting and filling machines, or even from product containers. This type of contamination may stem from either bulk materials used in the formulation or from the use of improperly cleaned equipment or containers, but is more likely to be the result of particles being shed through wear and tear of equipment. **Clarity of solution** for injections is particularly important.

An unusual example of such contamination is that of metal particles which have been found in eye ointments packed in metal tubes. In this case, the method of tube manufacture gives rise to metal splinters which cannot be detached from the inside of the tube by any of the usual washing and cleaning procedures. Some of the particles, however, are extruded with the ointment, the number of particles so extruded increasing with the viscosity of the ointment. Both tin and aluminium tubes yield metal particles, though particles from the former tend to be larger than those from the latter.

Non-particulate contamination

Gaseous atmospheric pollutants, such as sulphur dioxide, hydrogen sulphide, arsine and even carbon dioxide and water vapour, can constitute a hazard to products manufactured or stored under less than ideal conditions.

Cross-contamination

The handling of powders, granules and tablets in large bulk frequently creates a considerable amount of air-borne dust, which, if not controlled, can lead to cross-contamination of products. The danger is well known to experienced manufacturers, particularly of steroidal and other synthetic hormones. Precautions, such as the use of face-masks and special extraction equipment, used to protect operators from undesirable effects of certain drugs of this type, are also suitable for more general use to limit cross-contamination. Manufacturers of penicillin preparations in the United States are required by the Food and Drugs Administration (FDA) to institute adequate control of the manufacture, handling and storage of drugs and their preparations to limit cross-contamination of one by another. The application of special limit tests places a check on contamination by penicillin of other products manufactured on the same premises.

Microbial contamination

The pharmacopoeial requirement of sterility tests for all products intended for parenteral administration and ophthalmic preparations, irrespective of whether they are prepared by end-sterilisation processes or produced under aseptic conditions, provides an adequate level of control for such preparations. Many other products, especially liquid preparations and creams for topical application to severely injured skin, large open wounds or mucous membranes, which are liable to bacterial, mould, and fungal contamination from the atmosphere (or, less frequently, from contaminated equipment) during manufacture, should be controlled for microbial contamination. A few materials are self-sterilising, but many products capable of supporting microbial growth require the addition of suitable antibacterial or antifungal agents if microbiological spoilage of the product is to be completely avoided. Certain materials, which are particularly prone to microbial contamination, may constitute a health hazard unless they are carefully controlled. These are mainly substances of natural origin, which are known to be liable to contamination, usually with specific organisms. The most satisfactory control, therefore, is one which sets a requirement for freedom from specified microbial contaminants (Table 1.2). This is the preferred method of control in the United Kingdom. In some countries reliance is placed on a limit of **total viable count** of micro-organisms present. This method of control, however, is less satisfactory as growth of contaminating micro-organisms may occur on storage of the product.

Table 1.2 Recommended control of microbial contamination in natural pharmaceutical substances

Freedom from salmonellae
Acacia; Senna; Tragacanth
Freedom from Escherichia coli
Sterculia
Freedom from salmonellae and Escherichia coli
Cochineal; Prepared Digitalis; Gelatin; Pancreatin; Starch
Freedom from pseudomonads
Dried Aluminium Hydroxide; Dried Aluminium Phosphate

Process errors

Gross errors arising from incomplete solution of a solute in a liquid preparation should be detected readily by the normal analytical control procedures. Minor errors, however, could escape notice if the manufacturing tolerance for the quantity of active ingredient in the product is wide and analysis indicates that the lower limit has only just been reached. Whilst errors of this sort are undesirable, they are probably only of serious concern in the case of solutions of potent medicaments. The preparation of such solutions, therefore, calls for special precautions, such as filtration, to avoid the danger of undissolved solute contaminating part of the batch. Uneven distribution of suspended matter during manufacture can, similarly, become the source of batch variation, or even variation within a batch.

Failure to acknowledge the limits of mechanical efficiency of mixing, filling, tabletting, sterilising and other equipment can lead to minor variation and, very occasionally, gross error, in the content of both active ingredient and formulatory adjuncts in compounded products. The mixing of powders of significantly different particle size or density calls for special care, and still further care in transfer to avoid separation in later stages of manufacture. Special care is essential to avoid mixing and filling errors in the preparation of tablets and capsules of highly potent medicament present in low dosage (5 mg and less) or at high dilution; analytical standards and control procedures for such preparations should be appropriately designed to exclude all but minor variation in the individual dosage units.

Decomposition, and loss of potency, through insufficient attention to chemical stability of pharmaceutical chemicals or additives under the conditions of manufacture represent a further source of impurity. Careful attention to process design, and to the handling of compounds liable to decompose, for example, by hydrolysis, oxidation or, in the case of optically active drugs such as *Ergometrine Maleate* and *Adrenaline Acid Tartrate* (Epinephrine Bitartrate), by racemisation, is essential if products are to conform to the highest standards. The use of ionising radiations for sterilising pharmaceutical products may also lead to decomposition with consequent loss of the medicament and possible formation of toxic breakdown products.

Packing errors

Products of similar appearance, such as tablets of the same size, colour and shape, packed in similar containers can constitute a potential source of danger through mislabelling of either or both. The handling of two such products in proximity should be avoided and control procedures should guard against the possibility of such mishaps.

Inadequate checks on the issue of labels, on the filling of labelling machines, on the setting of ampoule or other printing machines, and the destruction or return to stock of unused labels also constitutes a major packaging hazard. Such misadventures can only be avoided by care in manufacture, with particular attention to detail and cross-checks in the matter of stock records, process dockets and batch-marking of both raw materials and finished products.

Standardisation of pharmaceutical chemicals and formulated products

Two types of specification are used to control the quality of pharmaceutical products:

(a) manufacturing product licence standards
(b) pharmacopoeial standards.

Manufacturing standards are laid down by licensing authorities on the basis of information supplied by the manufacturer on the quality of trial (scale-up) and typical manufacturing test batches of the material. These are release specifications that provide for control of the product at the time of manufacture and usually represent the best minimum quality that can be consistently guaranteed. Such specifications are usually confidential between the manufacturer and the licensing authority, and, for this reason, are not generally available to independent analysts who may wish to check that the product is up to standard. In the United Kingdom, however, independent checks are applied to new and recently prepared products sampled by the Medicines Inspectorate against the manufacturer's licence specification (*release specification*). This work is conducted on a confidential basis by the Pharmaceutical Society's Medicines Testing Laboratory in Edinburgh on behalf of the Medicines Division of the Department of Health and Social Security.

In contrast, the published standards of the British Pharmacopoeia (BP), European Pharmacopoeia (Pharm Eur) and the United States Pharmacopoeia (USP) are available to all users of the material. They are designed primarily to set permissive limits of tolerance for the product at the time it reaches the patient. They do not necessarily equate with manufacturing specifications, since they must also take into account possible degradation of the product throughout its shelf-life up to the time it is prescribed and ultimately used. In this sense, pharmacopoeial standards are check specifications. Official standards must also encompass all known methods of manufacture and safeguard against varying standards of purity and impurity patterns and varying degrees of stability arising from more than one method of manufacture. They are, therefore, a compromise and do not necessarily represent the highest standard attainable by a single efficient producer at the time the product is manufactured.

The prime consideration, undoubtedly, is that the product should be satisfactory clinically. Variation in biological response is such that it is doubtful whether the difference between one preparation containing 85% of the required amount of active ingredient and another containing 115% (±15% limit) would be detectable in practice, unless a very carefully controlled trial involving large numbers of such preparations and patients were undertaken. The standards, however, are essential to ensure that reasonably reproducible products can be prepared in different laboratories by different operators, and by the same operator on different occasions. This not only ensures a consistent therapeutic response but also ensures that the product retains an acceptable level of potency and freedom from toxicity during storage before use.

Pharmacopoeial standards, therefore, do not necessarily equate with manufacturing standards, a frequent misconception, since official standards must be framed to take account not only of the nature and purity of the raw materials and the methods and hazards of manufacture, but also the manner and hazards of storage and the conditions under which the product is likely to be used.

Manufacturing variations

The principal hazard is loss of active ingredient, and all official standards for drugs and formulated products must allow for unavoidable decomposition or loss in manufacture, and storage under reasonably adequate conditions for a reasonable period of time.

The limits for compounded preparations take account of such hazards. This is particularly important for small-scale and extemporaneous preparations, which, unlike large-scale manufacture, are not necessarily subject to rigorous analytical control and, hence, do not allow for adjustment of the product. Errors in weighing small quantities and in the measurement of small volumes of liquid may be significant and limits of error for dispensed mixtures are usually at least of the order of ±5%.

Storage and dating of products

Appreciable manufacturing losses or anticipated decomposition on storage may require the inclusion of a definite overage, so that the product will still comply with the official standards after a reasonable period of storage.

This method, which is practised extensively with certain unstable vitamin preparations, such as those containing *Vitamin A* and its esters, is only permissible provided it does not interfere with any precise dosage requirement and provided the breakdown products are not toxic and do not further catalyse decomposition. The method is, thus, undesirable on clinical grounds for use with products containing *Ergocalciferol* and *Cholecalciferol*, where consistent overdosage can lead to a condition of hypervitaminosis. Likewise, it is unsuitable for *Benzylpenicillin* (Penicillin G) products, in which the primary product of hydrolytic breakdown, penicilloic acid, is capable of accelerating the further decomposition of the parent antibiotic by functioning as an acid catalyst.

The alternative method of stipulating an expiry date is not normally used in official standards. Opinions are very much divided on the use of expiry dates and on the value of the alternative, a declaration of date of manufacture, on the label of products. The calculation of expiry dates is circumscribed by a number of factors, not the least of which is the precise control of storage conditions, particularly temperature and humidity, for sensitive products. Failure to comply with expressly stated storage conditions could completely invalidate a carefully calculated expiry date. A declaration of the date of manufacture, on the other hand, merely indicates the age of the product, and gives no guide in itself to the possible life of the product.

Conditions of use

Official standards must also have regard to the conditions under which the product is likely to be used. Thus, the geographical distribution for sale is important; semi-solid preparations, such as pastes and ointments prepared under temperate conditions but required for tropical use, must comply with standards which preclude liquefaction or cracking in tropical climes; similarly, the viscosity requirements of suspensions and emulsions for tropical use must be such that the product retains stability, despite the decrease in viscosity with increased temperature

Dosage forms

Consideration of dosage form and packs in relation to the intended use of the product is also important in relation to the setting of official standards. This is especially relevant to the preservation and prevention from contamination of such widely diverse products as multidose injections, eye-drops, and products for external application to broken skin surfaces, which may be subject to spoilage with repetitive use.

The presence of toxic impurities or toxic decomposition products constitutes a hazard, no less important than loss of active ingredient, which has long been appreciated. Official standards legislate for and seek to control such impurities by limit tests based on qualitative and semiquantitative procedures, which have become all the more sensitive with the refinements introduced by such techniques as gas and thin-layer chromatography (TLC).

Consistency of dosage units is important. For most medicaments, however, where precise dosage is not critical, general methods such as **uniformity of weight** for tablets and powders for injections or **uniformity of volume** for injection solutions suffice to provide a satisfactory measure of control of the variation between dosage units. Where the amount of active ingredient per dosage unit is small (2 mg or less), as, for example, in *Digoxin Tablets*, or where the proportion of active ingredient to diluents and excipients is very low (2% or less) and in a few special cases where very precise but somewhat larger dosage is required for diagnostic purposes (*Dexamethasone Tablets*; 5 mg), unit dosage variation should also be controlled by a specific **uniformity of content** requirement involving individual tablet assays.

The efficacy of certain solid dosage products (Tablets, capsules and suppositories) requires that standards be set to ensure release of the active ingredients *in vivo* in a form which will permit a reasonable rate of absorption. This safeguard may take the form of a simple disintegration test in which the tablet or capsule is required to break up into small fragments within a specified time. In the case, however, of compounds of relatively low water solubility, or where there is an established clinical need to delay or otherwise control the rate at which the medicament is available, a specific quantitative dissolution test may be applied to ensure batchwise uniformity of drug availability.

Pharmacopoeial standards

As might be expected the British, European and United States Pharmacopoeias each have their own distinctive format. Nonetheless, the monographs therein have much in common in that their content is both descriptive and informative in addition to prescribing standards and conditions of storage. As shown in the typical monograph structure based on Ampicillin (p.15), official monographs for pharmaceutical chemicals generally embrace the following:

(1) A definition of the nature of material.
(2) A statement of the minimum standard of purity as determined by assay.
(3) A description of its physical characteristics.
(4) Tests for use in verifying the identity of the product.
(5) Limit tests to exclude excessive contamination and/or decomposition.
(6) Official quantitative procedures for determination of the active ingredient, solvents and other constituents required to assess compliance with the standard.
(7) Physical constants and tests which supplement the standard.
(8) Other information on:
    (a) packaging and storage conditions
    (b) labelling and other regulatory requirements (e.g. FDA)
    (c) dosage
    (d) cautionary notices on cytotoxic and other such dangerous materials.

In general, statements under the headings 1, 2, 3 (except solubilities), 4, 5, 6 and 7 constitute the official standards and methods for the assessment of material bearing the name at the head of the monograph for use in human medicine and also, unless specified to the contrary, for veterinary use. Information on the general physical characteristics of the material, that is, particle size and odour normally constitutes part of the standard; solubility data does not. The role and limitations of the principal features of monograph specifications which constitute the standard for pharmacopoeial products are set out below.

A typical monograph structure based on Ampicillin

Title based on British Approved Name, United States Adopted Name or International Non-proprietary Name	AMPICILLIN
Structural formula	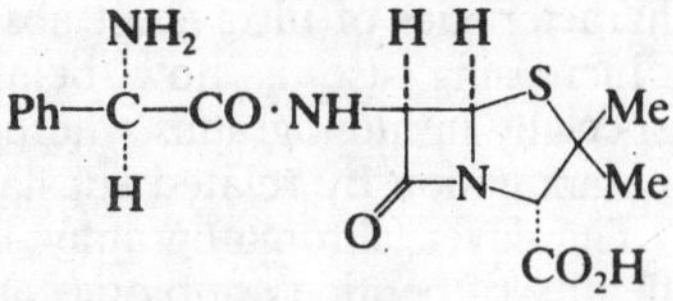
Molecular formula and weight, Chemical Abstracts Service Registry number	$C_{16}H_{19}N_3O_4S$ 349.4 7177-48-2
Definition of the nature of the material, by systematic chemical name or other suitable descriptive means	Ampicillin is (6*R*)-6(α-phenyl-d-glycylamino)-penicillanic acid
The standard of purity	It contains not less than 95.0% of $C_{16}H_{19}N_3O_4S$ calculated with reference to the anhydrous substance
Physical characteristics to aid recognition and assist handling	A white crystalline powder. Soluble in 170 parts of water; practically insoluble in ethanol (96%) chloroform, ether and fixed oils
Tests for verification of identity	(1) Infrared absorption spectrum (2) Specific colour tests to assist distinction from other penicillins (3) Specific test with penicillinase for β-lactam ring
Limit tests to control contamination by impurities arising from the method of manufacture or by degradation	Tests for: (1) Dimethylaniline by GLC (2) Heavy metals (3) Sulphated ash
Physical constants and tests which supplement the standard and help to confirm identity	(1) Specific optical rotation +280° to +305° (0.25% in water) (2) Acidity: pH of a 0.25% w/v solution, 3.5 to 5.5 (3) Clarity of solution
Quantitative evaluation of purity	Determination of: (1) Active ingredient by specific chemical *Assay* for β-lactam antibiotics (2) Water, by titration with Karl Fischer reagent
Additional information	(1) Storage. Keep in a well-closed container and store at temperatures ≯ 30° (2) Labelling requirements

Identification

The purpose of identification tests is to provide means of ensuring that materials have been correctly labelled.

Identification is usually achieved by a combination of simple chemical tests and measurement of appropriate physical constants. Because of the high structural specificity of infrared (IR) absorption spectroscopy, this technique now forms the basis of most identification procedures. For compounds with appropriate chromophores, infrared identification is usually supplemented by the somewhat less specific but nonetheless useful characteristics of ultraviolet absorption.

Increasing use is now being made of thin-layer chromatography, especially in monographs where this technique is also being used to limit contamination by related substances.

Thin-layer chromatography is particularly useful for confirming the identity of peptide antibiotics and hormones either by direct comparison against an authentic sample (*Framycetin Sulphate; Neomycin Sulphate; Calcitonin (Pork)*) or by acid hydrolysis and comparison of the chromatogram of the resulting amino acids with that derived from an authentic sample of the antibiotic (*Bacitracin Zinc; Polymixin B Sulphate*). It is also widely used for confirming the identity of active ingredients of injections (*Atropine Sulphate Injection*), eyedrops (*Gentamicin Eye Drops*) and other such preparations in which the active ingredient is present at relatively low concentration. Gas chromatography is used for identification in similar circumstances, when it is also used as the method of assay of *Homatropine Eye Drops* (Homatropine Ophthalmic Solution USP).

In a very few special cases, where complete characterisation by such physical tests is either inadequate or inappropriate, resort may occasionally be made to nuclear magnetic resonance (NMR) spectroscopy, as, for example, in the identification of *Gentamicin Sulphate.* In this case, however, the operator is not obliged to carry out all the prescribed physical tests, but is given the choice of identifying the material by NMR or by a combination of IR and TLC characteristics. The ratios of certain of the NMR signals do, however, specifically control the ratio of the components which make up this antibiotic mixture, though this is now more effectively achieved by high pressure liquid chromatography (HPLC).

Frequently, there is considerable overlap between identification tests, on the one hand, and limit tests, which are designed to limit undesirable amounts of impurity. Identification tests, no matter whether physical or chemical, can, if they are sufficiently specific, be used as the basis of a quantitative assay or in the design of specific limit tests and, in practice, a single chemical test or physical constant may contribute to both the identification and standardisation of the drug.

Chemical tests, which are used for identification, are included to establish as far as possible the presence of the required active groupings, to confirm the molecular structure. They may often be criticised on the grounds that they lack specificity but they are usually considered sufficiently specific when taken in conjunction with other requirements of the monograph. For inorganic substances, identification tests are usually based on those in use in general qualitative inorganic analysis. Organic substances are identified by the characteristic reactions of one or more of the functional groups present in the molecule.

Measurement of physical constants

Physical constants, such as melting point, boiling point, refractive index and solubility in water and organic solvents, are characteristic properties, useful for both identification and in the maintenance of standards of purity.

Melting point

The limitations of melting-point data as a criterion of purity are now generally recognised and, even when such factors as sample size, capillary dimensions, temperature of insertion into the heating bath or block and rate of temperature rise are standardised, it is difficult to ensure reproducibility from one laboratory to another. Melting range is, therefore, a more practical criterion of identity and the level of purity than melting point. Most organic substances have low melting points (i.e. below about 300°) and these melting points are more often than not depressed in the presence of an organic contaminant.

Solubility

Precise solubility requirements are seldom used as official standards for pharmaceutical chemicals. Statements included in official monographs of the British and United States Pharmacopoeias under the heading **solubility** are only approximate, and are intended solely for information (see British Pharmacopoeia General Notices). Statements under qualified headings, such as **solubility in ethanol**, however, do express an exact requirement and, hence, form part of the official standards for the substance. All are measured at 20°.

Descriptive terms, such as freely soluble, are frequently used to give an indication of the approximate solubility range of compounds at ambient temperature. The approximate solubilities implied by such terms when in the British, European and United States Pharmacopeias are set out in Table 1.3.

Table 1.3 Solubilities

Descriptive term	Approximate solubility in parts by volume of solvent for one part by weight of solute
Very soluble	< 1
Freely soluble	1 to 10
Soluble	10 to 30
Sparingly soluble	30 to 100
Slightly soluble	100 to 1000
Very slightly soluble	1000 to 10 000
Practically insoluble	$> 10\,000$

Weight per millilitre; specific gravity; relative density; refractive index

These constants are widely used as standards for liquids, including fixed oils (*Arachis Oil; Peanut Oil*), synthetic chemicals (*Dimethyl Sulphoxide, Glycerol, Propylene Glycol*) and solutions (*Syrup*). **Weight per ml**, the weight of 1 ml of liquid at 20°, is widely used in the British Pharmacopoeia. Because of Excise requirements, however, **specific gravity** (20°/20°), the ratio of the mass of a given volume of liquid to that of an equal volume of water (both at 20°), is still used for *Ethanol* and *Industrial Methylated Spirits*, though not for preparations (*Tinctures* and *Liquid Extracts*) made from them.

The term **relative density** (20°/20°), which is synonymous with specific gravity, is used in the European Pharmacopoeia. Similar standards for the weight of liquids and solutions, however, come under the heading of **specific gravity** in the United States Pharmacopoeia. Both specific gravity and refractive index are normally measured at 25° in the United States, in contrast to measurements at 20° in Britain and the rest of Europe. Exceptionally, some measurements are made at higher temperatures, as in the case of *Theobroma Oil* (Cocoa butter), which is a waxy solid melting at about 33°; the refractive index is measured at 40°.

Light absorption

Measurement of light absorption in the visible and ultraviolet range is now widely used as a means of identification (*Choline Theophyllinate*, λ_{max} 275 nm in 0.01M NaOH). It is especially useful when the observed spectrum shows more than one characteristic maximum (*Nitrofurantoin*, λ_{max} at 266 and 367 nm in buffer) or alternatively where the intensity and position of the maxima are influenced by pH (*Triprolidine Hydrochloride*, λ_{max} 290 nm in 0.05M H_2SO_4, and λ_{max} 230 nm and 276 nm in water). In some cases, the absorbance at specified wavelengths (not necessarily maxima) is used as the basis of limit tests to exclude undesirable impurities (*Hydrocortisone Acetate*). The ratio of the absorbances at two maxima also provides a useful criterion of identity (*Amphotericin*) or purity (*Viprynium Embonate*).

Infrared absorption

Infrared absorption spectroscopy is now firmly established as the method of choice for the verification of identity of pharmaceutical chemicals because of the high structural specificity of IR spectra in the so-called fingerprint region (1500 to 625 cm^{-1}). Comparison of the infrared absorption spectrum in the region 4000–625 cm^{-1} with that of an authentic reference compound (**BP Chemical Reference Substances, EP Chemical Reference Substances** and **USP Reference Standards**) has been widely used. The method, however, suffers from major disadvantages. These stem mainly from the excessively high cost of preparing, purifying, confirming, maintaining and distributing the reference materials. The risk, complications and general undesirability of making samples of narcotics and other controlled drugs readily available on request, even in small amounts and at high cost, has also deterred the introduction of infrared identification tests for derivatives of such important analgesics as morphine and pethidine. These problems however, have been overcome with the move in the British Pharmacopoeia and the British Pharmacopoeia (Veterinary) to replace IR identification tests based on reference compounds by comparable tests based on **reference spectra**.

In order to minimise unavoidable difficulties arising from instrument variation, reference spectra are produced on a spectrometer of medium resolving power and supplied with a reference spectrum of polystyrene film, produced on the same instrument to provide a baseline for comparison. The method of sample preparation is, of course, critical and is laid down in the monograph.

Infrared absorption spectroscopy is also useful for recognising differences in polymorphism (p.20), which may have therapeutic significance in some poorly soluble substances. Thus, *Tamoxifen Citrate* polymorphs have different solubility profiles. Compliance with the IR reference spectrum ensures that the material has the required crystal structure and solubility.

Optical rotation

Measurement of the optical rotation of chemically inhomogeneous liquids such as *Castor Oil* assists both identification and maintenance of quality. Additionally, the **specific optical rotation** of optically active compounds provides a valuable means of controlling the optical purity of pharmaceutical chemicals, in which pharmacological or physiological activity is highly correlated with molecular configuration, as, for example, with *Adrenaline Acid Tartrate* (Epinephrine Bitartrate), *Prednisolone Sodium Phosphate* and *Dextrapropoxyphene Hydrochloride* (Propoxyphene Hydrochloride). In *Levodopa*, for which the observed rotation at the D line of sodium is inconveniently small, measurement of the rotation of a suitable complex at a specific wavelength in the ultraviolet region provides a more effective means of ensuring that only the correct isomer is present. Circular dichroism spectrometry has also been shown to provide a more satisfactory means than simple optical rotation measurements for isomer control in *Phenthicillin* (Part 2).

The optical rotation has a special significance in the control of *Camphor* where the use of either natural $\{[\alpha]_D^{20} + 40°$ to $+43°$ (*c*, 10, 95% ethanol)$\}$ or synthetic camphor ($[\alpha]_D^{20}$ 0°), but not mixtures of the two, is permitted. This, although not specified explicitly, is definitely implied in the limits of specific rotation for synthetic camphor, which are $-1.5°$ to $+1.5°$.

Rotation measurements are also used as an assay to control the content of Dextrose in *Dextrose Intravenous Infusion* (Dextrose Injection USP) and also of dextrans in *Dextran Intravenous Infusions.*

Viscosity; jelly strength; swelling power

Viscosity measurements are used as a means of distinguishing various grades of liquid paraffin. They are also used to control the molecular size of dextrans in *Dextran Intravenous Infusion*, the composition of *Iron-Sorbitol Injection*, the extent of nitration of *Pyroxylin* and in the standardisation of *Methyicellulose, Povidone* and similar additives.

The properties of materials, such as *Gelatin*, which are capable of forming visco-elastic gels, are controlled by the measurement of **jelly strength** (Gel strength USP) of gels prepared from them in a Bloom gelometer. The quality of *Bentonite*, a suspending agent which forms gels by the absorption of water, is controlled by its **swelling power**, a measure of its increase in volume in water.

Polymorphism and particle size

The physical state of most water-soluble compounds is of no therapeutic significance, though their solid state properties, like those of other less soluble substances, may affect mixing and handling characteristics in the preparation of solid dosage forms. On the other hand, solid state properties of insoluble drugs, which are necessarily administered in solid form, can markedly influence the rate of both solution and absorption of the drug. Crystal form and particle size in particular are the important factors controlling surface area, the property which determines the level of activity of such drugs as *Griseofulvin*, *Spironolactone* and *Digoxin*. Thus, the official monograph for *Griseofulvin* provides for standardisation of the powder such that the bulk of the material consists of particles not more than 5 µm in maximum dimension, with the occasional particle up to 30 µm.

Control of **sediment volume**, as measured by the depth of sediment after standing for a period of hours in a vessel of specified dimension, provides a simple alternative means of controlling particle size in *Tetracosactrin Zinc Injection*. The suspendability of *Barium Sulphate* is similarly controlled by a test for **sedimentation** (Bulkiness USP).

Certain products may always be produced in one particular pharmaceutically more desirable polymorphic form, which, although not specifically described in the monograph, is controlled by an identity test using the Infrared Reference Spectrum of that particular polymorph. Polymorphism, the ability to crystallise in alternative crystal forms, is a widespread phenomenon. Like particle size, it, too, can be significant in compounds of relatively low solubility in determining the level and duration of the response. Thus, *Chloramphenicol Palmitate* exists in polymorphic forms, of which one is biologically inactive. Infrared spectroscopy is used to compare samples of the product recovered from *Chloramphenicol Palmitate Mixture* with an artificially prepared mixture of authentic material containing 10% of the inactive polymorph.

Assay

Ideally, assay methods should not only be specific for the chemical entity under examination, but also be stability-determining, as, for example, in the iodometric imidazole assays of β-lactam antibiotics such as *Benzylpenicillin Potassium* (Penicillin G) and *Ampicillin*. Nonetheless, non-specific assay methods are frequently used. This applies particularly in the case of acid-alkali titrations for bases and acids. Many inorganic salts, also, are determined simply on the content of one of the ions present; thus, *Sodium Sulphate* is assayed on its sulphate content by precipitation as barium sulphate. Even when modern physical methods are used, as for example the measurement of ultraviolet absorption, this may be by no means specific. For example, most simple aromatic substances show an absorption, which can form the basis of an assay, in the region 260-300 nm; this absorption is characteristic of the aromatic ring and, while differences exist in the spectra of different aromatic substances, these are seldom completely characteristic. The assays are, therefore, non-specific.

Despite elements of non-specificity, however, the assay process is, nevertheless, regarded as being sufficiently specific for the purpose when taken in conjunction with the other requirements of the monograph, as discussed above. Details of the methods used for assay are described in later chapters.

Assay tolerances

Assay tolerances are of considerable importance in fixing standards for pharmaceutical preparations. They include the limits of error of the actual assay process for the active ingredients, manufacturing tolerances for the particular dosage form, and sampling errors. The limits of error of the assay method itself depend upon the method adopted. Volumetric and gravimetric methods usually have quite narrow limits of error, e.g. *Sodium Chloride* ($\nless$99.5% calculated with reference to the dried substance). Non-aqueous titrations are usually less precise and titres are often marginally high. Typical limits are, therefore, somewhat wider, as in *Azathioprine* (98–101.5%) and *Atropine Sulphate* (98.5–100.5%). Spectrophotometric assays and assays involving extraction or other complex manipulations are usually subject to greater variation than volumetric and gravimetric methods. Thus, the limits applied to the ultraviolet absorption assay of *Testosterone Propionate* are 99–103% whilst those for *Phenobarbi-*

tone Sodium whose assay involves an extraction with ether, are 98–100.5%. Assay tolerances for formulated product are discussed in Chapter 10.

Sampling procedures and errors

Errors, due to sampling, arise in the selection of material for analysis where the material selected is not truly representative of the batch as a whole. The problem does not arise with homogeneous liquids or solutions unless these are distributed in separate containers which may contain material from different batches. Heterogeneous liquids, such as emulsions and suspensions, and solid samples are subject to mixing variations and to separation and the problem of obtaining a representative sample for analysis then becomes important. One of two methods is generally used for solid samples, depending upon whether or not the bulk material is reasonably homogeneous or not. Random sampling is used for products which are not subject to gross variation; the bulk is divided into real or imaginary units and a sample proportionate to the size of each unit is collected, the selection of individual samples being carried out in a random manner. For heterogeneous materials, such as crude drugs comprising various parts of the plant, representative samples of each type of material present are selected in a random manner in proportion to the amount of that material present in the whole.

The difference between the mean value of the active ingredient in the samples and the true value is known as the sampling error. The error due to sampling is inversely proportional to the square root of the number of samples taken. This error is additional to those introduced by the analytical method and by the fact that, in a formulated product, the active ingredient is itself subject to a permitted variation.

Where the material is packed in a number of separate containers, representative samples for analysis must be selected from a proportion of the containers taken at random. Emulsions, suspensions, pastes and ointments should be shaken or stirred to ensure the best possible mixing before sampling; fluid samples of emulsions and suspensions should be taken as far as possible from different parts of the container and/or from a suitable selection of containers. Pastes and ointments should be sampled with an auger to give cored samples from various parts of the container(s).

Limit tests

Limit tests are quantitative or semiquantitative tests designed to identify and control small quantities of impurity which are likely to be present in the substance. The quantity of any one impurity in an official substance is often small and, consequently, the visible response to any test for that impurity is also small. The design of individual tests is, therefore, important if errors are to be avoided in the hands of different operators. This is accomplished by giving attention to a number of factors, which are discussed below.

Specificity of the tests

Any test used as a limit test must, of necessity, give some form of selective reaction with the trace impurity. Many tests used for the detection of inorganic impurities in official inorganic chemicals are based upon the separations involved in inorganic qualitative analysis. A test may be demanded which will exclude one specific impurity, but highly specific tests are not always the best; a less specific test which limits several likely impurities at once is obviously advantageous, and in fact can often be accomplished. An example of such a test is the heavy metals test applied to *Alum*, which limits contamination by not only lead, but also other heavy metal contaminants precipitated by thioacetamide as sulphide at pH 3.5.

High specificity is readily achieved with limit tests based on thin-layer, gas and high pressure liquid chromatography, in which impurities are readily separated and are identified by their characteristic R_F values, retention or elution times.

Sensitivity

The degree of sensitivity required in a limit test varies enormously according to the standard of purity demanded by the monograph. The sensitivity of most tests is dependent upon a number of variable factors, all capable of strict definition and all favourable towards the production of reproducible results. Thus, the precipitation of an insoluble substance from solution is governed by such factors as concentration of the solute and of the precipitating reagent, duration of the reaction, reaction temperature and the nature and concentration of other substances unavoidably present in solution. As a general rule, cold dilute solutions give light precipitates, whereas more granular ones are obtained from hot concentrated solutions. Many limit tests, however, are concerned with very dilute solutions, which are often slow to react, and here sensitivity of the reaction can often be increased by extending the duration of the reaction or by raising the reaction temperature. Similar considerations apply in the design of colour and other tests.

Chromatographic tests are much more amenable than chemical tests to control sensitivity, which can be effected by variation of such factors as the support, solvent composition, temperature and method of detection.

Control of personal errors

It is essential to exclude all possible sources of ambiguity in the descriptions of a test. Vague terms, such as 'slight precipitate', should be avoided as far as possible. The extent of the visible reaction to be expected under the specified test conditions should be clearly and precisely defined. This is usually accomplished in one of three ways.

(a) **Tests in which there is no visible reaction.** A definite statement is incorporated in the wording of the test, which states that there shall be no colour, opalescence or precipitate, whichever is appropriate to the particular test. Such tests are rarely used since a negative result does not necessarily imply the complete absence of the impurity, the test as laid down merely indicating the absence of an undesirably large amount of the impurity.

(b) **Comparison methods.** Tests of this type require a standard containing a definite amount of impurity, to be set up at the same time and under the same conditions as the test experiment. In this way the extent of the reaction is readily determined by direct comparison of the test solution with a standard of known concentration. The official limit tests for chlorides, sulphates, iron and heavy metals and the majority of tests for specific impurities are based on this principle. The limit tests for lead and arsenic are, in practice, also comparison methods. They are, however, so designed that they can be readily applied as quantitative determinations.

Many thin-layer chromatographic tests are also based on direct comparison with reference samples of impurities applied at specific concentration.

(c) **Quantitative determinations.** Quantitative determination of impurities is only applied in special circumstances, usually in those cases where the limit is not readily susceptible to simple and more direct chemical determination. The method is used in many of the general limit tests for non-specific impurities, such as limits of insoluble matter, moisture, and residue on ignition.

General limit tests for non-specific impurity

Clarity of solution

The degree of clarity or opalescence of solutions is defined by the European Pharmacopoeia by one of the following terms:

(a) clear
(b) very slightly opalescent
(c) slightly opalescent
(d) opalescent
(e) very opalescent.

Measurement is made by direct comparison with a reference solution, either with the aid of a beam of light of standard luminosity or by viewing through the solution in diffused light against a black background. Solutions are **clear** if their clarity is the same as that of *water* or the solvent used. The various degrees of opalescence are measured against freshly prepared silver chloride opalescences prepared from standard sodium chloride solutions and silver nitrate.

Clarity of solutions for injection is particularly important to ensure reasonable freedom from **particulate matter**. Solutions for injection must appear to be free from insoluble matter when viewed by eye against a black background in upright, horizontal and vertical positions under a screened light source of specified intensity. Solutions should also be viewed in plane-polarised light to detect cellulose fibres. Large-volume intravenous injections with a nominal content >100 ml, such as *Dextrose Intravenous Infusion*, are further subject to a **limit test for particulate matter** using an instrument capable of counting particles within specific size ranges. This test requires the mean count of particles per ml from five containers to be $\not>$1000 greater than 2 μm and $\not>$100 greater than 5 μm in equivalent sphere diameters.

Tests for clarity of solution, often combined with a limit for colour, also constitute a means of limiting insoluble parent drugs in their more highly water-soluble derivatives (*Phenytoin Sodium; Warfarin Sodium*). Similar tests for **completeness and clarity of solution** and for **particulate matter** in the United States Pharmacopoeia are applied to sterile dosage forms from which reconstituted solutions are prepared for injection.

Insoluble matter

Limit tests for insoluble matter provide a useful means of controlling small amounts of solvent-insoluble contamination in otherwise readily soluble materials. A typical example is the determination of **ethanol-insoluble matter** to limit sodium chloride in *Chloramine*. Similarly, **water-soluble substances** in *Amaranth* exclude non-colorants and the requirement for **solubility in ethanol** in *Boric Acid* controls the presence of borax.

Colour of solution

Colour of solution is also frequently subject to control by the methods of the European and United States Pharmacopoeias. A solution may be described as **colourless** if it has the appearance of water when examined under specified conditions. Varying degrees of colour (**degree of colouration of liquids,** Pharm. Eur.) are permissible and controlled by direct comparison of solutions with specified standard colour solutions.

In the European Pharmacopoeia, colour standards are based on three primary solutions: yellow, red and blue prepared from iron (III) chloride, cobalt (II) chloride and copper (II) sulphate, respectively. These primary solutions are mixed in various combinations with or without 1% hydrochloric acid to give five **standard solutions**, which are Brown (B), Brownish Yellow (BY), Yellow (Y), Greenish Yellow (GY) and Red (R). The reference colours are produced from the standard solutions in six dilutions, denoted by subscripts 1 to 6 (e.g. B_1 to B_6), with 1% hydrochloric acid.

Similar primary solutions (Colorimetric Solutions) are used in the USP in tests for **color** and **achromicity**. The **matching fluids**, A to T, which are the comparison standards, are prepared by mixing the primary colorimetric solutions in various proportions with water.

Soluble matter

Tests are applied to limit soluble impurities in official substances which are themselves completely insoluble in a particular solvent, the object usually being to detect some specific impurity. The control of **water-soluble barium salts**, which are highly toxic and, therefore, stringently excluded from *Barium Sulphate* required for X-ray work is an excellent example of the value of this type of test. A general limit of matter soluble in dilute acid is applied to *Light Kaolin* by refluxing with 0.2M HCl, filtering and evaporating the filtrate. A limit of **acid-soluble matter** is applied to *Purified Talc* to control trace impurities of water-soluble salts, on the one hand, and of oxides, hydroxides and carbonates on the other. **Water-soluble acids** in *Undecenoic Acid* (Undecylenic Acid) are controlled by shaking a sample with warm water, filtering and titrating the filtrate with standard sodium hydroxide solution. **Matter soluble in petroleum spirit** is similarly controlled in *Griseofulvin*, while **free cyclobarbitone** in *Cyclobarbitone Calcium* is limited by extraction with toluene.

Moisture, volatile matter and residual solvents

Many substances absorb moisture on storage. Deterioration of this nature is readily limited by a requirement for the loss in weight **(loss on drying)** when the substance is dried under specified conditions. The amount of heat to which the substance is submitted varies considerably according to the nature of the substance. The temperature must be sufficiently high to produce the required result within a reasonable time, but not so high as to cause decomposition. If the substance is stable the test is usually applied by drying to constant weight at 105° as with *Ethynodiol Diacetate* and *Sodium Benzoate*.

Special modifications are adopted for thermolabile substances: in the case of *Hyoscine Hydrobromide* the sample is first dried at room temperature *in vacuo* for one hour, and then heated to constant weight at 105°, presumably because it is unstable only in the presence of moisture. Vacuum drying at various suitable temperatures is used for hydrated salts, such as *Tubocurarine Chloride*. Vacuum drying over phosphorus pentoxide is used where water is tenaciously held or the compound is heat sensitive, as in *Maleic Acid*, where higher temperatures might precipitate anhydride formation with release of water.

Inorganic salt hydrates, such as *Sodium Sulphate* ($Na_2SO_4,12H_2O$) and *Sodium Phosphate* ($Na_2HPO_4,12H_2O$ and $7H_2O$) lose all or part of their water of crystallisation on drying. The loss in weight is often considerable and, in fact, forms a secondary assay process. Such limits for loss in weight when a salt is dried are often necessarily wide (Sodium Sulphate loses between 52 and 57%) and may allow for either efflorescence or deliquescence.

In certain other substances, where the content of water is critical or where it is important to distinguish between moisture and other volatile matter, water may be determined titrimetrically with Karl Fischer Reagent. This method, which is described in detail on p.172, is used to determine **water** in antibiotics (*Erythromycin; Gentamicin Sulphate*) in certain steroids (*Hydrocortisone Sodium Phosphate*) and inorganic compounds (*Sodium Nitroprusside*). Determination of water by exchange with deuterium oxide (D_2O) can be monitored by NMR spectroscopy using a computer attachment in special cases where the amount of substance available for determination is very small (*Cloprostenol Sodium* and *Fluprostenol Sodium*).

Where an official moisture limit is specified, this is usually taken into account in calculating the results of the assay (Cyclizine Hydrochloride is required to contain 98.0 to 101% $C_{18}H_{22}N_2$,HCl calculated with reference to the substance dried to constant weight at 130°). This obviates the necessity of using dried samples for assay, which is advantageous because the assay and loss on drying procedure can be carried out simultaneously. Also, dried (or ignited) substances, because of their hygroscopic character, are more difficult to weigh than air-dried samples.

The introduction of gas-liquid chromatography provides a useful alternative method for limiting specific volatile substances. The method is used for the determination of **methanol** in *Streptomycin Sulphate*.

Determination of the loss on drying of Sodium Chloride

Method Heat a clean, dry, shallow, stoppered weighing-bottle in an air-oven at 130° for about 20 min. Transfer to a desiccator, cool for at least 20 min with the stopper on its side in the neck of the bottle, insert the stopper and reweigh. Reheat and reweigh until constant weight is attained. Introduce about 1 g of the sample into the bottle, insert the stopper and reweigh. Place in the drying oven with the stopper on its side in the mouth of the bottle and heat in the oven at 130° for 2 h, cool in a desiccator for at least 20 min, insert the stopper and weigh. Repeat the drying and cooling until constant weight is attained. Calculate the percentage loss in weight.

The following example illustrates the calculations involved in utilising the figures for moisture content in conjunction with the assay figures.

Calculation A sample of Sodium Chloride was found to contain 98.90% NaCl (from the assay figures) and upon drying lost 0.760% of its weight. Calculate the percentage of NaCl with reference to the sample.

$$\therefore\ 100\text{ g sample (undried)} \quad \text{contains} \quad 98.9\text{ g NaCl}$$

$$\text{But } 100\text{ g sample (undried)} \equiv 100.00 - 0.76 \equiv 99.24\text{ g dried sample}$$

$$\therefore\ 99.24\text{ g sample (dried)} \equiv 98.9\text{ g NaCl}$$

$$\therefore\ 100\text{ g sample (dried)} \equiv \frac{98.9 \times 100}{99.24}$$

$$\equiv 99.65\text{ g NaCl}$$

∴ *The sample of Sodium Chloride was found to contain* 99.65% NaCl *calculated with reference to the substance dried to constant weight at* 130°

Non-volatile matter

These limits are applied to both inorganic substances which are readily volatile. The Pharmacopoeia draws a distinction between substances which are readily volatile and those which are volatile only when ignited strongly. Limits of **non-volatile matter** are applied to the former group to control contamination by unspecified inorganic matter, such as other salts and dirt. *Ammonium Bicarbonate, Hydrogen Peroxide Solution, Water for Injections* and a number of readily volatile solvents and anaesthetics including *Ethanol, Isopropyl Alcohol, Chloroform, Halothane, Solvent and Anaesthetic Ether*, are examined in this way. The determination of non-volatile matter in *Hydrous Wool Fat* by heating to constant weight on a boiling water bath provides an approximate determination of the amount of water present in the sample.

Residue on ignition; loss on ignition

Limits of **residue on ignition** are generally applied to substances which undergo a major decomposition, leaving a residue of definite composition. Thus, *Calamine,* a basic zinc carbonate, decomposes when ignited to give carbon dioxide and water, leaving the oxide as residue. The process is, therefore, a non-specific assay. Ignition is also used as a method of assay to control zinc oxide in *Zinc Cream*.

Loss on ignition is applied to thermostable substances, which contain large amounts of tightly bound water (*Magnesium Trisilicate*), or thermolabile impurities, such as carbonate in *Light Magnesium Oxide*.

Determination of the loss on ignition of Zinc Oxide

Method Heat a clean crucible supported on a pipeclay triangle in a hot Bunsen flame for 10 min. Cool in a desiccator for about 20 min and weigh. Repeat until constant weight is attained. Introduce about 1 g of sample into the crucible, reweigh and then heat strongly in a hot Bunsen flame for about 1 h. Cool in a desiccator and reweigh. Repeat the heating and cooling until constant weight is attained.

Ash values

The ash content of a crude drug is the inorganic residue remaining after incineration. It represents not only the inorganic salts, such as calcium oxalate, occurring naturally in the drug, but also inorganic matter from external sources. The official ash values are chiefly of use in examining powdered drugs. Thus, they may have one or more of the following applications:

(*a*) to ensure the absence of an undue proportion of extraneous mineral matter introduced accidentally or by design at the time of collection or in subsequent treatment, e.g. earth, sand and floor sweepings
(*b*) to ensure absence of other parts of the plant, e.g. in *Cardamom Fruit*
(*c*) to detect adulteration with exhausted drug, e.g. in *Ginger*
(*d*) to detect adulteration with material containing stone cells or starch which would modify the ash values.

The four types of ash value used in the British Pharmacopoeia are **ash, acid-insoluble ash, water-soluble ash** and **sulphated ash**.

Ash

A figure for the total ash content is valuable for drugs in which little calcium oxalate is present (*Ginger*). If much calcium oxalate is present then the value for the acid-insoluble ash is a better criterion of purity. Ash in *Sterilisable Maize Starch* (Absorbable Dusting Powder) limits the amount of magnesium oxide which it is permitted to contain.

Method for the determination of ash in ginger Ignite a silica dish to constant weight at a dull red heat in a muffle furnace. Spread 2 to 3 g of the powdered drug evenly over the bottom of the dish and weigh. Ignite at low temperature until vapours have almost ceased to be evolved and then increase the temperature slightly, not exceeding 450°, to burn off the carbon (Note 1). Cool in a desiccator and weigh. Repeat the heating and cooling until constant weight is attained.

If a carbon-free ash is not obtained in this way, extract the residue with hot water, filter through an ashless paper, incinerate the residue and filter paper in the silica dish until all carbon is removed, using a higher temperature than that used previously if necessary (Note 2). Add the filtrate to the dish, evaporate to dryness and ignite to constant weight using not more than a dull red heat.

Calculate the percentage of ash with reference to the air-dried drug.

Note 1 Too high a temperature would cause a loss of such substances as alkali chlorides which are volatile at high temperatures.

Note 2 Sometimes heating at a low temperature does not give a carbon-free ash because carbon particles are trapped in fused alkali carbonates or phosphates. Treatment of the ash with hot water dissolves the salts. The insoluble residue can then be ignited rapidly at a higher temperature to remove carbon without loss of inorganic material; the filtrate is added and subsequently evaporated to dryness when the whole residue can be ignited at the lower temperatures, at which alkali chlorides are not volatile.

Acid-insoluble ash

Crude drugs containing calcium oxalate can give variable results upon ashing depending upon the conditions of ignition. Treatment of the ash with hydrochloric acid leaves virtually only silica. Hence **acid-insoluble ash** forms a better test to detect and limit excess of soil in the drug than does the total ash.

Method for the determination of acid-insoluble ash in Rhubarb Wash the total ash (obtained as above) from the silica dish into a 100 ml beaker using 25 ml of dilute hydrochloric acid. Boil for 5 min. Transfer the insoluble residue to a previously prepared, ignited and weighed Gooch crucible, wash well with hot water, allow suction to remove most of the water and then ignite at a temperature not exceeding 450° inside an ordinary crucible used as a 'jacket crucible'. Cool in a desiccator and weigh. Calculate the percentage of acid-insoluble ash with reference to the air-dried drug.

Water-soluble ash and water-soluble extractive

These determinations are only specified in the case of one official drug, *Ginger*, where it is helpful in detecting samples which have been extracted with water.

Method for the determination of water-soluble ash in Ginger As for acid-insoluble ash, but use 25 ml of water in place of the 25 ml of acid; subtract the weight of the insoluble residue from the weight of the total ash; the difference in weight represents the water-soluble ash. Calculate the percentage of water-soluble ash with reference to the air-dried drug.

Sulphated ash

The determination of **sulphated ash** is used in the case of unorganised drugs such as *Colophony, Podophyllum Resin, Wool Alcohols* and *Wool Fat*. It is also applied very widely to control the extent of contamination by non-volatile inorganic impurities in organic substances, as in *Aspirin, Benzyl Benzoate, Cyclizine Hydrochloride* and *Lignocaine Hydrochloride* (Lidocaine Hydrochloride). The test is also used to control traces of alkali metals in *Chlorbutol* resulting from the method of preparation by heating acetone and chloroform in the presence of potassium hydroxide.

The substance is ignited with concentrated sulphuric acid, which decomposes and oxidises organic matter, leaving a residue of inorganic sulphates. Reproducible results are more readily obtained in this determination than in a total ash determination because, in general, metal sulphates are stable unless heated very strongly.

Method for the determination of sulphated ash in Aspirin Ignite a large silica crucible to constant weight. Add approximately 5 g of sample, reweigh, moisten with sulphuric acid and heat gently at first and then more strongly as the volatile matter is removed. Ignite more strongly to remove the carbon, *cool*, remoisten with sulphuric acid and then reignite to constant weight. Calculate the percentage of sulphated ash.

Note This determination must be carried out in a fume-cupboard.

Precipitation methods

These are used when moderate amounts of impurities such as **iron, aluminium and silica** are permitted as, for example, in *Potassium Hydroxide*. A sample (5.0 g) is boiled with excess dilute hydrochloric acid and then made alkaline with ammonia. Any precipitate of metallic hydroxide or silica is collected, dried and after ignition should weigh $\not>$5 mg (limits of Fe, Al and matter insoluble in hydrochloric acid).

Similar tests are used to limit **iron, aluminium and phosphate** in *Calcium Hydroxide* and *Chalk*.

Limit tests for metallic impurities

Considerable emphasis is placed on the limitation of physiologically harmful impurities. Special attention is given to the control of such toxic elements as arsenic and antimony, and the heavy metals lead, cadmium and mercury, which are potent nerve poisons. Contamination by arsenic and, particularly, by lead is still widespread in industrialised countries as a result of atmospheric pollution. In consequence, quantitative tests for heavy metals and arsenic are applied particularly stringently to all compounds, such as, for example, antirheumatics, heart stimulants and antihypertensives, which are used for long-term administration.

Heavy metals

Contamination by **lead** and other **heavy metals** is most effectively controlled by precipitation of their relatively insoluble and characteristically coloured sulphides (Cd, yellow; Hg, black; Pb, brown), which occurs readily when aqueous solutions are treated with hydrogen sulphide or alkali metal sulphides (NaSH). The stability of metal sulphides to acid varies considerably with acid strength. Table 1.4 shows how the precipitation of heavy metal sulphides is affected by mineral acid, acetic acid, dilute ammonia and by potassium cyanide which forms stable water-soluble complexes unaffected by the precipitating agent (H_2S or NaSH) with the majority of heavy metals. Tests based on precipitation from ammoniacal solution by sodium sulphide in the presence of potassium cyanide, which are highly specific for lead (brown) and cadmium (yellow), have been widely used, but have now been largely superseded by the more general and less specific **heavy metals tests** of the European and United States Pharmacopoeias, which rely on sulphide precipitation in acid solution.

Table 1.4 Precipitation of heavy metal sulphides

Precipitated in the presence of mineral acid		Precipitated in the presence of a weak acid (acetic acid)	Precipitated by ammonia and hydrogen sulphide	
			KCN absent	KCN present
Ag, Hg, Pb, Bi, Cu, As, Sb, Sn, (Cd)	Qualitative analytical Groups I and II	Zn, Ni, Co	All the metals in the preceeding columns, except As, Sb, (Sn) *Note:* Al and Cr will also be precipitated as hydroxides	A large group of metals form complex salts with potassium cyanide, from which they are *not* precipitated as sulphides by hydrogen sulphide. The following give precipitates in the presence of potassium cyanide: Pb (black), Zn (white), Cd (yellow)

The British, European and United States Pharmacopoeias, however, still retain specific limit tests for **lead** where this has been established as a potential impurity in some compounds for long-term administration (*Chloroquine Phosphate* and *Sulphate*) and particularly in certain products for direct application to mucous membranes (*Sodium Cromoglycate*; Cromolyn Sodium) or large-dose administration by the intravenous route (*Dextrose*). A variety of test procedures are employed, ranging from precipitation with sodium sulphide to extraction with chloroform of the red complex which lead forms with dithizone (p.224) in cases where the parent substance would interfere with the standard method (*Ferrous Sulphate, Dextrose*). Atomic absorption spectroscopy is also used (*Lactose, Bismuth Subcarbonate*; Milk of Bismuth).

A specific test for **mercury** in a number of USP monographs for iron(II) salts employs a direct titration with dithizone in the presence of hydroxylamine and chloroform and comparison with a standard mercury solution.

The heavy metal tests of the European and United States Pharmacopoeias differ only in the actual precipitating reagent, which is either a freshly prepared saturated aqueous solution of H_2S (USP), or NaSH generated immediately prior to use by heating thioacetamide with sodium hydroxide solution (Pharm Eur). The tests are based on the formation of a brown colour due to precipitation of the metal sulphides at or about pH 3.5 in colloidal form stabilised by the presence of glycerol. The test solution is compared after a fixed time interval (2 min, Pharm Eur; 5 min, USP) with a standard preparation prepared in an identical manner with a Lead Standard Solution. Solutions are compared visually in matched, flat-bottomed tubes of specified dimension or volume, which are made of colourless, transparent, neutral glass. The liquid is examined by viewing vertically down through the solution against a white background.

A variety of test procedures are prescribed to enable the test to be applied to the widest possible range of compounds without interference from the parent substance. This is important as, not only may the colour produced be modified by the presence of the substance under test, but, in some cases, the drug substance itself may be precipitated or interfere with metal sulphide precipitation, as in the case of complexing agents such as *Disodium Edetate*. The procedures differ in the method of preparing the test solution. The general method is satisfactory for all substances which give clear solutions (*Acetic Acid, Citric Acid, Lactic Acid, Sodium Acid Phosphate*), if necessary after neutralisation, as in the case of water-insoluble substances soluble in sodium hydroxide (*Phenytoin Sodium, Probenecid* and sulphonamides). An alternative method widely used for organic compounds which do not readily yield clear aqueous solutions is applied to the sulphated ash obtained after ignition either with concentrated sulphuric acid alone (*Tolbutamide, Sodium Aminosalicylate*) or in admixture with magnesium sulphate (*Chlorthiazide, Hydroxyurea, Methylcellulose, Saccharin*). Other ignition methods using nitric acid followed by hydrogen peroxide or perchloric acid, nitric/sulphuric acid (*Glibencamide*) or magnesium oxide (*Levodopa, Methyldopa*) are used more rarely where the compounds are not amenable to the more general methods. Yet other compounds are submitted to the test in solutions requiring the use of such organic solvents as aqueous dioxan, aqueous acetone (*Paracetamol*), aqueous ethanol (*Sodium Salicyclate*) and methanolacetic acid (*Erythromycin*).

Standard test procedure for heavy metals

The test is complicated by the fact that many substances submitted to the test affect the colour of the colloidal metal sulphides produced. These effects are overcome by comparison of two solutions of the substance, test and standard, of which the former contains more of the sample than the latter. The standard solution, therefore, contains a small amount of sample in addition to the specified amount of lead which is normally at a concentration of either 1 or 2 ppm. The stringency of the test is varied from substance to substance, according to need, and is fixed by the weight of sample used in the preparation of the Test Solution.

Limit of heavy metals in lactic acid (Limit 10 ppm)

Lead Standard Solution (1 ppm) is a solution of lead(II) nitrate (0.000016 g $Pb(NO_3)_2$ equivalent to 0.00001 g Pb), containing a trace of nitric acid, in 10 ml.

Method Dissolve 5.0 g of sample in M NaOH (42 ml) and dilute to 50 ml with *water* (sample solution A). Prepare the test and standard solutions for colour comparison as follows:

	Test solution	Standard solution
(1)	Sample solution A (12 ml)	Sample solution A (2 ml)
(2)	–	Standard lead solution (1 ppm Pb; 10 ml)
(3)	Thioacetamide reagent (1.2 ml)	Thioacetamide reagent (1.2 ml)
(4)	Acetate buffer pH 3.5 (2 ml)	Acetate buffer pH 3.5 (2 ml)
(5)	Mix; allow to stand for 2 min.	Mix; allow to stand for 2 min.

Compare the colour of the two solutions. Any brown colour produced in the test solution should not be more intense than that of the standard solution.

As usually applied, the test simply records the fact that the sample merely complies or fails to comply with Official requirements for heavy metals, calculated as Pb. The test can be made more quantitative by observing the amount of Standard lead solution which must be added to the standard solution in order that the resulting colour shall equal that of the test solution. If, however, more than 15 ml of Lead standard solution is required, a smaller volume of Sample solution A must be taken. The following calculation demonstrates the way in which the Heavy Metal limit is quantified.

Calculation Lactic Acid is required to contain not more than 10 ppm of heavy metals, calculated as Pb.

Since the comparison is between 12 ml of sample solution A and 2 ml sample solution A plus 10 ml of standard lead solution (1 ppm Pb), if the colours are equal:

$$\text{10 ml (12 ml} - \text{2 ml) sample solution A} \equiv \text{10 ml standard lead solution (1 ppm Pb)}$$

$$\therefore \quad \text{1 g of sample} \equiv \text{0.000001 g Pb}$$

$$\therefore \quad \text{1 000 000 g of sample} \equiv \text{10 g Pb (i.e. 10 ppm).}$$

Iron

Two tests are widely used. One based on the formation of the red ferric thiocyanate after treatment with bromine or ammonium peroxydisulphate to ensure that all the iron is oxidised to the ferric (iron III) state, is official in the USP. The alternative test is based on the formation of a purple colour by reaction of the iron with mercaptoacetic acid in a solution buffered with ammonium citrate and comparison of the colour produced with a standard colour containing a known amount of iron. The purple colour is due to the formation of the coordination compound, ferrous mercaptoacetate, ferric iron being reduced to the ferrous state by the reagent:

$$2Fe^{3+} + 2HSCH_2.COOH \longrightarrow 2Fe^{2+} + HO.OC.CH_2S.S.CH_2.COOH + 2H^+$$

$Fe^{2+} + 2HSCH_2.COOH \longrightarrow$ 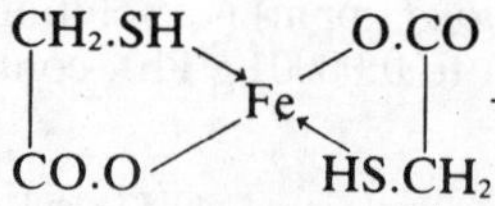 $+ 2H^+$

Limit of iron in Lactic Acid

Iron Standard Solution (1 ppm Fe) is an aqueous solution of ammonium iron(III) sulphate (0.01726 g), containing sulphuric acid in 2 l. It contains 0.01 mg Fe in 10 ml.

Method Dissolve 5.0 g of sample in M NaOH (42 ml) and dilute to 50 ml with *water* (Sample solution A). Prepare Test and Standard solutions for colour comparison as follows:

	Test solution	Standard solution
(1)	Sample solution A (10 ml)	
(2)	20% w/v aqueous citric acid (2 ml)	20 ml w/v aqueous citric acid (2 ml)
(3)	Add mercaptoacetic acid (0.1 ml)	Add mercaptoacetic acid (0.1 ml)
(4)	Mix; make alkaline with 10M ammonia and allow to stand for 5 min.	Mix; make alkaline with 10M ammonia and allow to stand for 5 min.

Compare the colour of the two solutions. The colour of the test solution should not be more intense than that of the standard solution.

Copper

Wool Alcohols may contain traces of copper arising from the use of copper catalysts during manufacture. **Copper** is controlled in the residue remaining on ignition by the specific reaction with sodium diethyldithiocarbamate.

Palladium, nickel and zinc

Traces of **palladium** in *Carbenicillin Sodium*, nickel in certain other compounds, and **zinc** in *Insulin* are controlled by the use of atomic absorption spectroscopy (Part 2).

Zinc

Zinc salts are used extensively in the dyestuffs industry to assist precipitation of organic dyes in a crystalline condition and, although zinc salts are not so used in the production of dyes for medicinal purposes, a test is always included to guard against accidental contamination in this way. The colour of the dye would mask any precipitation test, so the organic molecule is first destroyed by ignition with concentrated sulphuric acid. The residue which contains any zinc as sulphate is dissolved in dilute acid; the solution is made alkaline with ammonia and then treated with ammonium sulphide. Any zinc will appear as an opalescence or as a white precipitate. In water-soluble substances, such as *Sodium Sulphate*, zinc is controlled by precipitation as zinc hexacyanoferrate(II).

Alkaline earth and related impurities

Of these, **barium**, which is toxic, is a likely impurity, since barium salts are often used for the removal of large amounts of sulphate ions. Barium sulphate is insoluble in water and dilute acids and the metal is invariably tested for by the addition of dilute sulphuric acid. *Quinine Hydrochloride* and *Quinine Dihydrochloride*, which are prepared from the corresponding sulphate derivatives by treatment with barium chloride, may be contaminated by **barium** and this treatment is controlled by the addition of dilute sulphuric acid when no precipitate should be obtained.

Calcium sulphate is very much more soluble than barium sulphate and does not precipitate simultaneously with barium unless calcium impurity is present in very high concentration. Strontium is an unusual impurity, but would be precipitated under the conditions of the test for barium. It is, however, specifically controlled by atomic absorption spectrophotometry in *Calcium Acetate*, in which the levels of other ions are critical because of its use in the preparation of haemodialysis solutions. The test for **calcium** in *Sodium Sulphate* is precipitation as its insoluble oxalate by the action of ammonium oxalate in the presence of ethanol. This test is unsuitable for insoluble substances, such as *Magnesium Carbonate* and *Oxide*, and the sample is first dissolved in dilute acetic acid and ethanol then added. Whilst magnesium oxalate is readily soluble in dilute ethanol, calcium oxalate is quite insoluble and is, therefore, precipitated. The precipitation is compared with a standard calcium oxalate precipitate. In *Sodium Acid Phosphate,* **calcium** is controlled by precipitation with magnesium as phosphate in the presence of ammonia.

Magnesium

Magnesium impurity in calcium salts (*Calcium Carbonate, Calcium Chloride, Calcium Gluconate*) is controlled by precipitation as sulphate after removal of calcium as oxalate. Control of magnesium and other ions is more critical in *Calcium Acetate* and is effected by atomic absorption spectrophotometry. Formation of the magnesium 8-hydroxyquinoline (oxine) complex is also used as a test for magnesium in some monographs.

Alkali metals

In insoluble substances, such as *Magnesium Carbonate*, where control of alkali metals used in manufacture is important but not therapeutically critical, determination of **soluble substances** often suffices. In more criticial situations, such as *Calcium Acetate*, sodium and potassium are individually limited by atomic emission spectrophotometry. Wet chemical tests involving precipitation of potassium as its tetraphenylborate and **hexanitritocobaltate(III)** are also used for some compounds.

Ammonium salts

Ammonia is used in the preparation of a number of official organic substances and these may, therefore, be contaminated by ammonium salts. The standard test as applied to *Borax* and *Alum* is based on a comparison of the yellow colour produced by alkaline potassium mercuri-iodide against that formed by an Ammonium Standard Solution (1 ppm NH_3). Other substances such as *Dried Aluminium Hydroxide* which may be more heavily contaminated are treated quantitatively using the ammonia distillation method (p.162).

Aluminium

Aluminium can accumulate and reach toxic levels as a result of prolonged treatments with aluminium-containing materials. This is probably true of *Haemodialysis Solutions*, and stringent limits are imposed on such solutions to control impurity at levels not exceeding 15 ppm. Classical methods of analysis are not sufficiently sensitive, and control is effected by adding 8-hydroquinoline and measuring the fluorescence of the corresponding aluminium complex.

Arsenic

The test is a modification of the Gutzeit test in which all arsenic is converted into arsine (AsH_3) by reduction with zinc and hydrochloric acid. Reaction of the issuing gases with mercuric bromide paper produces a yellow stain, which can be compared with that produced from a known amount of arsenic. The apparatus used is of the type shown in Fig. 1.1. The capacity of the flask should be about 100 ml, and the dimensions of the tube are required to comply with certain definite specifications (length 200 mm, internal diameter 5 mm). The tube is open at both ends with a ground flanged surface at the upper end. A small hole in the side of the tube at the lower end is necessary to prevent condensed liquid from being forced up the tube by the pressure of hydrogen developed in the flask, thus blocking the tube. The tube is packed with cotton wool previously impregnated with lead acetate solution and dried. This serves to remove traces of hydrogen sulphide from the liberated gases, which would otherwise interfere with the test. A small extension tube of the same internal diameter and similarly flanged at one end is used to fix the mercuric bromide paper in position in such a way that all the arsine will pass through a circle of paper 5 mm in diameter. The two tubes with the mercuric bromide paper in place are held together by a spring clip.

Fig. 1.1 Arsenic limit test apparatus

Hydrogen gas is generated in the solution by the action

of stannated hydrochloric acid on arsine-free granulated zinc, which has been activated by treatment with chloroplatinic acid (Activated zinc). The presence of a small quantity of stannous chloride in the hydrochloric acid ensures a rapid reaction between the acid and the zinc and a steady evolution of hydrogen. Pure zinc is not very reactive toward hydrochloric acid and the presence of a small quantity of tin salt increases the reaction rate by the formation of localised spots of Sn/Zn electrolytic couples. Stannous chloride also acts as a reducing agent so that any pentavalent arsenic is reduced to the trivalent state. The hydrogen formed reduces any arsenic present to arsine, AsH_3, which is carried through the tube by a stream of hydrogen and out through the mercuric bromide paper. A reaction occurs between arsine and mercuric bromide which may be represented:

$$2AsH_3 + HgBr_2 \longrightarrow Hg(AsH_2)_2 + 2HBr$$

Other products, such as $AsH(HgBr)_2$, $As(HgBr)_3$ and As_2Hg_3, may also be formed. These reactions result in the formation of a yellow or brown stain on the mercuric bromide paper. Provided the diameter of the paper exposed to the issuing gases is constant, the depth of colour produced is proportional to the amount of arsenic present. Mercuric bromide paper becomes discoloured on exposure to light and should, therefore, be stored in the dark. Discoloured papers must not be used in the test. The test is comparative in that two solutions, one containing the sample under test and the other containing a known amount of arsenic, are submitted to the test at the same time. The stains are then compared by daylight.

The stringency with which the test is applied depends upon a number of factors, of which the dose of the substance under examination is probably the most important. Arsenic is a cumulative poison and substances likely to be administered in large doses, or repeatedly in small doses over a long period must be strictly controlled with respect to arsenic impurity. Other factors such as the origin of the medicament and the processes used in its manufacture will also have been taken into account in fixing the permitted limit for arsenic. In the test itself, stringency is determined by two factors, the weight of sample used in the preparation of the solution to be examined, and the volume of Standard arsenic (1 ppm As) solution used in the preparation of the standard stain. In practice, the latter is usually kept constant, comparison being made with what is known as a 1 ml Standard Stain. The stains fade on keeping and must be freshly prepared.

Limit of arsenic in Sodium Phosphate

Arsenic Standard Solution (1 ppm As) is a solution prepared from arsenic trioxide (1.32 g/l) in dilute aqueous hydroxide, further diluted (1/1000). It contains 0.001 mg As/ml.

Method The amount of substance to be examined and any preliminary treatment required vary considerably and are specified in each individual monograph. The general method is to dissolve the substance in a mixture of *water* and stannated hydrochloric acid. For the examination of *Sodium Phosphate*, prepare test and standard solutions as follows:

	Test solution	Standard solution
(1)	Dissolve 0.2 g of sample in *water* (25 ml) in the flask.	Dilute 1 ml of Arsenic standard solution (1 ppm) with *water* (25 ml) in the flask.

(2) Add hydrochloric acid (15 ml) and tin(II) chloride AsT to each solution.
(3) Add potassium iodide solution (M; 5 ml) to each solution and allow to stand for 15 min. This liberates hydriodic acid which assists in the reduction of pentavalent arsenic to the trivalent state from which arsine is then formed in the subsequent reduction at Step 4.
(4) Add activated zinc (5 g) to each flask, and immediately insert the tube already assembled with the mercuric bromide paper in position as shown in Fig. 1.1.
(5) Immerse both flasks in a water bath at a sufficient temperature to allow a uniform evolution of gas for 2 h.
(6) Compare the test and standard stains produced on the respective mercuric bromide papers.

The substance is said to comply with the requirements of the test if the colour of the test stain is not darker than that of the standard stain.

Calculation. Sodium Phosphate is required to contain not more than 5 ppm of As.

If the stains obtained from 0.2 g of Sample and 1 ml of arsenic standard solution are equal:

$$0.22\ \text{g of sample} \equiv 0.001\ \text{mg As}$$
$$\therefore\ 1\ \text{g of sample} \equiv 0.00005\ \text{g As}$$
$$\therefore\ 1\,000\,000\ \text{g of sample} \equiv 5\ \text{g As (i.e. 5 ppm)}$$

Modification of the general method for arsenic

Insoluble substances. In general, no special treatment is required for insoluble substances (*Light Kaolin* and *Magnesium Trisilicate*). Their insolubility does not interfere with the solution and reduction of arsenic to arsine. They are, therefore, simply suspended in water and treated with stannated hydrochloric acid by the general method. In *Barium Sulphate*, however, where the risk of arsenic contamination from the sulphuric acid used in its manufacture is much higher, the material is submitted to a preliminary digestion process with nitric acid to ensure all the arsenic is in solution.

Substances which evolve carbon dioxide with hydrochloric acid, or which react vigorously with hydrochloric acid. At this stage, arsenic is usually converted to arsenic trichloride which is volatile and, therefore, may be carried off with large volumes of carbon dioxide if these are produced at the same time. The use of brominated hydrochloric acid avoids this difficulty, the bromine oxidizing arsenic to the pentavalent state, in which it is non-volatile. *Chalk* and *Calcium Hydroxide* are treated in this way.

Metals which interfere with the normal reactions involved in the test. Iron will deposit on the surface of zinc, depressing the rate of reaction between the zinc and the acid.

(*a*) *Dried Ferrous Sulphate, Ferrous Succinate.* The sample is dissolved in water and stannated hydrochloric acid is added to convert all arsenic to the trivalent state as arsenic trichloride. Arsenic trichloride is volatile and may then be separated by distillation from the other metallic salts present and the distillate examined in the usual way.

(*b*) *Ferrous Fumarate.* Precipitation of fumaric acid on acidification is liable to cause frothing. The free acid is converted to its sodium salt with sodium carbonate and arsenic oxidised to the pentavalent state by evaporating the solution with bromine. The residue is gently ignited until carbonised to destroy organic matter, arsenic being retained as non-volatile sodium arsenate. The residue is dissolved in brominated hydrochloric acid and the test completed in the usual way. *Saccharin* and *Sodium Aminosalicylate* give rise to similar problems and are treated in the same way.

(*c*) *Antimony Sodium Tartrate.* Antimony compounds are reduced by zinc and hydrochloric acid to stibine (SbH_3) which reacts with the mercuric bromide paper to give a stain. The sample is distilled with hydrochloric acid to give a distillate containing all of the arsenic, but only a fraction of the antimony, as this is relatively non-volatile. A second distillation removes the last traces of antimony.

Potassium Nitrate. This substance produces an oxidising solution with hydrochloric acid and would yield oxides of nitrogen on the addition of zinc. Under these circumstances, no hydrogen would be produced for the reduction of arsenic. The sample is evaporated with concentrated sulphuric acid until the liquid commences to fume. Dilution with water decomposes nitrosylsulphuric acid which may have been formed, the decomposition being completed by a second evaporation until white fumes are again evolved. The usual test procedure may then be applied.

Sodium Metabisulphite. Acidification of this substance would give a precipitate of sulphur. To prevent this occurring it is oxidised to sodium sulphate by heating with potassium chlorate and hydrochloric acid. Most of the excess chlorine is removed by heating gently, and the final traces with stannous chloride.

Limit tests for acid radical impurities

The most common acid radical impurities are chloride and sulphate which generally arise from the use of tap water in manufacturing processes. Because of widespread contamination by these two impurities, the British Pharmacopoeia specifies general limit tests for them which are applicable, with minor modifications, to a larger number of medicinal substances.

The limit test for chlorides

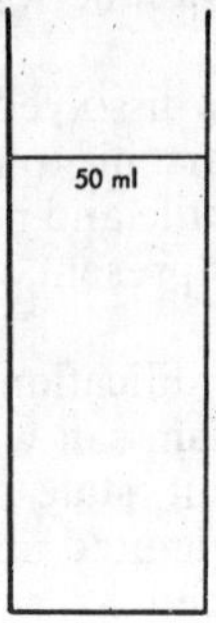

Fig. 1.2
Nessler Glass

This test, which is mainly used to control chloride impurity in inorganic substances, depends upon the precipitation of the chloride with silver nitrate in the presence of dilute nitric acid, and comparison of the opalescent solution so obtained with a standard opalescence containing a known amount of chloride ions. The opalescence in the two solutions is compared by examination in special vessels known as Nessler Glasses (Fig. 1.2; British Standard No. 612; 19), by viewing them transversely through the solution against a dark background. Borderline cases may often be decided by comparison over the printed page of a book, when the definition of the print is readily discernible and can be used as the criterion of the depth of opalescence.

Determination of Chloride in Sodium Sulphate

Chloride Standard Solution (5 ppm Cl) is prepared by diluting 1.9 ml of an 0.0824% w/v solution of sodium chloride with *water* to 100 ml

Method Prepare test and standard solutions as follows:

	Test solution	Standard solution
(1)	Dissolve 0.25 g of sample in *water* and dilute to 15 ml.	Dilute chloride standard solution (5 ppm Cl; 10 ml) with *water* (5 ml).

(2) Add 2M nitric acid (1 ml) to each solution.
(3) Add 0.1M silver nitrate (1 ml) to each solution.
(4) Stir immediately and allow to stand for 2 min protected from bright light.
(5) Compare the opalescence of the two solutions after 5 min. The opalescence of the test solution should not be greater than that of the standard.

Calculation *Sodium Sulphate* is required to contain not more than 200 ppm of chloride. If the opalescence of the test and standard solutions is the same, then:

$$0.25 \text{ g of sample} \equiv 0.00005 \text{ g Cl (10 ml of 5 ppm)}$$
$$\therefore \quad 1 \text{ g of sample} \equiv 0.0002 \text{ g Cl}$$
$$\therefore \quad 1\,000\,000 \text{ g sample} \equiv 200 \text{ g Cl (i.e. 200 ppm)}$$

The stringency of the test depends on the amount of material which is used. The greater the weight of sample specified, the more stringent does the test become. Apart from such variations as the weight of material examined, the test is always carried out as far as possible under the same conditions. Substances which react with nitric acid (*Magnesium Oxide* and *Carbonate, Calcium Hydroxide*) are dissolved in acetic acid. This solution is used in the general test procedure, so that the acidity of Test and Standard Solutions remain comparable.

Insoluble substances (*Light Kaolin, Magnesium Trisilicate*) are boiled with a mixture of water and dilute nitric acid, the solution filtered and the test applied to the filtrate. Insoluble organic substances (*Promethazine Theoclate, Propyl Gallate*) are merely shaken with cold water, filtered and the test applied to the filtrate. A similar modification is applied to metallic salts of aromatic acids (*Sodium Salicylate, Sodium Benzoate*) and also *Sodium Calciumedetate* after acidification with nitric acid. *Chlorbutol* is examined as a 5% solution in ethanol (95%), to suppress hydrolysis of the substance itself, which would yield chloride ions in aqueous solution.

Coloured substances, which would interfere, are specially treated before applying the usual test. Thus, *Potassium Permanganate* is decolorised by reduction with ethanol, filtered from the precipitated manganese dioxide and the test applied to the filtrate.

Chloride in Potassium Bromide

Chloride contamination of bromides can be appreciable and, in the case of *Potassium Bromide*, is determined in conjunction with the assay.

Chloride and bromide in Iodine

The necessity for this test arises because free iodine is sufficiently soluble in water to interfere with the general test. The difficulty is overcome by reducing the free iodine with zinc. Iodide is separated from chloride and bromide in the resulting solution by precipitation with silver nitrate in the presence of ammonia. Silver chloride and bromide are soluble. Acidification of the solution with nitric acid gives an opalescence due to chloride and bromide.

Chloride in Purified Water

A standard which closely approaches complete freedom from chloride impurity is required for *Purified Water*. A sample (10 ml) is tested with silver nitrate solution, acidified with nitric acid and should remain clear and colourless for 15 min. This test will also limit other acid radicals precipitated by silver nitrate in neutral solution.

Chloride contamination in organic compounds

The general limit test for chlorides is only directly applicable to those water-soluble organic compounds which are not precipitated under the conditions of the test. In a few cases (*Glycerol, Neostigmine Methylsulphate, Sodium Stibogluconate, Sulphobromophthalein Sodium*), a quantitative test depending upon a similar reaction with silver nitrate in the presence of nitric acid is used.

Special semiquantitative tests are used for substances which are heavily contaminated with chloride; *Liothyronine Sodium*, which precipitates in acid solution, is first dissolved in alkali and the solution is then acidified with nitric acid and titrated potentiometrically with silver nitrate solution.

Chloride in *Sodium Lauryl Sulphate* is controlled by Mohr titration (p.198) with silver nitrate after neutralisation with nitric acid.

The general test is not applicable to water-immiscible liquids. Chloride in *Trichloroethylene* and *Tetrachloroethylene*, and halides in *Halothane* and *Methoxyflurane*, are extracted by shaking the sample with freshly boiled and cooled water, separating and testing the aqueous layer with silver nitrate in the presence of dilute nitric acid. The tests are stringent and require that there shall be no opalescence.

Iodide contamination in organic compounds

Colorimetric methods involving oxidation of iodide to iodine and extraction of the latter into chloroform are applied to *Iodipamide Meglumine Injection, Iopanoic Acid, Liothyronine Sodium* and *Sodium Diatrizoate*.

The limit test for sulphates

This test is designed for the control of sulphate impurity primarily in inorganic substances. It depends upon the precipitation of the sulphate with barium chloride in the presence of hydrochloric acid and traces of barium sulphate. The latter, prepared *in situ* by precipitation from ethanolic potassium sulphate, assists rapid and complete precipitation by seeding. The opalescent solution so obtained is compared with a standard turbidity containing a known amount of sulphate ion.

Determination of sulphate in lactic acid

Ethanolic Sulphate Standard Solution (10 ppm SO_4) is prepared by diluting 1 ml of a 0.181% w/v solution of potassium sulphate in ethanol (30%) with ethanol (30%) to 100 ml.

Sulphate Standard Solution (100 ppm SO_4) is prepared by diluting 1 ml of a 0.181% w/v solution of potassium sulphate with *water* to 100 ml. It contains 0.00001 g SO_4/ml.

Method Dissolve 5.0 g of sample in M NaOH (42 ml) and dilute to 50 ml (sample solution A) Prepare the test and standard solutions as follows:

	Test solution	Standard solution
(1)	Barium chloride solution (25% w/v; 1 ml).	Barium chloride solution (25% w/v; 1 ml).
(2)	Add ethanolic sulphate standard solution (10 ppm SO_4; 1.5 ml) to each solution.	
(3)	Add sample solution A (7.5 ml) and *water* (7.5 ml).	Add sulphate standard solution (10 ppm SO_4; 15 ml).
(4)	Add 5M acetic acid (0.5 ml) to each solution and allow to stand for 5 min.	

Compare the opalescence of the two solutions. The opalescence of the test solution should be not greater than that of the standard solution.

Calculation *Lactic Acid* is required to contain not more than 200 ppm of SO_4.

If the degree of opalescence is the same in the test and standard solutions, then:

$$5 \times \frac{7.5}{50} \text{ g of sample} \equiv 0.75 \text{ g sample} \equiv 0.00015 \text{ g of } SO_4$$

$$\therefore \quad 1 \text{ g of sample} \equiv 0.0002 \text{ g of } SO_4$$

$$\therefore \quad 1\,000\,000 \text{ g of sample} \equiv 200 \text{ g of } SO_4 \text{ (i.e. 200 ppm)}.$$

The stringency of the tests is controlled in much the same way as in the limit test for chlorides. The acidity of the solution is controlled since the solubilities of barium sulphate precipitates are very much affected by the acid concentration.

Other acid radicals

Specific tests are also applied for a number of other acid radical contaminants.

Cyanide

Cyanide in *Disodium Edetate* is determined by titration with silver nitrate in neutral solution using an adsorption indicator.

Oxalate

Oxalate is a common impurity in organic acids and their salts, due to the use of oxalic acid to remove calcium during the manufacturing processes. Oxalic acid is also used in the isolation and purification of organic bases such as ephedrine, which form well defined crystalline oxalates. The standard reagent for the detection of these impurities is calcium chloride, with either ammonia or acetic acid, though colorimetric methods, such as that used, for example, with *Sodium Citrate* and *Sodium Acid Citrate*, are more sensitive.

Phosphate

A limit on **total phosphate** in *Sodium Phosphate* (^{32}P) *Injection* is imposed by means of a yellow colour reaction with ammonium vanadate and ammonium molybdate in the presence of perchloric acid. The exact composition of the molybdovanadophosphoric acid complex is uncertain. A similar test is used to control phosphate in *Tetracosactrin Zinc Injection*.

Silicate

Silicates and silica are insoluble in dilute hydrochloric acid and are usually limited by reference to solubility in this acid. They may also be limited, along with aluminium, iron and phosphate, by a precipitation test with ammonia. For example, *Chalk* is boiled with excess hydrochloric acid, ammonia added, the solution filtered and the residue ignited and weighed.

Limit tests for non-metallic impurities

Boron

Boron, which is capable of causing unwanted skin reactions, and is toxic to nervous tissue and muscle, is rigorously limited in compounds such as *Salbutamol*, where it arises as a result of the use of sodium borohydride in manufacture. Boron is converted to borate and organic matter destroyed by fusion with sodium and potassium carbonates. The boron is then determined colorimetrically.

Free Halogen

Some products containing organic iodo-compounds are liable to release traces of free iodine. This is controlled in *Iophendylate Injection* by measurement of the light absorbance at 485 nm due to the element. *Iodised Oil Fluid Injection* is shaken with starch mucilage containing cadmium iodide to extract iodine into the aqueous phase; a limit is placed on the intensity of the blue starch iodine complex formed. A similar test is used to control free chlorine in *Chloroform* and *Tetrachloroethylene* by displacement of iodine from cadmium iodide.

The control of organic impurity in organic medicinal substances

Physical methods

Contamination of organic medicinal substances occurs in exactly the same way as for inorganic materials. The wide range of chemical types, and the even more varied nature of the contaminating impurities covered by official substances, make the design of general tests for organic impurities difficult. Considerable reliance is now placed on the use of physical separation methods, particularly thin-layer chromatography, and on the measurement of appropriate physical characteristics, in combination with specific chemical tests designed to limit particular likely impurities.

The scope of such methods is set out in the following text. The fundamental principles and detailed applications of these and other physical methods of analysis are examained in detail in Part 2.

Solubility

Foreign alkaloids in *Pilocarpine* are controlled by the addition of 6M ammonia to an aqueous solution. Pilocarpine is not precipitated from aqueous solution at the concentration used in the test, whereas other foreign alkaloids are insoluble. Similarly, a limit of phenylbarbituric acid is imposed on *Phenobarbitone* by the requirement that a sample 20% w/v in boiling ethanol 96%) should give a clear solution. **Lumiflavine** in *Riboflavine* is controlled by its solubility in chloroform, the latter being virtually insoluble.

pH of Solution

Control of acidity and/or alkalinity provides a useful method of ensuring identity, stability and freedom from contamination. Water-soluble substances which are naturally acidic or alkaline usually have limits of **acidity** or **alkalinity** which are expressed as a pH range for a solution of specified concentration. Thus, the limits of acidity for *Adrenaline Acid Tartrate* (Epinephrine Bitartrate) ensure that it is essentially the acid tartrate. Likewise, limits of alkalinity for *Aminophylline Injection* help to control the amount of ethylenediamine which may be present. *Water for Injections*, which should have pH 7.0, has strict limits on both acidity or alkalinity which are quantified by titration with standard sodium hydroxide and hydrochloric acid, using phenol red as indicator.

Thin-layer chromotography

The speed and ease with which thin-layer chromatographic separations can be made and the high sensitivity of the method make this technique a particularly attractive one as the basis for a rapidly growing number of important specific limit tests. It is used in the control of **related substances and decomposition products** in *Chlordiazepoxide Hydrochloride*. The sensitivity of such tests as that for **free prednisolone** in *Prednisolone Sodium Phosphate* and **related foreign steroids** in *Dexamethasone, Hydrocortisone* and *Prednisolone* is enhanced by the use of alkaline triphenyltetrazolium chloride for colour development, which is highly specific for the C-17 α-ketol side-chain. Other important applications of thin-layer chromatography include the control of **4-chlorophenol** in *Dichlorophen*, other cinchona alkaloids in quinine and quinidine salts and **methimazole** in *Carbimazole*.

Problems arise where, for one reason or another, reference samples of the expected impurity are not available. In such cases, the usual device is to use two or more dilutions of the substance under examination in one of which the concentration is only a small fraction (0.5, 1.0 or 2.0%) of that in the other. Secondary spots observed in the chromatogram of the higher dilution are then required to be not more intense than the main spot due to the parent compound in the chromatogram of the lower dilution. This method has some disadvantages. Thus, impurity spots, which are referred in terms of intensity to those of the dilutions, may not only have different sensitivities to the visualising agent, but also may well appear at considerable distances from the reference spot, increasing the difficulties of comparing their relative intensities.

Thin-layer chromatography is sufficiently sensitive to provide a means of controlling impurity arising from epimers in optically active compounds.

Thus, it is used to control the inactive 4-*epi*-**tetracycline** as well as the closely related 4-*epi*-**anhydrotetracycline** in *Tetracycline Hydrochloride*. It has also proved valuable in providing a satisfactory method for limiting closely related, but unwanted, derivatives in *Pethidine Hydrochloride*, which otherwise were undetected, even when present in fairly substantial amounts.

Paper chromatography still finds some use in particular limit tests which do not yield to TLC techniques, such as that for the control of retinol in *Vitamin A Ester Concentrate*, and for controlling the composition of *Capreomycin Sulphate* ($\nless$90% Capreomycin I).

Gas–liquid chromatography

Gas–liquid chromatography is widely used in the control analysis of pharmaceutical chemicals, wherever its greater sensitivity and efficiency are demanded. A typical example of its use arises in the control of *N*-ethyl-α-methyl-4-trifluoromethylphenylethylamine in *Fenfluramine Hydrochloride*. The free fenfluramine base is isolated by basification and extraction with chloroform, and chromatographed on a suitable column using authentic fenfluramine hydrochloride for comparison against *N,N*-diethylaniline as internal standard. Similar GLC procedures are used both as assay and to control the same impurity in *Fenfluramine Tablets*.

Gas–liquid chromatography provides an even more sensitive method than thin-layer chromatography for separating and distinguishing between closely related materials, as, for example, the *cis*- and *trans*-isomers of *Tranylcypromine Sulphate*, in which it is used to limit the unwanted *cis*-isomer and cinnamylamine.

High pressure liquid chromatography

The advent of high pressure liquid chromatography has provided a further refinement of chromatographic analytical methods available for impurity and isomer control of pharmaceuticals which, because of instability or lack of volatility, may be unsuitable for gas–liquid chromatography.

Typical examples of the application of this method are seen in the control of z-isomer mix in *Clomiphene Citrate* and of *cis*-isomer in *Tranexamic Acid*.

High pressure liquid chromatography is now widely used in the control of a number of complex natural materials, such as blood products, antibiotics and hormones. For example, it is used as a specific assay of oxytetracycline in *Oxytetracycline Calcium*, thereby reducing the need for separate control of related substances. Similarly, it is used in control of the composition of antibiotics such as *Gentamicin* which is a mixture of several components (C_1, C_{1a}, C_2 and C_{2a}) of varying toxicity and potency. It is also used to ensure the homogeneity of pork and beef *Insulins*, to limit **liothyronine** and **thyroxine** in *Calcitonin (Pork)* and to determine the content of tetracosactrin peptide in *Tetracosactrin Acetate*.

Electrophoresis

Both paper and gel electrophoresis are used in certain limit tests, where TLC tests are ineffective. Paper electrophoresis in a formic acid–glacial acetic acid–acetone system is used to control **related substances** in *Cephaloridine* in a test which uses crystal violet to visualise the mobility of the test and control spots and cyanocobalamin to indicate the position of the true baseline. The behaviour of the sample under test is compared with that of authentic cephaloridine at 1% and 0.1% levels. Electrophoresis on polyacrylamide gel is used to control **related peptides** in *Glucagon*.

Gel permeation chromatography

The molecular fractionation which can be achieved with gels prepared from cross-linked dextrans is used to control **uniformity of molecular size** and, hence, homogeneity in the peptide hormone *Calcitonin (Pork)*. The product is required to elute under carefully controlled conditions in a single band with a defined relative elution volume in relation to two markers, bovine serum albumin and tyrosine. Gel filtration is also used as a test for denatured protein in *Albumin*. A dextran gel for fractionating proteins of molecular weight 5000 to 150 000 Daltons is used which will completely exclude denatured material (MW > 400 000 Daltons).

Gel permeation chromatography also provides a much more precise method than viscosity measurements for control of the average and range of molecular weight of polymers used in the preparation of *Dextran Intravenous Infusions*. This is particularly important in *Dextran 40*, which, because of its use in the reduction of blood viscosity, requires to be relatively free of the higher molecular weight fractions primarily used as blood expanders.

High performance size exclusion chromatography operated at pressures of 1000 p.s.i. is more effective and some twenty times as rapid as simple gel permeation chromatography and, hence, is set to replace the latter in many procedures.

Absorptiometric methods

A number of important tests limiting organic impurity in organic medicinal substances are based on the development of specific colours and comparing them photoelectrically with standard solutions. The Beer-Lambert Law, which applies to such measurements, gives the following relationships:

$$A_{\text{test}} = \log_{10} \frac{I_0}{I_{\text{test}}} = kc_1t_1$$

$$A_{\text{standard}} = \log_{10} \frac{I_0}{I_{\text{standard}}} = kc_2t_2$$

so that

$$\frac{A_{\text{test}}}{A_{\text{standard}}} = \frac{c_1t_1}{c_2t_2}$$

In photoelectric instruments, the thickness t is constant and, therefore,

$$\frac{c_1}{c_2} = \frac{A_{\text{test}}}{A_{\text{standard}}}$$

where the extinctions are measured with reference to a blank of reagents. For routine work, a calibration curve can be used so that, even though Beer's Law may not be obeyed, a particular method can still be of use.

The method used for the development of specific colours varies in both method and detail of procedure from compound to compound. The test for salicyclic acid in *Aspirin* by measurement of the violet colour which iron(III) chloride forms with the phenolic hydroxyl group of the impurity is a typical example of the method. A further example is the test for noradrenaline in *Adrenaline Acid Tartrate* (Epinephrine Bitartrate), in which the primary amino function of the impurity is condensed with sodium 1,2-naphthoquinone-4-sulphonate in an alkaline buffer:

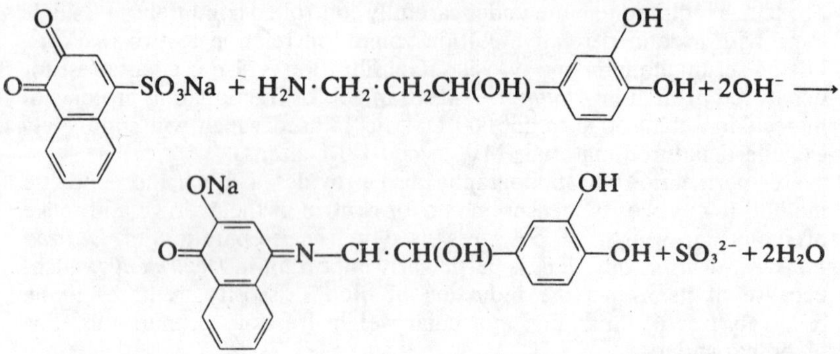

The coloured complex is extracted with toluene, in the presence of benzalkonium chloride. The colour of the organic phase is compared with that obtained by treating a known amount of noradrenaline-free adrenaline acid tartrate in the same way. The excess reagent remains in the aqueous phase and does not interfere.

Similarly, sulphathiazole and related aromatic amines in *Phthalylsulphathiazole* and *Succinylsulphathiazole* are controlled by diazotisation of the free primary amino group, coupling with *N*-(1-naphthyl)ethane-1,2-diammonium dichloride and measuring the intensity of the resulting colour against that of an appropriate standard. Chloroaniline in *Proguanil Hydrochloride* is controlled by a similar test.

Ultraviolet light absorption

Measurement of light absorption characteristics, particularly the wavelength and relative intensities of absorption maxima (or minima), provides a useful method of controlling related impurities. Thus, *Cycloserine* is required to show a single maximum at 219 nm with the absorbance lying between narrow limits. A separate light absorption measurement in alkaline solution at 219 nm provides for a limit of condensation products. *Demeclocycline*, likewise, is required to show a single absorption maximum in alkaline solution at 385 nm within closely defined limits. In other instances, such as *Cyanocobalamin*, precisely defined light absorption requirements form the basis of identification and freedom from absorption due to impurities.

Infrared spectroscopy

Infrared spectroscopy is used as a limit test for the biologically inactive **polymorph A** in *Chloramphenicol Palmitate Mixture*.

Chemical methods

Specific types of organic impurities are often controlled by specific or semi-specific functional group reagents in much the same way in different organic medicinal substances. Some of the more widely used tests are grouped together in the following sections.

Aldehydes, ketones, sugars and unspecified reducing substances (oxidisable matter)

Tests used to control contamination by aldehydes, ketones and similar reducing substances vary widely in their specificity and sensitivity. The highly sensitive alkaline potassium mercuri-iodide, which gives a yellow colour or precipitate with aldehydes and ketones, is applied to the anaesthetics *Chloroform* and *Ether*, whilst aldehydic substances in *Acetic Acid* are determined titrimetrically with iodine after treatment with sodium metabisulphite. **Aldehydes and ketones** in *Isopropyl Alcohol* and **acetaldehyde** in *Paraldehyde* are limited by adding hydroxyammonium chloride ($NH_2OH \cdot HCl$) and titrating the acid released (p.143). Fehling's Solution gives a red precipitate (Cu_2O) in the presence of **reducing sugars** and is used to control contamination by such substances (presumably lactose) in *Lactic Acid*.

Several monographs make miscellaneous provisions to guard against the presence of **reducing substances** and **oxidisable matter**. Decolorization of potassium permanganate is a suitable test similar to that used for **oxidisable matter** in *Purified Water*, provided that the substance to which the test is applied is stable to the reagent, as, for example, *Nikethamide*. However, the very much more sensitive ammoniacal silver nitrate is used to control **reducing substances** in the anaesthetic *Nitrous Oxide*, the sensititivity being enhanced by the requirement that **no** cloudiness or darkening be observed on passing a litre of gas through the reagent solution. The same reagent is used to control reducing substances in *Glycerol*; the test, however, is much less sensitive; a brown or grey colour may be formed, though neither a precipitate nor a silver mirror should be produced.

Where a less sensitive test is required, potassium dichromate and sulphuric acid followed by addition of potassium iodide is used, as in the control of **formic acid and oxidisable impurities** in *Acetic Acid*. In these tests, a limited amount of potassium dichromate is added and the test solution allowed to stand for a specified time. Iodine liberated from unused potassium dichromate by the addition of potassium iodide is titrated with standard sodium thiosulphate solution.

Free bases in neutral compounds

Pyridine in *Cetylpyridinium Chloride* and **ammonia** in *Benzalkonium Chloride Solution*, which are instantly recognised by odour, are controlled by an olfactory test, after treating the drugs with sodium hydroxide solution. **Free amines** in *Crotamiton* are limited by dissolving in ether, extracting with hydrochloric acid and evaporating the acid extract. The weight of residue is limited.

Methanol in Solvent Ether

Such contamination is likely to arise in practice through the use of Industrial Methylated Spirits in place of ethanol during manufacture. The test, which is used to limit **methanol** in *Solvent Ether*, depends upon a selective oxidation of methanol to formaldehyde by potassium permanganate in the presence of phosphoric acid. Excess potassium permanganate is destroyed by the addition of a mixture of oxalic acid and sulphuric acid. Formaldehyde so produced can be detected by the addition of Schiff's reagent to give a pink colour. Thirty minutes is allowed for development of the colour with this reagent, which normally gives a rapid reaction, except in very dilute solutions. The absence of colour after this time has elapsed, therefore, gives a strict control over possible contamination. Ether is not miscible with potassium permanganate solution and is shaken in a separator with 10% ethanol to transfer any methanol to the aqueous phase.

In the determination of alcohol in galenicals in the preparation of which Industrial Methylated Spirits has been used, the distillate obtained must comply not only with the necessary limits for refractive index and specific gravity, but also with a GLC test for methanol, which must not be present in greater proportion than that permitted in the spirit which was used.

Neutral and basic substances in Barbiturates

Neutral and basic substances in *Amylobarbitone* and a number of other barbiturates are controlled by dissolving in aqueous sodium hydroxide, extracting with ether and limiting the weight of residue obtained from the ether on evaporation.

Contamination by organic halogen compounds

Combined halogen is usually converted to halide ion before examination in the usual way. **Halides** in *Dimercaprol* would represent contamination by 1,2-dibromopropanol, an intermediate in its synthesis. The bromine is obtained in an ionised form by refluxing the sample with ethanolic potassium hydroxide. Oxidation of the dimercaprol to the corresponding disulphonic acid with hydrogen peroxide prevents any interference in the subsequent Volhard titration for bromide. **Chlorinated compounds** (dichlorodiethyl ether) in *Vinyl Ether* are converted to chloride by refluxing with sodium in amyl alcohol and determined by Volhard's method. **Halogen-containing substances** (1,3-dibromopropane and 1-bromo-3-chloropropane) in *Cyclopropane*, arising from the processes of manufacture, are decomposed by oxidation over platinised quartz, and then examined by a method similar to the general limit test for chloride, using a standard solution prepared from potassium bromide.

Contamination by peroxides

Peroxides may be present in both *Solvent* and *Anaesthetic Ether* as a result of light-catalysed air oxidation. Peroxides are toxic and, moreover, can give rise to mixtures which are explosive when distilled. This danger can be largely removed by distillation in the presence of a small quantity of ferrous sulphate. Removal of peroxide impurity during preparation does not, however, guarantee freedom from peroxides after a period of storage and careful tests are essential to guard against the harmful effects of these impurities. The test is applied by shaking solvent ether with twice its volume of potassium iodide in a stoppered tube. Air must be completely excluded from the tube by filling it to the brim. After standing for 30 min in the dark the aqueous phase must not be more yellow, due to liberation of iodine from the potassium iodide (by peroxides), than the same volume of aqueous potassium iodide containing 0.5 ml of 0.001M iodine. A more sensitive test is obtained for *Anaesthetic Ether* by shaking under similar conditions with aqueous potassium iodide containing starch mucilage. Iodine liberated must not impart a brown colour to the solution.

Ethyl Oleate, which is used as a solvent for injections of certain steroid hormones, may contain toxic **peroxides** formed by air oxidation at the double bond of the oleate radical. The reaction with potassium iodide must be carried out in a mixed solvent of chloroform and glacial acetic acid, since ethyl oleate is insoluble in water. Under these conditions, peroxides liberate iodine from the potassium iodide, which, after dilution with water, can be titrated with standard sodium thiosulphate solution. A similar test is used in the heavily unsaturated *Synthetic Vitamin A Concentrate*.

Related substances in alkaloids, glycosides and synthetic compounds

Plants usually produce not a single alkaloid or glycoside, but rather a whole group of chemically related substances. The relationship is often quite close, e.g. quinine and the related cinchona alkaloids; morphine and certain opium alkaloids. Consequently, separation may be a tedious and lengthy process. Contamination by other alkaloids or glycosides is, therefore, something which is extremely difficult to exclude. Nevertheless, it must be kept under control and most monographs contain some provision for the exclusion (or limitation) of such impurity.

Likewise, synthetic organic medicinal compounds are liable to be contaminated by trace impurities arising from starting materials, from intermediates, or from decomposition products. A small selection of typical chemical tests for the control of related substances in such compounds is given in Table 1.5.

Unsaturated substances

Potassium permanganate is also used to test for **cinnamylcocaine** in *Cocaine* and *Cocaine Hydrochloride*. Cinnamylcocaine is oxidised at the carbon-carbon double bond by the reagent giving first benzaldehyde and then benzoic acid. Cocaine itself is stable towards potassium permanganate.

A much more delicate test is necessary in the control of **unsaturated substances** in *Cyclopropane*, which is used as a general anaesthetic. The gas is passed through a specified volume of a solution of iodine monochloride (in glacial acetic acid) and then through a solution of potassium iodide. The two solutions are mixed with the result that unreacted iodine monochloride gives free iodine which can be titrated with 0.1M sodium thiosulphate. Iodine monochloride adds readily to organic substances which contain a carbon-carbon double bond in the molecule.

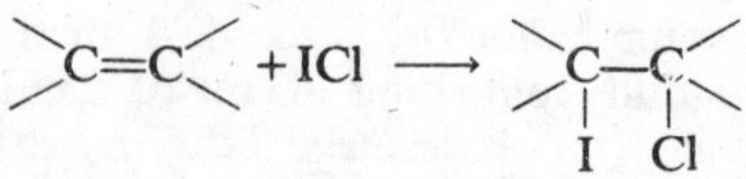

Acetylenic compounds in *Trichlorethylene* and *Tetrachloroethylene* are limited by shaking the sample with an ammoniacal solution of copper nitrate and hydroxylamine. The latter acts as a reducing agent to give cuprous salts, which would then precipitate acetylene in the form of red cuprous acetylide.

Control of fixed oils, fats and waxes

The close similarity between the physical constants (refractive index and wt/ml) for fixed oils in common use lays the more expensive ones open to sophistication by admixture with cheaper oils. Adulteration of this kind is not always readily detectable by chemical determination of acid value (p.146), saponification value (p.158), iodine value (p.188) or, where applicable, ester value (p.159), ratio number (p.160) and unsaponifiable matter (p.240). Appropriate combinations of physical tests are, therefore, required, together with a number of specific tests for the exclusion or limitation of particular oils which may be present as contaminants or adulterants. The more important of these tests are described below.

Table 1.5 Some chemical tests for related substances

Official Substance	*Related Substance*	*Test*	*Remarks*
Carbimazole	Methimazole	Thin-layer chromatography in comparison with authentic methimazole	
Codeine Phosphate	Morphine	A solution in dilute hydrochloric acid treated with sodium nitrite followed by ammonia should give a yellow colour not deeper than that similarly obtained from a stipulated amount of morphine	Codeine does not give a yellow colour under these conditions
Digitoxin	Digitonin	A solution in ethanol (95%) does not give a precipitate with an ethanolic solution of cholesterol	Digitonin forms an insoluble complex with 3-β-hydroxysteroids such as cholesterol
Homatropine Hydrobromide	Atropine, hyoscyamine and hyoscine	A small quantity treated with fuming nitric acid and evaporated to dryness on a water bath does not give a violet colour on further treatment with acetone and methanolic potassium hydroxide	The test is a specific colour reaction for atropine and a few closely related compounds, and homatropine does not interfere. A positive reaction is associated with the tropic acid portion of the molecule
Mepacrine Hydrochloride	3-Chloro-7-methoxyacridone	A sample extracted with anaesthetic ether shows only a limited fluorescence	3-Chloro-7-methoxyacridone may be produced by decomposition of mepacrine. The former is soluble in ether whereas the latter, being present as a salt, is insoluble
Morphine Sulphate	Other alkaloids	A solution in sodium hydroxide yields only a limited residue to chloroform	Morphine, which is a phenolic alkaloid, gives a water-soluble sodium derivative; other non-phenolic alkaloids (e.g. codeine) are extractable by chloroform

Test for the absence of arachis oil in other oils

This test depends upon the very much lower solubility and higher melting point of arachidic acid, the product of hydrolysis of arachis oil, than of the acids obtained from the oil under examination (*Almond Oil*). The conditions of experiment must be adhered to rigidly. The oil is boiled with 1.5M potassium hydroxide under reflux for 10 min and then 70% ethanol containing concentrated hydrochloric acid added. With a thermometer immersed in the liquid, it is cooled slowly with continuous stirring (rate of cooling 1°/min). No turbidity should appear above 4° for *Almond Oil* or above 9° for *Olive Oil*. Should a precipitate be obtained, further detailed tests may be applied, involving multiple recrystallization of the precipitate, which should not melt above a stated temperature.

Test for the absence of cottonseed oil in other oils

Mix equal volumes of the oil, amyl alcohol and carbon disulphide containing 1% of precipitated sulphur. Place in a glass-stoppered boiling tube and tie in the stopper. Immerse in a boiling water-bath to one third of its depth and heat for 30 min. There should be no pink or red colour. It is best to carry out the test on an authentic specimen at the same time. The test is not sensitive to much less than 10% of cottonseed oil in the sample. It is applied to *Almond* and *Arachis Oils*.

Test for the absence of sesame oil in other oils

Mix 2 ml of oil with 1 ml of concentrated hydrochloric acid containing 1% w/v of sucrose and set aside for 5 min. The acid layer should not be coloured pink. The test is sensitive only for concentrations greater than about 5% and is applied to *Almond* and *Arachis Oils*.

Tests specifically applied to individual oils

A number of other, less general tests are also applied in the same way, similarly to guard against adulteration of particular individual oils.

Almond Oil. Tests are included to guard against adulteration with cheap oils such as **apricot-kernel and peach-kernel oils**. Shake 5 ml of oil vigorously with a mixture (freshly prepared) of equal parts by weight of sulphuric acid, fuming nitric acid and water, keeping cool. The whitish mixture should show no pink colour after 15 min.

Cetostearyl Alcohol. Hydrocarbons are limited by chromatographing a solution in light petroleum (b.p. 40–60°) on a column of alumina under specified conditions and weighing the residue obtained from the first portion of eluate.

Oleic Acid. Tests are included to exclude adulteration by **mineral acids, neutral fats** and **mineral oils**.

Mineral acids. On shaking with water, the aqueous phase, after filtration, is not acid to methyl orange.

Neutral fats and mineral oils. Boil 1 ml with 0.5M Na_2CO_3 (5 ml) and water (25 ml); oleic acid is soluble. The solution, while hot, is clear or, at the most, opalescent. Neutral fats and mineral oils are insoluble in dilute aqueous sodium carbonate.

Congealing point. Dry a sample by heating at 110°, with constant stirring. Cool a small sample in a tube (20 mm in diameter), immerse in a suitable water-bath at 15° and cool steadily at a rate of 2°/min, stirring continuously. It does not become cloudy above 10° and congeals to a white solid mass at about 4°.

Undecanoic Acid. A limit of **fixed** and **mineral oils** is imposed by the requirement that a sample boiled with aqueous sodium carbonate should give a clear solution whilst hot.

Wool Fat. Paraffins are limited by chromatography in light petroleum on an alumina column and elution of paraffin hydrocarbons in the same solvent.

2 Registration and assessment of medicines

Introduction

Legislation and its implementation in the United Kingdom

Under the United Kingdom Medicines Act 1968, **Product Licences** are required for the manufacture of all medicines, and for surgical and dental materials containing them, for sale or supply within the United Kingdom, for importation and, in some cases, for export. Similar provisions relate to the issue of **clinical trial certificates** for the trial of medicinal products by doctors and dentists in human patients, and to the issue of **Animal Test Certificates** for comparable trials of veterinary products.

All licences and certificates are issued by the appropriate licensing authority. This is the Department of Health and Social Security Medicines Division, for the issue of product licences for human medicines and clinical trial certificates, and the Ministry of Agriculture, Fisheries and Food for veterinary product licences and animal test certificates. The processing of applications for licences and certificates is conducted through advisory committees. The organisational relationship of these committees and of the parallel committees concerned with published pharmacopoeial standards for medicines is shown in Fig. 2.1. Product licences for human medicines and clinical trial certificates are issued by the licensing authority, normally on the recommendations of the **Committee on Safety of Medicines** (CSM), whose task it is to advise on the **safety, quality** and **efficacy** of the products concerned. Additionally, the Committee on Safety of Medicines is also charged with the collection, investigation and dissemination of information on **adverse reactions** of drugs and medicines, so that effective advice on matters of safety may be given.

Under Section 20 of the Medicines Act, the licensing authority, before refusing to issue a product licence or clinical trial certificate, must consult the appropriate committee. In practice, therefore, all applications are referred to the Committee on Safety of Medicines for its advice. Before offering its advice, the Committee, acting through the licensing authority under Section 44 of the Act, may request the applicant to supply further information. After considering all the relevant information, the Committee may advise that the licence or certificate be issued. If, however, the Committee has in mind to advise a refusal, or wishes to place substantial conditions under which a licence may be issued, then, under Section 21 of the Act, the applicant has the right to be heard and/or to make representations in writing to the Committee. In the event of a refusal being sustained after a hearing and/or the submission of written representations,

Fig. 2.1 Organizational chart of Licensing and Standards Authorities and Committees

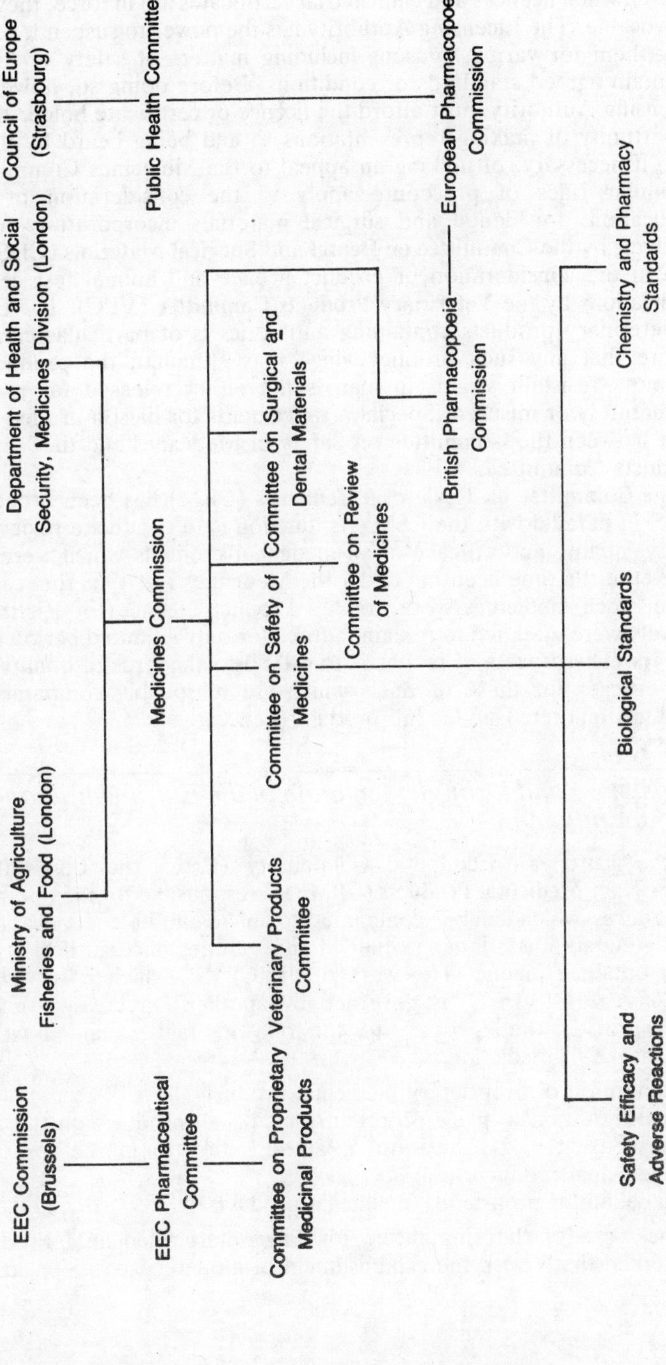

the applicant has the right of appeal to the **Medicines Commission** against the decision.

Even when licences and clinical trial certificates are in force, they are not irrevocable. The Licensing Authority has the power to suspend, revoke or vary them for various reasons including matters of safety or failure to maintain agreed standards or conditions. Before doing so, however, the Licensing Authority must afford the licence or certificate holder the same opportunity of making representations to and being heard by the CSM and, if necessary, of making an appeal to the Medicines Commission.

Similar rules of procedure apply to the consideration of licence applications for dental and surgical materials incorporating medicinal products by the **Committee on Dental and Surgical Materials** (CDSM) and also in the consideration of product licence and animal test certificate applications by the **Veterinary Products Committee** (VPC). The licensing of veterinary products containing antibiotics is of particular concern to ensure that any such product which may stimulate the production of antibiotic-resistant strains in man is not either released for marketing prematurely or misused. Special arrangements for liaison in such matters exist between the Committee on Safety of Medicines and the Veterinary Products Committee.

The Committee on Review of Medicines (CRM) has been brought into effect in parallel with the CSM. Its function is to conduct a review of the safety, quality and efficacy of all medicinal products which were on the market at the time licensing under the Medicines Act 1968 first came into force. Such medicines were granted Product Licences of Right, which initially were intended to remain in force for only a limited period of time, 5 years. The process of review is to establish standards of quality, safety and efficacy for these products which are reasonably comparable with products marketed under full product licences.

Legislation and its implementation in the European Economic Community

In the European Economic Community (EEC) the **Committee for Proprietary Medicinal Products** (CPMP) is responsible for the operation of procedures which enable companies to make multiple applications for marketing authorisations in other Member States once authorisation has been obtained in one. The work of the CPMP, which is set up under Directive 75/319/EEC, is governed by another Directive, 65/65/EEC, which provides the legal basis for the following matters falling within the Committee's jurisdiction:

(a) definition of proprietary medicinal products
(b) authorisation to place proprietary medicinal products on the market
(c) suspension and revocation of authorisation to market proprietary medicinal products
(d) labelling of proprietary medicinal products.

The work of the Committee for Proprietary Medicinal Products is concerned firstly with the establishment of an agreed basis for achieving comparable standards of quality, safety and efficacy in Member States and, secondly, with the establishment and implementation of procedures whereby a manufacturer who has received marketing approval in one country can apply for similar approval in at least five others. The general basis for comparable standards rests in the so-called 'norms and protocols' directive (75/318/EEC) which specifies in general terms the essential items of information which should normally be included in an application to market a medicinal product. These are set out in the Annex to the Directive covering:

Part 1 Physico-chemical, biological or microbiological tests:
(a) Qualitative and quantitative particulars of the constituents
(b) Description of method of preparation
(c) Control of starting materials
(d) Control of tests carried out at an intermediate stage of the manufacturing process
(e) Control tests on the finished product
(f) Stability tests.

Part 2 Toxicological and pharmacological tests:
(a) Objectives
(b) Toxicity
(c) Fetal toxicity
(d) Examination of reproductive function
(e) Carcinogenicity
(f) Pharmacodynamics
(g) Pharmacokinetics
(h) Products for topical use
(i) Presentation of particulars and documents.

Part 3 Clinical trials:
(a) Conduct of trials
(b) Presentation of particulars and documents
(c) Examination of applications for authorisation.

Notwithstanding the detail in each of these sections, the norms and protocols directive only gives general guidance. This has, therefore, been amplified in **notes for guidance** to applicants which have been drawn up by the CPMP to give more specific advice on the nature of the information that manufacturers should supply. Member States are still free to produce their own notes for guidance, within the general terms of the norms and protocols directive, but these must incorporate the notes for guidance on specific topics, produced by the CPMP, which require a wide concensus of opinion. Current CPMP notes for guidance include:

(a) efficacy requirements for drugs intended for long-term use, e.g. non-steroidal anti-inflammatory agents, antihypertensives, anticonvulsants, diuretics, anti-arrhythmics, oral contraceptives
(b) safety requirements for drugs intended for long-term use covering:
 (i) carcinogenicity testing
 (ii) reproduction studies

(iii) chronic toxicity studies
(iv) mutagenicity testing.

The directive (75/319/EEC) which sets up the CPMP also makes a number of other important provisions. Thus, it lays down that marketing applications are drawn up and signed by experts with the necessary technical or professional qualifications, defines the basis of their qualifications and places certain obligations on such experts. It sets out a basis for the verification of particulars and quality control procedures, makes regulations for importation from non-Member States, and provides for the inspection of manufacturing premises.

Specific time limits for the processing of marketing applications are set down in Article 7 of Directive 65/65/EEC. These require that applications be processed within 120 days, but allow for an extension for a further 90 days in exceptional cases. If, however, the Licensing Authority requests further information in order to facilitate a decision, then there is provision for these time limits to be suspended whilst this information is awaited. The mechanism for processing multiple applications in more than one Member State requires that the State which has issued marketing authorisation submit a copy of the authorisation with all the manufacturer's application documents to the CPMP, for forwarding on to the States named in the multiple application. Decisions to grant marketing authorisations must be reached and notified within 120 days. In the event of a refusal, reasoned objections must, likewise, be lodged with the CPMP within 120 days.

It should, perhaps, be noted that, whilst the norms and protocols directive (75/318/EEC) sets guidelines for the conduct of clinical trials and the manner of their reporting, there is no agreement on the legal certification of trials. It is only in the United Kingdom that authorisation to conduct clinical trials of drugs and medicines is mandatory. Somewhat similar, though in some respects less exacting, provisions exist in the United States of America under the Federal Food, Drug and Cosmetic Act.

Legislation and its implementation in the United States of America

Testing and licensing of human medicines in the USA are controlled by the provisions of the Federal, Food, Drug and Cosmetic Act. The regulatory procedures are administered by the Food and Drugs Administration (FDA), a branch of the Department of Health, Education, and Welfare, which lays down requirements for the testing of new drugs and medicines and issues authorisations for marketing. Regulations provide for two stages in the registration procedure corresponding to the issue in the UK of the clinical trial certificate and product licence, respectively. The first stage requires that, before commencing human studies, the sponsor (either a pharmaceutical company or a physician) should complete and lodge with the FDA a **Notice of Claimed Investigational Exemption for a New Drug** in the prescribed form. This is required to contain the following:

(a) the composition, source and manufacture of the drug and its dosage form
(b) results of all preclinical animal studies demonstrating both safety and potential usefulness in human medicine
(c) a plan (protocol) of the proposed investigation
(d) the qualifications and experience of the investigator
(e) agreement to obtain the informed consent of patients taking part in the trial
(f) agreement that the protocol of the trial will be subject to the approval of an ethical committee in the institution in which it is to be tested
(g) agreement to notify the FDA and all investigators of any adverse effects which arise during either further animal or human tests
(h) agreement to submit annual progress reports.

The sponsor must wait 30 days before commencing any clinical study of the **Investigational New Drugs** (IND), to allow the FDA time to review the trial protocol and the supporting data. The 30-day delay may be extended by the FDA if time is required to correct deficiencies in either the protocol or the animal data.

Provision is made for three phases in the clinical evaluation of an IND.

Phase I Human pharmacology primarily to establish basic safety; routes of administration and dose-ranging studies; pharmacokinetics and metabolism. These studies are usually conducted in normal human subjects.

Phase II Initial trials on a limited number of patients for the specific treatment under examination to establish efficacy. Once this is established and parallel long-term animal tests have shown that the drug is safe, these investigations may proceed to the third stage.

Phase III Extended trials to confirm efficacy and safety; also, to establish that the drug is essentially free from hazardous side-effects and that non-hazardous side-effects are minimal.

In Phase III, the drug is used in the way it is intended to be marketed. Once this stage is satisfactorily completed, the sponsor may proceed to apply for approval to market the product, by making a **New Drug Application** (NDA). This is a compilation of all the available information on the product and is reviewed by one of the six divisions within the FDA's Bureau of Drugs. These divisions, namely cardiopulmonary-renal, neuropharmacological, metabolic-endocrine, anti-infective, oncology-radiopharmaceutical and surgical-dental, are staffed by physicians, pharmacists, chemists and other experts who, with the help of advisory committees of independent experts, reach a decision on each application. There is a legal time limit for review of 180 days but, not surprisingly, there are powers for its extension if the information submitted is considered to be deficient in any respect. Once marketing permission is obtained, the company is required to keep production and other records and make periodic reports to the FDA.

Codes of good laboratory practice

The validity of data submitted in support of licensing applications depends on the quality of the laboratory practices on which they are based. Some of the more important aspects of **Good laboratory practice** for nonclinical laboratory studies have been set out by the FDA (Federal Register, 1976, **41**(225), 51206–30) and have been the subject of comment in a Report by W. Horwitz (*Analytical Chemistry*, 1978, **50**, 521A). The proposed standards and protocols set out in the Federal Register are intended primarily to regulate toxicological studies but, in so doing, also lay down standards for controlling the identity, purity, stability, strength and storage of products under test. The maintenance of adequate records, covering the number, nature and detail of experiments, batch records of materials and animal diets used; source, strain and weights of animals used, calibration of reference standards and even calculation checks, is essential. Similar guidelines on good laboratory practice are also set out in an appendix to the so-called orange guide to good pharmaceutical manufacturing practice (1977)

Codes of good manufacturing practice

The quality of medicinal products is dependent just as much on the conditions under which they are manufactured as on the requirements of the product licences granting authority for their manufacture. The development within the industry over many years of practices designed to ensure reliability and safety in manufacture have resulted in the establishment of guidelines to good manufacturing practice which all manufacturers of repute would wish to emulate.

These now widely accepted practices, which stem from the high ideals of dedicated production teams within the industry, have been codified and issued in written form by such international authorities as the World Health Organization and various other national authorities. Sources such as **Quality Assurance on Pharmaceutical Supply Systems** (WHO/Pharm/79.497) and the **Guide to Good Pharmaceutical Manufacturing Practice** published by HMSO (1977) on behalf of the Department of Health and Social Security provide guidelines on such matters as:

(a) the location, construction and adaptation of buildings
(b) the design, construction, location and maintenance of equipment
(c) personnel and training
(d) cleanliness and general and personal hygiene
(e) production procedures, documentation and records
(f) sampling, quality control and assurance
(g) storage and transportation
(h) product recall.

Special provision for safeguarding the manufacture and control of sterile medicinal products, which apply not only to injectable products but also to certain other products specially liable to microbial contamination, are set down in an appendix to the Guide. Many of these provisions are based on the **Report on the Prevention of Microbial Contamination of Medicinal Products** (HMSO, 1973). It covers the requirements for clean and aseptic areas, the design and maintenance of equipment, sterilisation methods and their control.

Quality control should be the responsibility of a specifically designated **Quality Controller** appointed by the management. His independence and authority are critical to the success of any policy of good manufacturing practice. He must be completely free of all responsibility for actual production processes, so that objective criticism of production methods can be made should this be necessary. The Quality Controller should also be equally independent of other divisions of the firm, so that all decisions on matters of quality and safety may be reached on their merit without pressure or the threat of being overruled on grounds of commercial expediency.

Satisfactory laboratory facilities must be provided under the responsibility and supervision of the Quality Controller to permit such testing as he may deem necessary to decide on the acceptance or rejection of raw materials, materials in-process, finished products and packaging materials. Raw materials may only be released for use in manufacture when the batch has been passed as satisfactory by the Quality Controller. Where in-process control is in effect, material at each stage must have the approval of the Quality Controller before being passed on to the next stage of the process. Similarly, the release or rejection of each batch of finished product for packaging and for distribution must have the agreement of the Quality Controller.

The duties of the Quality Controller must also include the compilation and approval of **specifications** for all active ingredients and excipients used in the process, for in-process controls, and for finished products. Duties also extend to examination and evaluation of the stability of each product, the retention of representative samples from production batches and the determination of shelf-life and expiry dates. In order to ensure true independence of control, samples for analysis should be taken by the quality control staff using approved sampling methods. The only allowable exception to this rule is in respect of samples for in-process control, which may be taken by the production staff, provided laid-down sampling procedures are followed.

The preservation of analytical reports relevant to each batch, and of the signed and dated authorisations for each batch of product released for sale, is imperative.

Clinical trial applications

Clinical trial objectives

The purpose of a clinical trial is primarily to establish efficacy and demonstrate freedom from unwanted side-effects in man. The obvious aspects of quality and safety should already have been demonstrated in the pre-trial laboratory studies and the results of these studies presented in the application to authorise the trial.

Applicants for product licence and clinical trial authorisations are required to present individual applications in respect of each product or trial giving appropriate data on the lines set out in **notes for guidance** issued by the Licensing Authority. In general, the information required on chemistry, pharmacy, standards and animal studies for the issue of a clinical trial certificate or for exemption therefrom, is similar to that for a product licence to cover the manufacture, sale or supply of a product, the clinical effectiveness of which has already been supplied. In both instances, the product is for administration to human subjects, and safety is all important. There can be no question of double standards, as is sometimes argued on the grounds that fewer patients are at risk in a clinical trial than when the product is generally available on licence. The risk of any wholly unexpected and life-threatening event is, indeed, small. The real risk lies in the exposure of patients to the possibility of unforeseen long-term effects and this can only be minimised by adequate requirements for pre-trial laboratory studies. Adequate safeguards in matters of safety are, therefore, every bit as important at the stage of clinical trial, if not more so, as at the time of issue of a product licence.

Data relating to safety

There are no fixed rules on what information should be supplied in a clinical trial application. Notes for guidance, although intended to be helpful, are often over-complex because they cater for every type of product imaginable. A flexible approach based on a shrewd combination of scientific evaluation of appropriate data and common sense is essential. The nature and volume of data required is obviously a matter of judgement, but should relate sensibly to the following parameters:

(a) the duration of the trial
(b) the number of patients in the trial at each centre
(c) the number of centres
(d) the dose and frequency of dosage
Small, single or infrequent doses might require only minimal toxicological studies. In contrast, multiple dosage over short periods or continuous medication, irrespective of dose level, over long periods demand much more exhaustive safety evaluation. This should cover quality and stability, metabolism, retention and excretion, mutagenicity, carcinogenicity, teratogenicity and effects on fertility.
(e) the route of administration
Toxicological studies must cover effects observed using the route of administration which it is proposed to use in man.
(f) the intended use of the drug and the type of patient involved in the trial. Thus, teratogenicity studies would be relatively unimportant in a trial involving geriatrics. Likewise, exclusion of women of child-bearing potential might remove the need for teratogenicity and fertility studies at that stage. It would not, however, avoid the need for such

checks prior to marketing of any drug likely to be widely used by the public at large.

Data relating to quality

Quality must be safeguarded for the purposes of a clinical trial just as much as for the manufacture and sale of a product of proven efficacy, since, in the absence of adequate safeguards, the trial may well be rendered uninformative, or in an extreme case be completely vitiated if a product of doubtful composition or stability were used. Some relaxation on the complete characterisation of potential impurities or in the matter of complete stability studies on tentative dosage forms, which may well require modification in the course of a clinical trial, may be conceded provided there are reasonable safeguards. Thus, the method by which a particular compound is synthesised for the purposes of a clinical trial may not necessarily be the method ultimately used for the production of much larger batches for sale when a product licence is ultimately issued. In these circumstances, the impurity patterns may be completely different. Extensive investigation and characterisation of all significant impurities in compounds for clinical trial might, therefore, be obviated by the presentation of more limited data, provided this is adequately supported by sufficient additional information showing the product to be reasonably free from toxicity under the maximum dosage regimens proposed for the trial. Nevertheless, specifications must be such as to ensure that the essential characteristics of both the drug substance and its dosage form remain reasonably constant throughout the trial and are reproducible in any product proposed in subsequent marketing applications. Otherwise, the proposed trial ceases to be a meaningful exercise as a guide to the safety, quality and efficacy of the final marketable product.

On similar grounds, stability testing of the drug substance, which may, on occasion, be related to its particular impurity pattern, and stability studies of the provisional dosage form may, in certain circumstances, reasonably be less extensive than those required for the issue of a product licence. The essential criterion is that they should relate appropriately to the method of handling and storage proposed and to the detailed arrangements to control the life of the product and the disposal of unused remainders during the course of the trial. Apart from any such provisos limiting the stringency of requirements for products on clinical trial, requirements for submissions both for clinical trial certificates and for product licences are essentially the same in respect of basic information on active ingredients (drug substances) and other components of medicines (dosage forms). Product licence submissions must, however, provide appropriate evidence of clinical efficacy, and safety in use in patients.

Phasing of clinical trials

In practice, all clinical trials are arranged to commence with a primary study limiting risk at the time when least is known of human response to the product. These, Stage 1, clinical pharmacological studies, which are heavily monitored, are followed first by single centre (Stage 2) and then multicentre trials (Stage 3).

Special arrangements are available in the UK (and USA, p.61) to speed entry into Stage 1 clinical trials by means of a scheme for exemption from clinical trial certification. This still requires the presentation of all the usual essential data on chemistry, pharmacy, pharmacology and toxicology of the product and the usual safeguard of approval of the clinical trial protocol by the appropriate ethical committee. Approval to proceed with the trial is automatic in absence of a stop-order from the Licensing Authority within 35 days (30 days in the USA). In the event of a stop-order being issued, application for a Clinical Trial Certificate may still be made in the usual way.

The exemption scheme is designed to minimise delays in drug development by facilitating early clinical trials, thus enabling companies to determine whether further investment in development is justified at the earliest possible point in the development programme. So far as data requirements are concerned, these are basically the same for exemption application as for a full clinical trial certificate. Toxicity testing requirements, however, are generally more limited reflecting the number of patients, scale and frequency of dosing. Teratology and carcinogenicity studies are not required, but mutagenicity testing is a requirement, as are the essential chemistry and pharmacy requirements for characterisation of the product and stability.

Product licence applications

Applicants for product licences are required to submit particulars of the product according to the official notes for guidance under the following headings:

(a) Summary of particulars
(b) Chemistry and pharmacy
 (i) Active ingredients (Drug Substances)
 (ii) Dosage form
(c) Reports of experimental and biological studies
(d) Reports of clinical trials

The division of information is largely one of convenience for assembly, presentation and submission of data for assessment, but it cannot be emphasised too strongly that it is the sum total of all information presented which must be considered in terms of safety, quality and efficacy. The work of the pharmaceutical analyst is concerned primarily with quality control, but the extent and effectiveness with which controls are applied also depends very much on the need for the product to meet requirements for safety and efficacy. Accordingly, whilst the emphasis in the sequel rests primarily on the requirements in respect of the chemistry and pharmacy section of submissions, its relationship to information presented under the other sections must always be kept continuously in mind. Applicants should, therefore, take care to ensure that a reasonable and meaningful correlation exists between data recorded in different laboratories and presented in separate sections of the submission. This can only be achieved by careful appraisal and editing of all the information by one person, who should be an experienced scientist with a broad overall view of the submission's requirements.

Whilst general guidelines are issued, the precise information to be supplied is a matter of judgement on the part of the applicant. Clearly, it varies according to the type of product, depending upon whether the application is for an entirely new drug, a new presentation of an existing drug, a new combination of existing drugs, or merely a reformulation for the purpose of updating and improving the product, or meeting some change in source or supply of a particular excipient. The precise format is, therefore, a matter for decision on the part of the applicant, but whatever its form and content, it should be, if nothing else, accurate, scientifically sound, strictly relevant, concise and free from major inconsistencies.

The following passages relate specifically to product licence applications in the United Kingdom. Whilst application procedures elsewhere in the world may differ in matters of detail, the same general approach applies.

Summary of particulars

General information

The purpose of this section of a submission is to give certain general information relating to the applicant, the licencee and the product. Thus, it should clearly indicate where the product is to be manufactured and, in the case of importation or assembly of the final product, the precise origin of the imported material. Constituents of medicines or complete products, manufactured overseas to national pharmacopoeial or other foreign standards, must be shown to conform either to the standards in force in the United Kingdom for use of such substances in medicine (i.e. British Pharmacopoeia or, where appropriate, European Pharmacopoeia standards) or, if they do not precisely meet the appropriate standards, to deviate from them only in trivial respects which do not affect their safety or stability. In this connection, it cannot be assumed that the licensing authority and its advisers are *au fait* with standards published in languages which are unfamiliar and little used in the United Kingdom. The onus is, therefore, on the applicant to ensure that his application is complete in such details including authenticated translations of relevant monographs, if these are appropriate.

Description and name of product

The summary of particulars should provide a brief description of the pharmaceutical form of the medicine (i.e. tablets, capsules, suspension, cream, injection etc.), including size, shape, colour and markings of solid dosage forms (tablets and capsules), a statement of the active ingredients and an indication of the manner in which these will be stated on the label of the medicine and in any associated descriptive material. The British

Approved Name (BAN) should be used for the specification of all materials but, in the absence of an Approved Name, the International Non-Proprietary Name or, failing that, the Trade Name will do. In the USA, United States Adopted Names (USAN) take precedence.

Physical properties

The physical characteristics of the active constituents should be reported. These may range from such basic characteristics as colour, odour, taste, melting point, specific gravity, refractive index, viscosity and light absorption, which, taken together, provide a proper basis for identification, to more specialised characters, such as pK_a, crystal habit, polymorphism, particle size, specific surface area and bulk density. The latter are important criteria in determining the batch to batch consistency of both drug substances (active ingredients) and finished products and the availability in terms of release rate, blood levels and clinical efficacy of active ingredients from the product (Chapter 1).

Information on solubilities and pK_a relate in some degree to the choice of dosage form and proposed route of administration. For solid dosage forms, and suspensions administered orally or as depôt injections, drug solubility provides a useful criterion by which to assess the need for controlled release rates and uptake of the medicament from the product (i.e. bio-availability). Solubilities in water and chloroform provide the best indication of likely behaviour in terms of absorption, distribution and excretion in animals and human subjects, since solubility in chloroform is a reasonably good guide to solubility in biolipids.

In submitting data on acids, bases and their salts, it is important to bear in mind the effects of pH on dissociation, both in dissolution tests and biological media. The pK_a and both water and chloroform solubilities of the parent acids and bases as well as solubilities of the salt actually used in compounding the medicine, provide valuable data against which to assess the potential efficacy of a particular dosage form in terms of its bio-availability. For similar reasons, an inappropriate choice of solvent in an *in vitro* dissolution test can give misleading indications. Thus, *Chloropropamide*, which has a weakly acidic sulphonamido group, has been known to show good release rates from tablets measured in Tris buffer (triethanolamine), due to the formation of a water-soluble salt, despite the fact that its bio-availability from the same batch of tablets, measured in terms of actual blood levels, may be poor. The latter correlates with its low water-solubility, and the limiting factor in uptake from the acidic milieu of the upper gastrointestinal tract is, in fact, the specific surface area of the powder.

Recommended clinical use and dosage

Statements on the recommended clinical use, the dosage and route of administration are of material concern in deciding acceptable levels of permitted impurities. Levels of trace elements and other impurities, which would be acceptable in a product for administration at milligram level on a single occasion, may be wholly unacceptable in comparable treatments requiring dosage in grams, or for repeated administration, even at intermediary dose levels over long periods of time. Other things being equal, control of impurity patterns might reasonably be less stringent in a highly potent medicament which is only required for occasional administration than for, say, an antibiotic used in the treatment of a deep-seated infection, such as tuberculosis, or a symptomatic drug for the relief of rheumatism, hay fever, catarrh, or mental disorders where long-term therapy is the order of the day.

In determining the maximum acceptable level of any permitted impurity, it is essential to consider the nature of the hazard involved. Clearly, inactive but relatively non-toxic isomers or by-products of the manufacturing process cause less concern than contaminants which show adverse clinical reactions such as nausea, vomiting, skin reactions, liver or kidney damage, teratogenicity or carcinogenicity. Each drug substance and impurity pattern requires individual consideration in the light of the proposed dosage form and route of administration. Trace elemental contamination is, however, widespread due in part to atmospheric pollution, but mainly as a result of the use of particular catalysts or reagents in organic synthesis. Certain materials are prone to concentrate in particular tissues, irrespective of the route by which the contaminated drug is administered, but it is evident, for example, that contaminants capable of causing undesirable skin reactions should be particularly stringently controlled in products for application directly to the skin. Similar contaminants in products administered internally, although equally undesirable, may well be tolerable at higher levels due to more efficient metabolic detoxification and excretion. By the same criteria, contaminants known to cause lung or eye irritation and damage should be more stringently controlled in inhalant aerosols and eye preparations, respectively. It is, however, difficult to get quantitative data on the toxicity of particular trace element impurities. Table 2.1 shows some qualitative data on tissue deposition, vulnerable tissues, organs and enzymes and indicates potential areas of hazard arising from trace elemental contamination.

Place of manufacture; sale and supply

In addition to these aspects, which are of direct relevance to the scientific assessment of the product, the summary of particulars is also required to state the place of manufacture to establish that the premises are properly licensed under the Act. They must also indicate whether or not quality control will be exercised over the process and product, the type of container to be used for the product, including details of its size, shape and other distinguishing features or markings. Labelling particulars must be given, and the method proposed for sale and supply, i.e. whether the product is proposed for general sale, supply through registered pharmacies only, either over the counter or on prescription only, or for distribution via alternative outlets, such as hospitals or herbalists.

Active ingredients (Drug Substances)

This section of the application is intended to provide the Licensing Authority with detailed information on source, manufacture, identity, standards and stability of the active constituents and other ingredients of the medicine which forms the subject of the submission. It should also provide information on the rationale underlying the choice of the proposed dosage form, the method of manufacture, the stability of the product, and control standards and methods.

Nomenclature and composition

In addition to the approved, non-proprietary, or other name of each active ingredient, the submission must give for each substance, which is not the subject of a pharmacopoeial monograph, the systematic chemical name, its molecular and structural formulae and state its molecular weight. In some cases, it may not be possible to meet all these requirements precisely. Thus, it may not be possible to state the precise composition of certain hydrates, solvates, clathrates and inclusion complexes, but merely to indicate the limits within which the composition of the active ingredient will be controlled. Similar criteria also apply to the composition of certain inorganic materials (e.g. *Hydrotalcite*, an aluminium magnesium hydroxide carbonate hydrate), to organic intermolecular complexes (e.g. *Dichloralphenazone, Tetracycline Phosphate Complex*) and, also, to polymeric products (e.g. *Polyglactin, Malethamer, Polysorbates*) in which only an indication of average molecular weight can be given.

This information is necessary to define the composition of the product and, also, for labelling purposes. To the expert, it provides a basis for predicting physical, chemical and biological properties and, hence, for critical appraisal of data supplied on these matters, on stability, and on the nature and properties of the proposed dosage form.

Manufacturing process

Details of the method of manufacture are essential as evidence of identity and chemical structure. Specifications for starting materials, and purification and control of intermediates, are required to provide a firm basis for the assessment of impurity patterns. The complete synthetic route from appropriate intermediates which are readily available items of commerce must be specified in sufficient detail to ensure satisfactory identification by physical, chemical or sensory characters, adequate quality by assay and satisfactory control of contaminants which may be deleterious either to the process or the final product.

It is frequently argued, where several stages are involved in the synthesis of the active ingredient, that the quality of the starting materials is immaterial, as impurities will be removed in the course of the process itself. This is frequently true in linear processes in which the final product is built up in successive steps, with intermediate purification. Convergent syntheses in which major fragments are linked at a late stage in the process offer less opportunity for the elimination of impurity derived from the starting materials or early-stage intermediates. Much also depends on the contaminant and the process since, in certain circumstances, the latter may serve to concentrate rather than eliminate undesirable impurities arising from impure source materials. Thus, ethyl iodide may contain small amounts of methyl iodide, which can arise from the use of industrial methylated spirits as source material. If used in the quaternisation of tertiary bases such ethyl iodide can give rise to undesirably high levels of the methiodide as a contaminant of the product ethiodide. This is because of the much higher rate of reaction with methyl iodide compared to that with ethyl iodide. Similarly, steric effects which inhibit the rate of reaction of aromatic compounds in the *ortho* position compared with that in the *para* position can lead to concentration of *p*-substituted intermediates and products in reactions conceived essentially as an attack on the *ortho* position.

Reagents, catalysts and solvents used in manufacture, which are likely to be retained in the final product, must also be considered as potential impurities, from the point of view of both toxicity and their effects on stability. In an ideal process, these will be eliminated or reduced to innocuous levels, but traces of nickel and palladium hydrogenation catalysts, also boron, arising from the use of boron trifluoride or sodium borohydride, should be particularly carefully controlled on account of their potential for producing toxic effects (Table 2.1). Retained solvents are seldom likely to be present in quantities which present serious toxic hazards. Significant levels of methanol have, however, been found in some antibiotics, such as *Streptomycin Sulphate*, which may well be administered in substantial doses for several weeks or months at a time. Such contamination is clearly undesirable and strict upper limits for solvent contamination should be imposed in the finished product specification for any material administered on this scale. Contamination with hydrocarbon and other solvents which can cause liver damage, should be even more strictly controlled.

In-process control

Chemical syntheses carried out on a production scale should be subject to in-process checks on intermediate stage products, in which the material is positively identified, checked in some way for quality and, if critical, for freedom from particular contaminants. Some processes, such as, for example, the Merrifield type synthesis of peptides, in which the product is built up unit by unit on an ion-exchange resin support without isolation of intermediates, are not amenable to control in this way. In such cases, the success of the process depends on rigidly defined and strictly controlled operating conditions, backed by tight specifications designed to limit any inadequacies in the process.

Potential impurities and their detection

Submissions should include details of methods used to detect likely

Table 2.1 Potential toxic hazards arising from trace element contaminants

Element	*Specific tissue deposition*	*General systemic poisons*	*Vulnerable tissues and organs* Skin	Lung (Mucous membranes)	Eye	Nerve and brain	Muscle and heart	Liver	Kidney	*Potential carcinogens*	*Vulnerable enzymes*
Antimony (organ)			+	+				+			Phosphoglucomutase
Arsenic (organo)	Skin, hair, nails, liver, bone	+	+	+	+					+	Phosphatases and other sulphydryl enzymes
Barium	Bone, lungs			+			+				
Beryllium	Bone			+						+	Alkaline phosphatases
Boron	Brain, nerve		+			+	+				
Cadmium		+		+		+				+	
Chromium			+	+						+	
Cobalt			+	+						+	
Gallium	Bone Malignant tissue					+					
Indium	Calcification at injection site							+	+		
Lanthanum	Nucleic acids (complex formation)										
Lead and organo lead	Bone Brain	+				+					
Mercury (organo)	Nerve					+					Catalase
Nickel	Nerve, lungs, heart		+	+						+	
Palladium	Heart	±					+	+	+		
Platinum			+	+							
Selenium			+		+						
Thallium		+									
Tin (organo)		+	+								
Tungsten				+							
Vanadium			+	+							

impurities in the final product. Thin-layer chromatography is widely used for this purpose, but gas-liquid chromatography, paper chromatography and high pressure liquid chromatography provide useful supplementary and alternative techniques for the identification and quantification of impurities which are often closely related chemically to the parent compound. A critical approach to these techniques is essential. A thin-layer method which shows excessive tailing or one in which one of the principal impurities has an R_F value barely distinguishable from that of the parent compound is unlikely to suffice unless it can be quite clearly shown that this is the best that can be attained after several likely alternative solvent systems and supports have been tried. Systems in which likely impurities sit on the baseline or run very close to the solvent front are equally unlikely to be acceptable. A clear indication of the sensitivity of the method, including evaluation of the technique used for detection of the spots, together with the presentation of visual evidence in the form of copies of relevant chromatograms and, where appropriate, photographic records, are the hallmark of a good submission.

Similarly, GLC systems with grossly asymmetric peaks, inadequate separation of solvent, main and impurity peaks, or with either inordinately long or widely separated retention times are undesirable. But, whatever the technique employed, whether it be, for example, chromatographic, spectroscopic, or enzymatic, it should give a clear distinction on a quantitative or semi-quantitative basis between individual impurities and the parent compound. In particular, the development work must establish that the product is free from harmful amounts of occluded solvents, and trace elements derived from the process, special attention being paid to known potential carcinogens (arsenic, beryllium, cadmium, chromium, cobalt, nickel) and other toxic elements, including boron, mercury, lead, barium and palladium, capable of causing nerve, muscle, liver or skin damage (Table 2.1).

Evidence of chemical structure

Unequivocal evidence for the structural formula of the active ingredients must be presented. This should rely primarily on the synthetic route, but should be supplemented by spectroscopic data derived from ultraviolet, infrared, NMR, mass, optical rotatary and circular dichroic spectrometry, as appropriate to the structure concerned (see Part 2 for examples of these methods in structure elucidation). Sufficient evidence should be provided to demonstrate that the structure is unequivocal. Examples will arise from time to time where, for solubility or other reasons, none of these techniques is capable of providing the required information. In such an event, the product at the subterminal stage or a suitable derivative may prove more amenable to spectroscopic examination, or useful information may be available from thermal analysis or even X-ray crystallography.

Evaluation of evidence relating to the structure of intermolecular complexes should take account of the solvents used in spectroscopic studies. Thus, unequivocal evidence of a complex, formed by two compounds present in stoichiometric proportions and obtained solely in non-aqueous media, may have little relevance to the physical state of the complex in essentially aqueous biological media. For example, the well-defined hydrogen bonded complexes of chloral hydrate with phenazone and acetylglycinamide, which are physically distinguishable from chloral hydrate by melting point and by the complete masking of the latter's characteristically unpleasant taste, are almost certainly decomposed in aqueous solution and merely provide an alternative means of administering chloral hydrate. Such lack of stability in biological media is relevant and the identity of the pharmacologically effective compound should be clearly apparent to the reader of submissions relating to complexes of this sort.

Specifications and batch analyses

Proposed specifications of the drug substance should take into account all aspects of the manufacturing process. They should provide for the proper identification of the product, to avoid mistakes in handling, and for the maintenance of adequate standards for content of the drug substance. Specifications should also give reasonable control of impurities actually found to occur in batches of product at the time of manufacture, or arising subsequently as a result of instability on storage or mishandling. Standards for content and impurity should take account of the precision of the methods employed and should be reasonably related to the levels encountered in actual batch analyses. They should provide for the maintenance of reasonable standards whilst allowing latitude for acceptable operator and instrument errors of particular techniques. It would not be reasonable, however, to submit specifications allowing assay tolerances down to a minimum active ingredient content of, say, 94% by non-aqueous perchloric acid titration of a base, a technique which would normally be expected to give results within 99 to 101.5%, when a set of six batch analyses for the material indicate levels of 98.2, 99.5, 100.3, 99.7, 99.5 and 98.6%. Likewise, basic impurities in such a compound, which would also be titrated by perchloric acid, should also be controlled by a separate limit test. Similarly, limits of trace impurities detectable at 0.1% level by TLC and batch analyses indicating maximum and minimum levels found as 0.3% and 0.1% would not call for a specification limit of 1% without good reasons being given.

The specifications must also take into account any fall in potency on storage during the expected life of the product and consequent increase in decomposition products. They should, however, be sufficiently strict to exclude products which have been subjected to mishandling, with the attendant possibility of an unnecessarily high rate of decomposition. Evidence of excessive decomposition on storage of the bulk drug cannot be made the excuse for a lax specification. Exceptionally, where products with important medical indications can only be produced in a form susceptible to significant decomposition on storage, it may be necessary to have a realistic **release specification**, applicable at the time of manufacture, and a separate **check specification** permitting somewhat lower minimum standards to apply at some later date when the product is incorporated into a particular dosage form.

Specifications for such materials as antibiotics, hormones, dextrans, enzymes and enzyme inhibitors, which are subject to control by biological assay, should conform to the requirements of the **Compendium of Licensing Requirements for the Manufacture of Biological Medicinal Products** (HMSO, 1977).

Stability studies

Stability data used to determine shelf-life, and co-related specifications, must take full account of the chemistry of the active ingredient and its likely vulnerability to degradation by oxidation, carbon dioxide, moisture, heat and light and container materials. Product analyses conducted in connection with stability must, therefore, be reasonably comprehensive. Assay figures alone, particularly where the assay process is relatively non-specific and unlikely to distinguish the drug substance from anticipated decomposition products, are not sufficient. Any properly conducted stability study must also include an examination of specific decomposition products by appropriate techniques to establish the identity and relative toxicity of the decomposition products and the concentrations in which they are formed.

Stability studies should not only take account of the physical state in which the compound is likely to be used, but also the immediate biological environment likely to be met on administration. Thus, substances for tabletting, encapsulation and the preparation of inhalant cartridges or suspensions should be examined primarily in the solid state. Substances for injection, which must on this account be subjected to some form of sterilisation procedure, must be examined particularly for stability at elevated temperatures, for possible hydrolysis or rearrangement in aqueous media and the effects of exposure to carbon dioxide and light. Similarly, all substances intended for oral administration must be chemically stable to the pH and enzymic conditions likely to be met in the gastrointestinal tract. Exceptions clearly apply where, for example, the drug substance is specifically designed to release the active ingredient by such decomposition, but if this is the case the submission should make this evident and include those conditions under which the compound is expected to be stable.

To encompass all these requirements, therefore, stability studies must be conducted on the drug substance in the solid state over a range of temperatures, at varying degrees of humidity, in both light and dark, in air and with air excluded. Also, if the product is such that it is likely to be subjected in use to widely varying temperature fluctuations, i.e. a product to be used in multiple dose form in the tropics, which should be stored, ideally, in cool or refrigerated conditions, then stability tests should include a study of the effects of fluctuating temperatures.

Pharmacology and toxicity

The pharmacology of the active ingredient must be clearly established and reported, including all effects relevant to the proposed use of the medicine. All other actions must be reported, including possible drug interactions which could provide a key to undesirable side-effects or otherwise affect its safety in use.

Animal toxicology, covering acute (single-dose) and chronic (long-term) toxicity, carcinogenicity, teratogenic and fertility studies in appropriate species, are essential. Reports must be presented in detail, giving particulars of the strain, diet, age, sex, weight and number of animals used in each experiment. In all repeat dose studies, the route of administration, method of dosage and food consumption must be relevant to that proposed for use in man. Reports are expected to include the results of haematological studies, biochemical investigations and urinalysis. Post-mortems, including histopathological studies, must be carried out in all animals dying in the course of experiments and the cause of death established. Where internal effects or significant lesions are found, the extent of reversibility on withdrawal of the drug should be examined.

Dosage forms

Formulation

The formulation of the finished product or dosage form must be declared so far as is practicable in such a way as to show *either* the amount of each of the active ingredients and excipients per dose *or*, if this is not possible, the proportion of each on a percentage basis. The formulation may in some cases contain an *overage* to compensate for loss of potency due to decomposition of active ingredients either during manufacture or on storage. Certain preparations, notably single-dose injections, may, additionally, contain an *overfill* in order to permit easy withdrawal of the correct dosage volume.

All constituents, active ingredients and excipients alike, must conform to a fully stated specification, pharmacopoeial or otherwise, appropriate to their use in human medicine. Certain exemptions from licensing of medicines are allowed under the Medicines Act, 1968 in respect of herbal remedies, provided the process of manufacture consists only of *drying, crushing* or *comminuting*, and provided the product is sold under a designation which only specifies the plants used and the process by which it is compounded. No other name may be applied, nor must there be any written recommendation for use, otherwise the product ceases to be exempt and is classifiable as a medicine.

Any formulation may be proposed for licensing a medicine, but combinations of active ingredients which are not already recognised for use in medicine may require justification in terms of both safety and efficacy. The possibilities which exist for drug interactions, whether these be in a biological sense leading to modified clinical response or alterations in the pattern of metabolism, or merely physico-chemical interactions affecting absorption, distribution and excretion kinetics, suggest a need for caution in proposing the use of particular unproven drug combinations.

Any formulation proposed for a medicine in a marketing application must either be identical with that used during clinical trials of the active ingredient or, failing that, must be capable of being meaningfully equated with the trial formulation(s). Thus, product licence submissions which propose major changes from clinical trial methodology concerning the route of administration, type of product, excipients, physical properties or manufacturing process, should be backed up by adequate data which clearly demonstrate bio-availability, safety and stability.

The choice of a new route for administration, for example, by metered-dose aerosol directly into the lungs, when the same compound has previously been administered orally for absorption from the gastrointestinal tract, may pose new and important questions. These concern absorption, and ultimate fate, which can perhaps only be answered by fresh pathological studies and clinical pharmacology relevant to local toxicity, carcinogenicity or even teratogenicity. Actual experimental findings are essential since, although drugs administered other than by the oral route may by-pass the liver and, hence, be subjected to early metabolism by non-hepatic pathways, the amount of drug actually reaching the bronchi from a single puff of a metered aerosol suspension is very much dependent on the particle size and other physical characteristics of the active ingredient, and can be as low as 7% of the average dose per puff. Thus, part of the drug may be absorbed from the buccal mucous, the stomach, the intestine or a combination of these, depending on its physico-chemical characteristics.

Excipients

The choice of excipients should be carefully considered and be capable of rationalisation. Ideally, these should be limited to the minimum necessary to ensure uniformity of dosage and stability throughout the period of the proposed shelf-life of the product. There may be no legal barrier to the marketing of products with multitudinous unnecessary excipients, but there is little scientific or technical justification for products which might be colloquially described as containing 'the kitchen sink'. Ideally, the excipients used in a preparation should be limited to the minimum necessary to achieve correct and uniform dosage and stability throughout the shelf-life of the product. However, some latitude is allowable in the proportions of excipients to provide reasonable manufacturing tolerances, but these should be clearly stated.

It is equally important that there should be no chemical reaction between excipients and the active ingredients, though the manufacturing process may be permitted to incorporate such reactions, as, for example, the in-process formation of a particular salt of an organic acid or base, provided the true nature and form of the product active ingredient so formed is properly declared on the label. On the other hand, physical interactions may often provide the *raison d'être* for a particular excipient, such as a wetting agent used in tabletting of a relatively water-insoluble drug to aid its solution and more rapid absorption, or the incorporation of a particular solid support, such as silicon dioxide to enhance the surface area of a silicone used as an antiflatulent.

The manufacturing process

The method of manufacture must be set out in sufficient detail to ensure its reproducibility if placed in the hands of a suitably qualified operator without special knowledge or experience of the process. This is particularly important in respect of processes leading to the production of products which are essentially physical mixtures, whether these be solid (pro-injection solids, tablets and capsules), semisolid (creams and ointments) or liquid products (emulsions and suspensions), and especially so when the fine detail of the process may well provide the principal means of controlling physico-chemical characteristics. Thus, milling processes and micronisation of active ingredients which are relatively insoluble must be so specified as to ensure reasonable batch to batch reproducibility of particle size or better still specific surface area, and to ensure that any changes in solid state characteristics which occur in the process are consistent batchwise. Similarly, solid products and suspensions incorporating wetting and dispersing agents can show marked differences in the rate and extent of bio-availability of insoluble components from them depending on the manufacturing process. The influence of formulation and process on polymorphism of active ingredients, on colours and on 'Ostwald ripening' leading to crystal growth in suspensions are important aspects of manufacture requiring close control (A. L. Smith (ed.), 'Particle growth in suspensions', *Soc. Chem. Ind.,* Monograph 28).

Dosage form specifications

Specifications for dosage forms must not only embrace those of the individual constituents, but also be such as to provide for a reasonable measure of control over the product as a whole. It is essential, therefore, that they should provide for the identification and control of the content of active ingredients, to ensure that they are present within reasonable working tolerances in amounts which accord with the declared label strength. Tolerances must, therefore, take into account not only the precision of the assay, covering inherent errors of methodology, instrumentation and operator, but also the precision of the actual manufacturing process.

Allowing for the normal variation of patient response, it is generally considered that unit dosage variations of up to ±10% are unlikely to produce observable differences in the clinical efficacy of the great majority of drugs, particularly if they are administered in repetitive doses. For a few compounds, however, which are employed in microgram or milligram doses in clinically critical situations, uniformity of dosage is essential and must be strictly controlled. Typical examples include the heart stimulant, digoxin, for which the normal adult dose is one or two 250 μg tablets, and the contraceptive pill in which the small oestrogen content requires careful control at 50 μg per tablet, for both efficacy and safety. These quantities are very small in relation to the total weight of the tablets and special care in manufacture is essential to ensure even distribution throughout the batch. Moreover, the normal tablet assay procedure based on examination of a group of 20 tablets would not reveal gross inter-tablet variation. Specifications for such products must, therefore, additionally incorporate control procedures based on the assay of individual tablets to ensure uniformity of content.

Single-dose injection solutions containing small amounts of potent medicaments seldom present problems, since it is easy to ensure uniformity of content with a solution. Uniformity of dosage may, however, need to be monitored for single-dose injectables where the active ingredient is present at microgram level together with a diluent, such as lactose, to visualise the contents of the ampoule, as for example in *Vincristine Injection*.

Metered-dose pressurised aerosols and powders for insufflation directly into the lungs also present problems in the control of unitary dosage. Aerosol specifications should control the total weight of active ingredient per can, measured as an average of the weight found in samples of at least ten cans drawn from each batch. The total number of puffs per can should also be specified as should the average weights of aerosol and active ingredient per puff. The latter, too, should remain reasonably constant throughout the life of the can and specifications should, therefore, provide for checks on the active ingredient per dose, both when cans are full and when almost empty. Even with this close control of the dosage form, the effective dose in the lungs depends very much on particle size of the drug substance, and the characteristics of the can adapter used during administrations. Specifications must, therefore, provide for strict control of these factors.

Micronised powders for insufflation into the lungs, such as *Sodium Cromoglycate* (Cromolyn Sodium), are packed in hard-gelatin capsules. They are usually used where the weight of powder to be administered is rather greater than can be conveniently projected into the lungs by an aerosol. The capsules are used in special inhalers, which incorporate a mechanism to puncture the capsule and allow its contents to be drawn into the lungs by a miniature breath-actuated turbo-fan or other appropriate device. As with aerosol preparations, control of particle size, cartridge content and inhaler and adapter characteristics are criticial factors in standardisation of dosage and must be tightly specified.

Bio-availability

Evidence that the formulation is capable of releasing the active constituent at a clinically effective rate is an essential aspect of all product licence applications. Presentation of correlated dissolution, absorption and excretion kinetics both in animals and in man is, therefore, imperative. The metabolic pathway and extent of metabolism should be determined, but if this work is based on the use of radioactive tracers, the position of the label within the molecule must be indicated so that it is clearly apparent that the label has been incorporated in a metabolically stable position.

Ready and consistent bio-availability may well be self-evident in the case of water-soluble injections and even solid dosage forms of readily water-soluble compounds, but formulations of insoluble compounds require careful evaluation. Tablets, capsules and suppositories must be capable of rapid disintegration and should show evidence of appropriate release rates into solution in suitably designed *in vitro* tests. Plasma levels, as measured by the **area under the plasma concentration-time curve** (AUC), provided they correlate reasonably with clinical response, however, are more likely to give a meaningful indication that the active principle is effectively absorbed from the point of administration, whether this be mouth, skin, lung, or other body cavity. Relatively few preparations require compilation of standards for batchwise monitoring of correlated release rates. They fall into two categories: tablets of insoluble, highly potent medicaments, administered in small doses for the treatment of vital medical conditions, and tablets specially designed to give controlled release over a prolonged period, from which too rapid or too slow release could be hazardous to the patient.

Stability studies

Product stability of material produced in full-scale production batches and stored in the final containers is also of paramount importance. This is obviously related in some degree to the inherent stability or otherwise of the active ingredients, but applications for product licences must, additionally, include evidence to show that the product retains an acceptable level of potency on storage and, equally importantly, that toxic decomposition products are not produced in significant amount. Accelerated storage tests, conducted at elevated temperatures and under other conditions likely to give a clear indication of all readily conceivable hazards, should be used to determine an acceptable shelf-life for each product.

The product should be examined under all conditions likely to be met during its storage in practice. For example, a large-volume pack of a syrup containing an oxidisable phenothiazine such as chloropromazine, which may be kept in broken bulk as stock on the dispensary shelf, should be examined for sulphoxide formation in partially filled containers, even if the product is to be issued in well-filled containers. Fluctuating temperatures with partly filled containers should be used to simulate conditions arising in patient use. Similarly, the unusually large pack of a sterile cream for the treatment of burns, which the patient or nurse might be tempted to retain for use on a second or third occasion, even if these be within a matter of hours, should be tested under appropriate conditions to demonstrate the efficacy of any preservatives it may contain.

If stability risks are shown to exist, it is preferable that they even be avoided by the use of unit dosage forms. The choice of container and pack should, therefore, reflect the stability characteristics of the product. Light-sensitive materials should be packed to exclude light—e.g. tablets in metal foil strip packs; amber glass is no longer considered effective and injectables or other liquid preparations should be packed in clear glass vessels, if necessary wrapped individually in foil, but otherwise in packages capable of excluding light.

All product stability data should be referable to a particular production batch, so that information is on file giving the batch number, size of batch and date of manufacture. This information is essential to enable the origin of particular ingredients to be traced if a query should arise.

Tests relating to the use of the product are equally important. Thus, a liophilised dispersion for reconstitution immediately prior to use should be checked for resuspendability and stability within the permitted period of use. Injection solutions, which may be added to particular large volume intravenous infusions should, likewise, be checked for freedom from interaction.

Containers

The type of container and the nature of the container material must be so specified to ensure that it is adequate for the purpose, for example, giving light and moisture protection if this is required.

Package insertions, whether these be space fillers to prevent breakage of friable tablets in the course of normal container handling or desiccant packs to exclude moisture, have been recognised as potential hazards to unsuspecting patients. Any such insertion must be clearly distinguishable, by a suitable combination of size, shape, weight, colour and texture, from the product itself, to ensure immediate distinction by sight and touch, particularly in the hands of handicapped patients such as blind persons. The pack should be labelled with instructions that the insertion is not to be eaten. The need for inclusion of a desiccant pack should be clearly demonstrated by presentation of evidence of moisture-induced physico-chemical changes. The compatibility of desiccant and product must be established, and the desiccant must be such as to ensure satisfactory desiccant action without leakage from the pack. The quantity of desiccant should also be related to the shelf-life of the product.

Reports of experimental and biological studies

Whilst it is generally considered sufficient for animal toxicity and related biological studies to be carried out on the individual active ingredients and excipients, it is important to consider the possible influence of formulation and route of administration on the significance of results so obtained. Since both these parameters can markedly influence the extent of absorption, the pattern of distribution and, consequently, the kinetics of metabolism and excretion, it is clearly important that chronic toxicity, carcinogenicity, teratogenicity and fertility studies should be carried out on either the dosage form actually proposed or something very close to it administered by the same route as that proposed for clinical use.

Reports of clinical trials

Reports of clinical trials must be of such nature and sufficiently well documented to provide adequate evidence of efficacy and safety of the agent when administered to patients by the proposed route in the dosage indicated for the treatment of the indications proposed. The results must be reported so as to indicate clearly the number of patients at the commencement and on completion of each trial, the range and mean dosage employed, the results obtained and any adverse reactions which have been observed. In all applications for product licences, it is essential that some evidence of clinical efficacy in relation to the proposed *indication* for that medicine be demonstrated. Failing such evidence, the only reasonable course of action on the part of an applicant is an appropriate modification in the claims made for the product.

On matters of safety, product licence applications should take due note of established hazards and recommended practice in products containing substances for which particular hazards have been shown to exist. Well known examples, which have received considerable publicity, include products containing *Aspirin, Halothane, Hexachlorophane, Practolol,* Monoamine Oxidase Inhibitors and the question of the oestrogen content of oral contraceptives. The formulation of such products and recommendation for their use should conform to established safety standards.

Long-term safety

No testing or licensing procedure can ever ensure complete safety of any new medicine. Only marketing and widespread use can establish the degree of safety and the extent to which it is free from undesirable effects. The detection of hazardous effects still remains difficult. It depends, essentially, on the recognition of unexpected events by doctors and others and the speed with which they can be reported to a central data bank. It is the collation of reports on unusual drug-related events culled from many different sources which can most rapidly spotlight the occurrence of a serious **adverse reaction**.

The Committee on Safety of Medicines compiles and maintains a Register of Adverse Reactions based primarily on reports from prescribers. This runs to several volumes, of which the first group consists of notifications by drug used and the second group, notifications by adverse reaction observed. Entries in the first group list all the various adverse reactions reported for each compound, giving the number of reports received and the number of deaths for each such reaction. Entries in the second group list all the compounds which have been reported to give rise to a particular reaction. All entries also indicate the number of National Health Service Prescriptions issued per year.

Experiments with monitored release procedures in cases where some degree of caution seems appropriate have proved difficult to operate and, even when release has restricted use of the medicine to hospitals, this has not proved particularly effective. Moves towards post-marketing surveillance, using procedures which allow prescribers of new drugs to be identified to the manufacturer for ongoing transmission of information on drug effects, are expected to be more profitable.

3
The theoretical basis of quantitative analysis

Acid-base titrations

Electrolytic dissociation

Certain substances known as electrolytes dissolve in water to yield solutions which will conduct electricity. In 1887, Arrhenius suggested that this ability to conduct electricity was due to the fact that, in solution, electrolytes undergo dissociation into positively and negatively charged fragments which are called ions. Positive ions move towards a negative electrode and negative ions towards a positive electrode. The passage of ions, and the subsequent neutralisation of the ionic charge at the electrode, brings about conduction of electric current through the solution.

Ions are usually solvated but, for simplicity, the effects of ion solvolysis have been ignored in the subsequent discussion of the behaviour of electrolytes in solution.

The law of mass action

This law, which was first stated by Guldberg and Wage in 1867, may be expressed in the following form:

'The rate of a chemical reaction is proportional to the active masses of the reacting substances.'

In dilute solution where conditions approach the ideal state, 'active mass' may be represented by the concentration of the reacting species, i.e. gram-molecules or gram-ions per litre. The constant of proportionality is known as the velocity constant, so that in the simple reaction, $A \rightarrow B$, the rate of reaction $= k[A]$, where $[A]$ is the concentration of A and k is the velocity constant.

Consider now the homogeneous, reversible reaction:

$$A + B \rightleftharpoons C + D$$

According to the law of mass action

$$v_f = k_1[A]\cdot[B] \quad \text{and} \quad v_b = k_2[C]\cdot[D]$$

where v_f = velocity of the forward reaction; v_b = velocity of the backward reaction; $[A]$ denotes molar concentration of A; and k_1 and k_2 are constants.

At equilibrium, $v_f = v_b$

$$\therefore \quad k_2[\mathrm{C}]\cdot[\mathrm{D}] = k_1[\mathrm{A}]\cdot[\mathrm{B}]$$

$$\frac{k_1}{k_2} = \frac{[\mathrm{C}]\cdot[\mathrm{D}]}{[\mathrm{A}]\cdot[\mathrm{B}]}$$

Since k_1 and k_2 are both constants, the fraction k_1/k_2 must also be a constant.

Hence

$$K = \frac{[\mathrm{C}]\cdot[\mathrm{D}]}{[\mathrm{A}]\cdot[\mathrm{B}]}$$

where K = the **equilibrium constant** of the reaction (constant at a given temperature).

In extension, the equilibrium constant for the general reversible reaction:

$$a\mathrm{A} + b\mathrm{B} + c\mathrm{C} + \ldots \rightleftharpoons p\mathrm{P} + q\mathrm{Q} + r\mathrm{R} + \ldots$$

$$\text{is} \quad K = \frac{[\mathrm{P}]^p\cdot[\mathrm{Q}]^q\cdot[\mathrm{R}]^r}{[\mathrm{A}]^a\cdot[\mathrm{B}]^b\cdot[\mathrm{C}]^c}$$

where a, b, c, and p, q, r are the number of molecules of the reacting species.

Application of the law of mass action to solutions of weak electrolytes

Electrolytes may be classified as either strong or weak electrolytes, depending upon the extent to which they are dissociated into ions in solution. Strong electrolytes are almost completely dissociated even in moderately concentrated solutions and, hence, do not constitute equilibrium systems. Weak electrolytes, on the other hand, are only incompletely dissociated even in the favourable ionisation conditions of dilute solution; therefore, an equilibrium, which can be considered in terms of the law of mass action, is reached between undissociated molecules and ions.

The dissociation of water

Water is an extremely weak electrolyte and is only very slightly dissociated into its ions:

$$H_2O \rightleftharpoons H^+ + OH^-$$

From the law of mass action:

$$K = \frac{[H^+]\cdot[OH^-]}{[H_2O]}$$

In pure water and in dilute aqueous solutions the concentration of free water may be considered constant and hence

$$K_w = [H^+]\cdot[OH^-]$$

where K_w is known as the **ionic product of water**. The latter varies with temperature, but under ordinary experimental conditions (about 25°), its value may be taken as 1×10^{-14}, when the concentrations of hydrogen and hydroxyl ions are expressed in gram-ions per litre.

In pure water,

$$[H^+] = [OH^-]$$

and hence

$$[H^+] = \sqrt{K_w}$$
$$= 10^{-7} \text{ gram-ions per litre.}$$

Solutions in which the hydrogen ion concentration is greater than 10^{-7} are **acidic**; when it is less than 10^{-7} the solution is **alkaline**.

The hydrogen ion exponent (pH)

Because of the very great variations in hydrogen ion concentration met with in practice, it is often convenient to adopt the pH notation first introduced by Sörensen. pH is defined as the negative logarithm (to base 10) of the concentration of hydrogen ions in solution:

$$\mathrm{pH} = -\log_{10}[H^+] = \log_{10}\frac{1}{[H^+]}$$

This method of stating hydrogen ion concentration has the advantage that all degrees of acidity and alkalinity between that of a solution molar (or normal) with respect to hydrogen and hydroxyl ions can be expressed by a series of positive numbers between 0 and 14. A neutral solution is one in which pH = 7, an acid solution one in which pH < 7 and an alkaline solution one in which pH > 7.

The dissociation of weak acids and bases

Consider an aqueous solution of a weak acid, HA, in which the following equilibrium between ions and undissociated molecules obtains:

$$HA \rightleftharpoons H^+ + A^-$$

From the law of mass action:

$$K_a = \frac{a_{H^+} \times a_{A^-}}{a_{HA}} \tag{1}$$

where K_a = the ionization constant or dissociation constant (at constant temperature);* a = the activity of the various species present.

In dilute solution the activity terms can be equated to concentrations, so that (1) may be rewritten in the form:

$$k_a = \frac{[H^+]\cdot[A^-]}{[HA]} \tag{2}$$

where k_a is the approximate dissociation constant.

Now if v is the volume of solution (in litres) which contains 1 gram equivalent of HA ($v = 1/c$, where c is the concentration in gram

*K is used (as opposed to k) in this text to denote thermodynamic constants.

equivalents per litre) and α is the degree of dissociation, then there will be present at equilibrium $(1 - \alpha)$ gram equivalents of unionized acid, HA, and α gram equivalents of each of the ions H^+ and A^-. The corresponding concentrations of unionized acid HA will then be $(1 - \alpha)/v$ and of the ions H^+ and A^- will be α/v, so that equation (2) may be expressed in the form:

$$k_a = \frac{\alpha^2}{(1-\alpha)v} \qquad (3)$$

For very weak electrolytes, α may be neglected in comparison with unity; the expression (3) then reduces to

$$k_a = \frac{\alpha^2}{v} = \alpha^2 c$$

$$\therefore \quad \alpha^2 = k_a/c \quad \text{and} \quad \alpha = \sqrt{(k_a/c)}$$

This relationship can now be used to derive an expression from which the pH of the solution can be determined as follows:

$$[H^+] = \alpha c$$

$$\therefore \quad [H^+] = c\sqrt{(k_a/c)}$$

$$= \sqrt{(k_a/c)}$$

Taking logarithms

$$\log [H^+] = \tfrac{1}{2}\log k_a + \tfrac{1}{2}\log c$$

$$\therefore \quad -\log[H^+] = -\tfrac{1}{2}\log k_a - \tfrac{1}{2}\log c$$

$$\therefore \quad pH = \tfrac{1}{2}(pk_a - \log c)$$

Similar equations may be derived for aqueous solutions of a weak base, the equilibrium system in this case being:

$$B + H_2O \rightleftharpoons BH^+ + OH^-$$

B represents the weak base and BH^+ the conjugate acid, whence

$$k_b = \frac{[BH^+]\cdot[OH^-]}{[B]}$$

$$= \frac{\alpha^2}{(1-\alpha)v} \simeq \frac{\alpha^2}{v}$$

where k_b = dissociation constant of the base at constant temperature
α = degree of ionisation
v = volume in litres containing one gram equivalent of the weak base.

Since

$$k_b = \alpha^2/v = \alpha^2 c$$

$$\alpha^2 = k_b/c$$

and

$$\alpha = \sqrt{(k_b/c)}$$

Now

$$[OH^-] = c\alpha$$

$$\therefore \quad [OH^-] = c\sqrt{(k_b/c)}$$
$$= \sqrt{(k_b/c}$$

But

$$[OH^-] = K_w/[H^+]$$
$$\therefore \quad K_w/[H^+] = \sqrt{(k_b/c)}$$
$$\therefore \quad [H^+] = K_w/\sqrt{(k_b c)}$$

Taking logarithms

$$\log [H^+] = \log K_w - \tfrac{1}{2}\log k_b - \tfrac{1}{2}\log c$$
$$\therefore \quad pH = pK_w - \tfrac{1}{2}(pk_b - \log c)$$

The strength of acids and bases

An acid may be defined as a substance which ionises to yield hydrogen ions or protons and a base as a substance which combines with hydrogen ions. An acid is, accordingly, a proton donor and a base a proton acceptor. Both can be defined by the expression

$$A \rightleftharpoons H^+ + B$$

where A is the acid and B is the base.

The strength of an acid is related to the concentration of hydrogen ions which it yields upon ionisation and will depend upon the value of the degree of dissociation, α, at any given concentration. The acid dissociation constant, k_a, gives a relationship between α and the concentration and, accordingly, is a measure of the acid strength. The strength of a base is likewise related to its dissociation constant.

The relationship between the dissociation constants of an acid (HA) and its conjugate base (A^-) can be derived from the equilibria concerned, i.e.

$$HA \rightleftharpoons H^+ + A^-$$
$$A^- + H_2O \rightleftharpoons HA + OH^-$$

Since

$$k_a = \frac{[H^+]\cdot[A^-]}{[HA]} \quad \text{and} \quad k_b = \frac{[HA]\cdot[OH^-]}{[A^-]}$$

$$k_a \cdot k_b = \frac{[H^+]\cdot[A^-]}{[HA]} \cdot \frac{[HA]\cdot[OH^-]}{[A^-]}$$
$$\therefore \quad k_a \cdot k_b = [H^+]\cdot[OH^-]$$
$$\therefore \quad k_a \cdot k_b = K_w$$

It follows from the expression that the dissociation constants of a weak acid and its conjugate base are complementary, i.e. the stronger the acid the weaker its conjugate base, and vice versa. For convenience in practice, the strength of a base is often expressed in terms of the dissociation constant of its conjugate acid.

The dissociation constant exponent (p*K*)

Dissociation constants of weak acids and bases are numerically small and a logarithmic notation is, therefore, convenient. The dissociation constant exponent, pK, is derived from the dissociation constant in a manner analogous to the derivation of pH from hydrogen ion concentration. Hence

$$pK_a = -\log_{10} K_a$$

It follows from this expression that the higher the value of K_a the smaller the value of pK_a, so that the stronger the acid the smaller the pK_a value.

The hydrolysis of salts

When salts are dissolved in water, interaction may occur with the ions of water and the resultant solution may become acid or alkaline, according to the nature of the salt. Such interaction is termed hydrolysis.

With an aqueous solution of a salt of the strong acid–strong base type, e.g. sodium chloride, neither do the anions have any tendency to combine with the hydrogen ions nor do the cations with the hydroxyl ions of water, since the related acids and bases are strong electrolytes and are, themselves, completely dissociated. The equilibrium between hydrogen and hydroxyl ions in water is, therefore, not disturbed and the solution remains neutral.

Weak base – strong acid salts

Consider the salt of a weak base and a strong acid, e.g. ammonium chloride. The chloride ions do not react significantly with hydrogen ions. The ammonium ions, on the other hand, are cations of a weak base and have a tendency to combine with hydroxyl ions to form undissociated ammonium hydroxide. Consequently, the hydroxyl ion concentration of the water will be decreased and the hydrogen ion concentration increased since the product $[H^+]\cdot[OH^-]$ must remain constant.

$$NH_4Cl \rightleftharpoons NH_4^+ + Cl^-$$

$$NH_4^+ + H_2O \rightleftharpoons NH_3 + H_3O^+$$

Its pH can be calculated from the following expression, which relates the pH of a solution of a weak acid with its dissociation constant, namely:

$$pH = \tfrac{1}{2}(pk_a - \log c)$$

by substituting in this expression for pk_a which equals pK_w – pk_b

$$pH = \tfrac{1}{2}(pK_w - pk_b - \log c)$$

Weak acid – strong base salts

Similarly, the salts of a strong base and a weak acid, e.g. potassium cyanide, will have an alkaline reaction because of hydrolysis represented by the equilibria:

$$KCN \rightleftharpoons K^+ + CN^-$$
$$CN^- + H_2O \rightleftharpoons HCN + OH^-$$

The pH of the solution can be calculated from the expression which relates the pH of a solution of a weak base to its dissociation constant, namely

$$pH = pK_w - \tfrac{1}{2}(pk_b - \log c)$$

Substituting in this expression for pk_b which equals $pK_w - pk_a$ we obtain the expression:

$$pH = pK_w - \tfrac{1}{2}pK_w + \tfrac{1}{2}pk_a + \tfrac{1}{2}\log c$$
$$\therefore \quad pH = \tfrac{1}{2}(pK_w + pk_a + \log c)$$

Weak acid – weak base salts

Both ions derived from the salt of a weak acid and a weak base, such as ammonium acetate, undergo hydrolysis in aqueous solution according to the equilibria:

$$NH_4^+ \rightleftharpoons H^+ + NH_3$$
$$CH_3COO^- + H_2O \rightleftharpoons CH_3COOH + OH^-$$

Summation of these equilibria gives the expression:

$$NH_4^+ + CH_3COO^- + H_2O \rightleftharpoons NH_3 + CH_3COOH + H^+ + OH^-$$

Provided the dissociation constants of the acid and base are not widely different, hydroxyl and hydrogen ions will be produced in approximately equal amounts. It is, therefore, permissible to subtract the water equilibrium $H_2O \rightleftharpoons H^+ + OH^-$ from the overall expression, to give the simplified expression:

$$\underset{(1-x)c}{NH_4^+} + \underset{(1-x)c}{CH_3COO^-} \rightleftharpoons \underset{xc}{NH_3} + \underset{xc}{CH_3COOH}$$

$$\therefore \quad k_h = \frac{[NH_3]\cdot[CH_3COOH]}{[NH_4^+]\cdot[CH_3COO^-]}$$

where k_h is the hydrolysis constant.

Introduction of the ionic product of water into this expression gives

$$k_h = \frac{[CH_3COOH]}{[H^+]\cdot[CH_3COO]^-} \times \frac{[NH_3]}{[NH_4^+]\cdot[OH^-]} \times [H^+]\cdot[OH^-]$$

$$\therefore \quad k_h = 1/k_a \cdot 1/k_b \cdot K_W$$
$$= K_w/k_a k_b$$

Consider a solution of ammonium acetate containing c mol l^{-1}. If the degree of hydrolysis is x, then

$$[CH_2COOH] = [NH_3] = xc$$

and

$$[CH_3COO^-] = [NH_4^+] = (1 - x)c$$

Substituting in the expression

$$k_h = \frac{[NH_3]\cdot[CH_3COOH]}{[NH_4^+]\cdot[CH_3COO^-]}$$

$$k_h = \frac{xc\cdot xc}{(1-x)c\cdot(1-x)c}$$

$$\therefore\ k_h = \frac{x^2}{(1-x)^2}$$

The pH of the solution may be calculated from a knowledge of the concentration of hydrogen ions in equilibrium with acetic acid as follows:

$$k_a = \frac{[H^+]\cdot[CH_3COO]}{[CH_3COOH]}$$

$$\therefore\ H^+ = \frac{k_a\cdot[CH_3COOH]}{[CH_3COO^-]}$$

$$= \frac{k_a\cdot xc}{(1-x)c}$$

Substituting for $x/(1-x)$ which equals $\sqrt{k_h}$:

$$[H^+] = k_a\cdot\sqrt{k_h}$$

and substituting for k_h which equals K_w/k_ak_b:

$$[H^+ = k_a\sqrt{(K_w/k_ak_b)}$$
$$= \sqrt{(K_wk_a/k_b)}$$

whence:

$$\log[H^+] = \tfrac{1}{2}\log K_w + \tfrac{1}{2}\log k_a - \tfrac{1}{2}\log k_b$$
$$\therefore\ pH = \tfrac{1}{2}(pK_w + pk_a - pk_b)$$

Buffer solutions

The resistance of a solution to changes in hydrogen ion concentration upon addition of small amounts of acid or alkali is termed **buffer action**; a solution which possesses such properties is known as a buffer solution. Buffer solutions usually consist of solutions containing a mixture of a weak acid or base and its salt. Buffer action in a solution of a weak acid and its salt is explained by the fact that hydrogen ions are removed by the anions of a weak acid to form unionised molecules, thus:

$$H^+ + A^- \rightarrow HA$$

Hydroxyl ions are also removed by neutralisation, according to the equation:

$$OH^- + HA \rightleftharpoons H_2O + A^-$$

The concentrations of hydrogen ion (relative to those of the weak acid and its salt) in such a buffer solution will be determined by the dissociation constant of the acid according to the expression:

$$k_a = \frac{[H^+]\cdot[A^-]}{[HA]}$$

$$\therefore \quad \log k_a = \log [H^+] + \log \frac{[A^-]}{[HA]}$$

$$\therefore \quad -\log k_a = -\log [H^+] - \log \frac{[A^-]}{[HA]}$$

$$\therefore \quad pk_a = pH - \log \frac{[A^-]}{[HA]}$$

$$\therefore \quad pH = pk_a + \log \frac{[A^-]}{[HA]}$$

Now, the acid HA is weak and only slightly ionised. Moreover, its ionisation is repressed by the relatively large concentration of anions A^- from the fully dissociated salt.

Hence, [HA] is numerically equal to the initial concentration of acid, and $[A^-]$ is numerically equal to the initial concentration of salt.

$$pH = pk_a + \log \frac{[salt]}{[acid]}$$

This is known as the Henderson equation.

If [salt] = [acid] the expression becomes

$$pH = pk_a + \log 1$$

$$\text{i.e.} \quad pH = pk_a$$

A tenfold increase or decrease of the ratio [salt]/[acid] would raise or lower the pH of the solution by one pH unit. The resistance of a buffer solution to such a pH change is a measure of the **buffer capacity**. This is defined as the number of gram equivalents of strong acid or strong alkali necessary to produce a change of 1 pH unit in 1 litre of the solution.

Consider a solution containing 0.5 mol l^{-1} of a monobasic acid and 0.5 mol l^{-1} of its salt. An increase of 1 pH unit will be brought about when the salt concentration has been raised to approximately 0.91 mol l^{-1} and the acid concentration reduced to approximately 0.09 mol l^{-1}. This change would require the addition of 0.41 mol l^{-1} of strong base. Hence the buffer capacity is 0.41 mol l^{-1}.

In a buffer solution which consists of a mixture of a weak base and its salt, hydroxyl ions are removed by the salt cations (BH^+) to form unionised molecules:

$$OH^- + BH^+ \rightarrow B + H_2O$$

Hydrogen ions are also removed by neutralisation,

$$H^+ + B \rightarrow BH^+$$

The concentrations of weak base and its salt relative to that of hydroxyl ion in the solution will be determined by the dissociation constant of the base, i.e.

$$k_b = \frac{[BH^+]\cdot[OH^-]}{[B]}$$

$$\therefore \quad \log k_b = \log [OH^-] + \log \frac{[BH^+]}{[B]}$$

$$\therefore \quad -\log k_b = -\log [OH^-] - \log \frac{[BH^+]}{[B]}$$

$$\therefore \quad pk_b = pOH - \log \frac{[BH^+]}{[B]}$$

but

$$pOH = pK_w - pH$$

$$\therefore \quad pk_b = pK_w - pH - \log \frac{[BH^+]}{[B]}$$

Hence,

$$pH = pK_w - pk_b - \log \frac{[BH^+]}{[B]}$$

Now, the base B is weak and only slightly ionised. Also its ionisation is repressed by the relatively large concentration of cations BH^+ from the fully dissociated salt. Hence, [B] is numerically equal to the initial concentration of base and $[BH^+]$ is numerically equal to the initial concentration of salt,

$$\therefore \quad pH = pK_w - pk_b - \log \frac{[\text{salt}]}{[\text{base}]} \qquad (4)$$

That this is merely a restatement of the Henderson equation is shown by substitution for $pK_w - pk_b = pk_a$ in eq. (4) when

$$pH = pk_a - \log \frac{[\text{acid}]}{[\text{salt}]}$$

and

$$pH = pk_a - \frac{[\text{salt}]}{[\text{acid}]}$$

This follows from the fact that BH^+ is the conjugate acid of base B, and that the latter, as before, can be equated to the concentration of salt.

Neutralisation indicators

It follows from the foregoing discussion of the behaviour of weak acids and bases, that the equivalence point in the titration of standard acids and alkalis will not always be the point of exact neutrality (pH 7.0). Coincidence of equivalence point and exact neutrality is attained only in strong acid–strong base titrations, since if either base or acid is weak the resulting salt will be hydrolysed and the solution will become either acidic or alkaline, respectively. The actual pH of the solution at the end point can be determined potentiometrically, or by means of a neutralisation indicator which changes colour according to the hydrogen ion concentration of the solution.

Indicators are weak acids or weak bases which have different colours in their conjugate base and acid forms (two-colour indicators); others are one-colour indicators and have one form coloured with a colourless conjugate form. Most indicators in common use are intensely coloured, and can be used in dilute solution in such small quantities that the acid-base equilibrium which is under examination is not disturbed by the addition of the indicator. As weak acids or weak bases, they are able to reach instantaneous equilibrium with the system and the colour of the solution will range between the extreme colours of the two forms as the proportion of acidic and basic forms automatically adjusts itself to the pH of the solution.

The following equilibrium will apply for an indicator functioning as a weak acid:

$$\underset{\text{unionised colour}}{HIn} \rightleftharpoons H^+ + \underset{\text{ionised colour}}{In^-}$$

In acid solution, the excess of H^+ ions will depress the ionisation of the indicator. The concentration of In^- will be small and that of HIn large and the colour will be that of the unionised form. Alkali will promote removal of hydrogen ions from the system with an increase in the concentration of the ionised form (In^-), so that the solution acquires the ionised colour.

Then

$$k_{In_a} = \frac{[H^+]\cdot[In^-]}{[HIn]}$$

$$[H^+] = k_{In_a}\frac{[HIn]}{[In^-]}$$

$$\therefore -\log [H^+] = -\log k_{In_a} - \log\frac{[HIn]}{[In^-]}$$

$$\therefore \quad pH = pk_{In_a} + \log\frac{[In^-]}{[HIn]}$$

Similarly, for an indicator functioning as a weak base, the following equilibrium will apply:

$$\underset{\text{unionized colour}}{In} + H_2O \rightleftharpoons \underset{\text{ionized colour}}{InH^+} + OH^-$$

and

$$k_{In_b} = \frac{[InH^+]\cdot[OH^-]}{[In]}$$

$$\therefore \quad pH = pK_w - pk_{In_b} - \log\frac{[InH^+]}{[In]}$$

Tautomeric neutralization indicators

Although the behaviour of indicators can be explained in terms of ionization of weak acids and bases, as above, the equilibrium is actually more complex, the colour changes being brought about by tautomeric changes in the structure of the molecule. This is illustrated by the behaviour of phenolphthalein in solution:

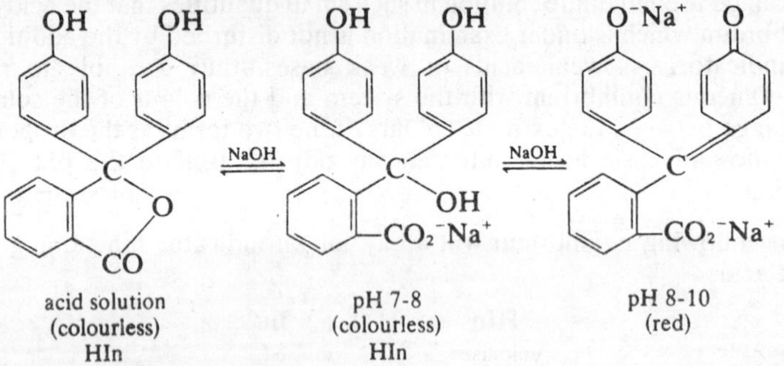

The red colour in alkaline solution is due to the quinonoid structure, with the resulting increased possibilities for resonance between the various ionic forms, as:

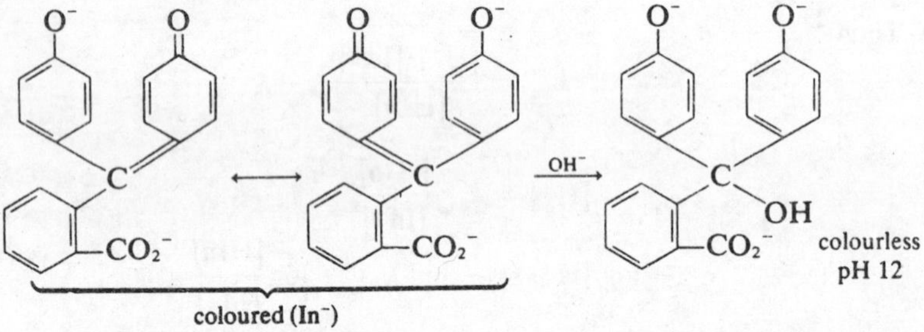

However, despite the complexity of equilibria such as these, for most practical purposes the expression

$$pH = pk_{In_a} + \log \frac{[In^-]}{[HIn]}$$

can be adopted.

The range of indicators

The observed colour of a two-colour indicator is determined by the ratio of the concentrations of ionized and unionized forms. Observed colour changes are, however, limited by the ability of the human eye to detect changes of colour in mixtures. This is particularly difficult where one colour predominates and, in practice, is almost impossssible when the ratio of the two forms exceeds 10 to 1. Thus the limit of visible colour change will be represented by the introduction of the term ± log 10 for + log $[In^-]/[HIn]$ in the above expression so that

$$pH = pk_{In_a} \pm 1$$

The average colour-change interval (range) of an indicator is, therefore, about two pH units. The observed colour changes within the indicator range are seen as a gradual change of tint or shade which ranges from one extreme colour to the other. The shade of colour is independent of the amount of indicator present, but the use of too much indicator should be avoided as slight changes are then more difficult to detect.

With a single-colour indicator, such as phenolphthalein, the intensity of colour is important, not shade difference. The actual concentration of indicator is, therefore, significant and should be carefully controlled.

Since the useful range of an indicator only extends over approximately two pH units, it is essential to have a series of indicators available to cover the complete pH scale. A list of such indicators in common use, together with their colour changes, is given in Table 3.1.

Precipitation and complex formation

Solubility product

Consider a solution of a slightly soluble salt, BA, which is in equilibrium with the solid phase BA:

$$\underset{\text{(solid)}}{BA} \rightleftharpoons \underset{\text{(solution)}}{B^+ + A^-}$$

Applying the law of mass action to this system,

$$K = \frac{[B^+]\cdot[A^-]}{[BA]}$$

The concentration of BA in solution will be constant in the presence of undissolved BA.

$$\therefore \quad K \times \text{constant} = [B^+]\cdot[A^-] = S_{BA}$$

where S_{BA} is a constant (at constant temperature) and is called the solubility product of the salt BA.

If the sparingly soluble salt has the general formula B_mA_n each molecule will furnish m cations and n anions:

$$B_mA_n \rightleftharpoons mB^+ + nA^-$$

and

$$S_{B_mA_n} = [B^+]^m\cdot[A^-]^n$$

In a saturated solution of the slightly soluble salt BA in water:

$$[B^+] = [A^-] = S$$

where S is the molar solubility of the salt, hence

$$S = \sqrt{S_{BA}}$$

Table 3.1 Indicator pH ranges and colour changes

Indicator	pH 1 2 3 4 5 6 7 8 9 10 11 12 13		
Alizarin Yellow G	Yellow	Orange	Orange-red
Alizarin Yellow GG	Colourless	Pale Yellow	Yellow
Bromocresol Red	Yellow	Green	Blue
Bromocresol Purple	Yellow	Grey	Purple
Bromophenol Blue	Yellow	Grey	Blue
Bromothymol Blue	Yellow	Green	Blue
Congo Red	Blue	Violet	Red
Cresol Red	Red	Orange	Yellow
	Yellow	Pink	Red-violet
Dimethyl Yellow	Red	Orange	Yellow
Litmus	Red	Violet	Blue
Metacresol Purple	Red	Orange	Yellow
	Yellow	Grey	Violet
Metanil Yellow	Red	Orange	Yellow
Methyl Orange	Red	Orange	Yellow
Methyl Orange – Xylene Cyanol FF	Violet	Grey	Green
Methyl Red	Red	Orange	Yellow
1-Naphtholphthalein	Pale Red	Violet	Blue
Naphthol Yellow	Colourless	Pale Yellow	Yellow
Neutral Red	Red	Orange-Red	Orange
Phenolphthalein	Colourless	Pink	Red
Phenol Red	Yellow	Pink	Red
Thymol Blue	Red	Orange	Yellow
	Yellow	Grey	Blue
Thymolphthalein	Colourless	Pale Blue	Blue
Titan Yellow	Yellow	Orange	Red
Tropaeolin O	Yellow	Orange-yellow	Orange
Tropaeolin OO	Red	Orange	Yellow

Common-ion effect

In any system in which solid is in equilibrium with its solution, the product of the ion concentrations is determined by the solubility product. Thus, if an excess of silver ions is added to a saturated solution of silver chloride in water, the solubility product [Ag][Cl] is exceeded and, consequently, some silver chloride will be precipitated, equilibrium being reached when the product of the silver and chloride ion concentrations becomes equal to the solubility product. A similar effect will occur if an excess of chloride ions is added to a saturated silver chloride solution. A compound having an ion in common with a slightly soluble salt decreases the solubility of the latter (common-ion effect). The extent of the depression of the solubility can be calculated if the excess of the common ion is known.

Example Calculate the solubilities of silver chloride in 0.001M, 0.01M and 0.1M potassium chloride, respectively.

In 0.001M potassium chloride $[Cl^-] = 10^{-3}$.

$$\therefore \quad [Ag^+] = \frac{S_{AgCl}}{[Cl^-]} = \frac{10^{-10}}{10^{-3}} = 10^{-7}$$

In 0.01M potassium chloride

$$[Ag^+] = \frac{10^{-10}}{10^{-2}} = 10^{-8}$$

In 0.1M potassium chloride

$$[Ag^+] = \frac{10^{-10}}{10^{-1}} = 10^{-9}$$

In the above equations the concentration of Cl^- ions furnished by the silver chloride itself is neglected, since it is very small compared with the concentration of the excess Cl^- ions added.

Depression of solubility by the common-ion effects is of fundamental importance in gravimetric analysis. Addition of a suitable excess of a precipitating agent usually decreases the solubility of a precipitate to such an extent that the loss by washing is neglible. On the other hand, a large excess of precipitant must be avoided, one reason being that the salt effect counteracts the common-ion effect.

Influence of temperature upon solubility

The solubility of the precipitates encountered in quantitative analysis increases in varying degrees with rises of temperature. In many instances the common-ion effect reduces the solubility to so small a value that the temperature effect, which might otherwise be appreciable, becomes very small. It is, however, significant with magnesium ammonium phosphate hexahydrate and silver chloride, which are usually filtered at room temperature to avoid appreciable solubility loss.

Complex ions

Increase in solubility of a precipitate upon the addition of excess of the precipitating agent can also occur. This is frequently due to the formation of a complex ion. A complex ion is formed by the union of a simple ion either with ions of opposite charge or with neutral molecules. Thus, when potassium cyanide is added to a solution of silver nitrate, a white precipitate of silver cyanide is first formed because the solubility product of silver cyanide is exceeded. The precipitate dissolves on the addition of excess of potassium cyanide due to the formation of the complex ion $[Ag(CN)_2]^-$

$$AgCN \text{ (solid)} + CN^- \text{ (excess)} \rightleftharpoons [Ag(CN)_2]^-$$

That the complex ion itself dissociates is shown by the fact that silver sulphide can be precipitated from the solution with hydrogen sulphide, dissociation occurring as follows:

$$[Ag(CN)_2]^- \rightleftharpoons Ag^+ + 2CN^-$$

Application of the law of mass action to the equilibrium gives:

$$K = \frac{[Ag^+]\cdot[CN^-]^2}{[Ag(CN)_2]^-}$$

in which K is the dissociation or instability constant of the complex ion. The experimentally determined value of K is very small ($K = 1.0 \times 10^{-21}$); also, the dissociation of the complex ion is repressed by the excess of cyanide ions present in the solution. Hence, the silver ion concentration is reduced to such a low level that the solubility product of silver cyanide is not exceeded.

Indicators in argentimetric titrations

Potassium chromate (Mohr method)

Potassium chromate is used as an indicator in the titration of chloride ions with standard silver nitrate in neutral solution, giving a precipitate of red silver chromate at the end point. The process is one of fractional precipitation of a pair of sparingly soluble salts — silver chloride ($S_{AgCl} = 1.56 \times 10^{-10}$ at 25°) and silver chromate ($S_{Ag_2CrO_4} = 9 \times 10^{-12}$ at 25°).

As the titration proceeds, silver chloride, the least soluble salt, will be precipitated so long as the chloride ion concentration is significant. The end point is reached when chloride ions are still present ($[Cl^-] = \sqrt{S_{AgCl}} = 1.249 \times 10^{-5}$) since silver chromate will commence to precipitate as soon as its solubility product is exceeded. This occurs when:

$$[Ag^+]\cdot[Cl^-] = S_{AgCl} = 1.56 \times 10^{-10}$$

$$[Ag^+]^2\cdot[CrO_4{}^{2-}] = S_{Ag_2CrO_4} = 9 \times 10^{-12}$$

$$\therefore \quad [Ag^+] = S_{AgCl}/[Cl^-] = \sqrt{\{S_{Ag_2CrO_4}/[CrO_4{}^{2-}]\}}$$

$$\therefore \quad \frac{[Cl^-]}{\sqrt{[CrO_4^{2-}]}} = \frac{S_{AgCl}}{\sqrt{S_{Ag_2CrO_4}}}$$

$$= \frac{1.56 \times 10^{-10}}{\sqrt{(9 \times 10^{-12})}}$$

$$= 5.2 \times 10^{-5}$$

$$\therefore \quad [CrO_4^{2-}] = \left(\frac{[Cl^-]}{5.2 \times 10^{-5}}\right)^2 = \left(\frac{1.249 \times 10^{-5}}{5.2 \times 10^{-5}}\right)^2$$

$$= 5.77 \times 10^{-2}$$

Hence potassium chromate must be present at a concentration of at least 0.0058M if precipitation is to occur at the end point. This calculation is based on solubility product figures determined at 25°; if the laboratory temperature is lower these will be significantly smaller, the corresponding potassium chromate concentration to ensure precipitation at 15° being 0.015M.

In practice, two other factors also operate. High concentrations of potassium chromate tend to obscure the end point and it is also necessary to add a small excess of silver nitrate (*c.* 0.05 ml 0.1M $AgNO_3$) before the eye can detect a change. The errors involved can be corrected by a blank determination carried out with the same volume of indicator (usually 1 ml of 5% solution) in sufficient water to reproduce the indicator concentration which obtains at the end point of the actual determination. The blank titration volume is subtracted from the determination titre.

Iron(III) thiocyanate (Volhard method)

Ferric alum (iron(III) ammonium sulphate) is used as an indicator for the titration of silver ions with ammonium thiocyanate solution in the presence of nitric acid. This method forms the basis of the Volhard determination of chlorides, bromides and iodides in acid solution. When excess of standard silver nitrate solution is added to a solution of the halide, precipitation of silver halide occurs:

$$MX + AgNO_3 \rightarrow MNO_3 + AgX \downarrow$$

The excess of silver nitrate is then back-titrated with ammonium thiocyanate:

$$AgNO_3 + NH_4SCN \rightarrow AgSCN \downarrow + NH_4NO_3$$

At the end point, excess thiocyanate reacts with ferric iron to give the red colour due to formation of iron(III) hexathiocyanoferrate(III):

$$2Fe^{3+} + 6SCN^- \rightleftharpoons Fe^{3+}[Fe(SCN)_6]^{3-}$$

In the determination of a chloride by this method, it is necessary to consider only the equilibria:

$$Ag^+ Cl^- \rightleftharpoons AgCl \downarrow$$

$$Ag^+ + SCN^- \rightleftharpoons AgSCN \downarrow$$

Both 'insoluble' salts are in equilibrium with the solution, so that

$$[Ag^+]\cdot[Cl^-] = S_{AgCl} = 1.56 \times 10^{-10}$$
$$[Ag^+]\cdot[SCN^-] = S_{AgSCN} = 1.16 \times 10^{-12}$$
$$\therefore \quad [Cl^-]/[SCN^-] = S_{AgCl}/S_{AgSCN} = 1.56 \times 10^{-10}/1.16 \times 10^{-12} = 134$$

Because silver thiocyanate is less soluble than silver chloride, it follows that when the equivalence point is reached, further addition of thiocyanate would result in reaction with the precipitated silver chloride and the reaction would proceed until the $[Cl^-]/[SCN^-]$ ratio reached 134, before any reaction occurred with the indicator. This is prevented, in practice, either by filtering out the precipitated chloride ions before back-titration or, alternatively, by addition of a small quantity of nitrobenzene (*c.* 1 ml) which assists coagulation of the precipitate and coats the coagulated particles with a film of oil, thus preventing reaction with the thiocyanate ions.

The corresponding ratio of $[Br^-]/[SCN^-]$ which controls the end point in the Volhard titration of bromides is very much smaller, being only 0.66 at 25°. There is little or no titration error in practice and there is, in consequence, no need for removal or treatment of the silver bromide precipitate. Similarly, the titration error with iodides is negligible.

Adsorption indicators

Adsorption indicators are acidic or basic dyes which change colour on adsorption on to the precipitate at the end point. Silver chloride precipitated during the titration of sodium chloride by silver nitrate adsorbs chloride ions on to the surface of the precipitate to form a layer of adsorbed ions. This layer of negatively charged ions in turn promotes secondary adsorption of oppositely charged ions (cations) present in solution, as shown in Fig. 3.1(*a*). At the end point some of the silver ions, now present in excess, will also be adsorbed on other particles of precipitate and will also constitute a primary adsorption layer, as in Fig. 3.1(*b*). The formation of the secondary adsorption layer involves competition between the anions present in the solution and the dye (usually fluorescein or dichlorofluoroscein) which is adsorbed preferentially. Combination of the dye and Ag^+ ions on the surface of the precipitate gives the characteristic orange end point colour.

Recommended adsorption indicators. Fluorescein (adsorbed colour pink) is suitable for the titration of chlorides, but since it is a very weak acid ($pK_a = 8.0$) it cannot be used in acid solution. Dichlorofluorescein (adsorbed colour orange), which is a stronger acid, is suitable for use in weakly acid solution ($pH > 4.5$).

Phenosafranine is suitable for the titration of both chlorides and bromides. It acts by a mechanism different from that of the other adsorption indicators, in that the indicator is adsorbed throughout the titration, the adsorbed colour changing at the end point from pink to pale lilac (chlorides) and to deep lilac (bromides).

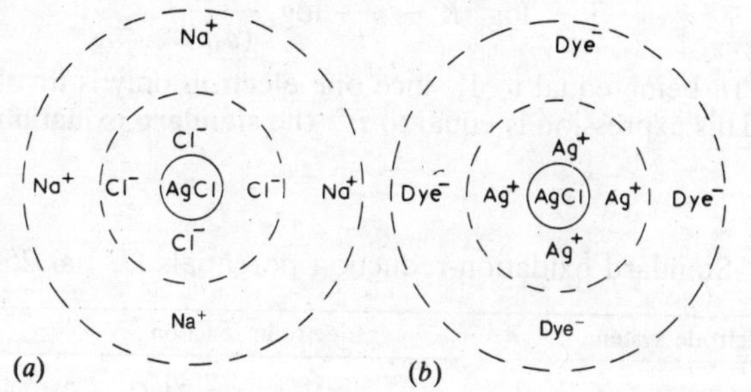

Fig. 3.1 Adsorption of indicators

Oxidation-reduction titrations

Oxidation and reduction can be defined in terms of loss or gain of electrons. A definite equilibrium exists in any oxidation-reduction system if both oxidised and reduced forms are present. This is shown in the tendency of the system to gain or lose electrons from an inert metal (Pt) electrode in contact with the system. Such an electrode would assume a definite potential, losing electrons to become positively charged in an oxidising system and gaining electrons to become negatively charged in a reducing system.

Standard oxidation-reduction potentials

Oxidation-reduction potentials may be calculated by measuring the potential difference of a cell in which the oxidation-reduction half cell is coupled with a standard reference electrode. This is usually the normal hydrogen electrode, the potential of which is taken as zero. Thus, for the ferrous-ferric system, the cell would be:

$$\mathrm{Pt}\left|\begin{matrix}\mathrm{Fe}^{3+}\\ \mathrm{Fe}^{2+}\end{matrix}\right\|\mathrm{H}^+\left|\mathrm{Pt},\ \mathrm{H}_2\right.$$

In this system the cell reaction is $Fe^{3+} + \frac{1}{2}H_2 \rightleftharpoons Fe^{2+} + H^+$ and application of the van't Hoff isotherm gives the expression:

$$-\Delta G = nFE = RT \log_e K - RT \log_e \frac{a_{Fe^{2+}} \cdot a_{H^+}}{a_{Fe^{3+}} \cdot (a_{H_2})^{1/2}}$$

$$\therefore \quad E = \frac{RT}{nF} \log_e K - \frac{RT}{nF} \log_e \frac{a_{Fe^{2+}} \cdot a_{H^+}}{a_{Fe^{3+}} \cdot (a_{H_2})^{1/2}}$$

If the potential of the hydrogen electrode is taken as zero. then the e.m.f. of the cell is equal to the oxidation-reduction potential of the particular ferrous-ferric system under examination.

The expression

$$\frac{RT}{F}\cdot\log_e K - \frac{RT}{F}\log_e\frac{a_{H^+}}{(a_{H_2})^{1/2}}$$

is constant (n being equal to 1, since one electron only is involved in the reaction). This expression is equal to $E°$, the **standard oxidation-reduction potential**.

Table 3.2 Standard oxidation-reduction potentials ($E°$) at 25°

Electrode system	Electrode reaction	$E°$ (volts)
MnO_4^-, MnO_2/Pt	$MnO_4^- + 4H + 3e \rightleftharpoons MnO_2 + 2H_2O$	+1.59
MnO_4^-, Mn^{2+}/Pt	$MnO_4^- + 8H^+ + 5e \rightleftharpoons Mn^{2+} + 4H_2O$	+1.52
Ce^{4+}, Ce^{3+}/Pt	$Ce^{4+} + e \rightleftharpoons Ce^{3+}$	+1.45
$Cr_2O_7^{2-}$, Cr^{3+}/Pt	$Cr_2O_7^{2-} + 14H^+ + 6e \rightleftharpoons 2Cr^{3+} + 7H_2O$	+1.36
IO_3^-, I_2/Pt	$IO_3^- + 6H^+ + 5e \rightleftharpoons I_2 + 3H_2O$	+1.20
Fe^{3+}, Fe^{2+}/Pt	$Fe^{3+} + e \rightleftharpoons Fe^{2+}$	+0.77
H_3AsO_4, H_3AsO_3/Pt	$H_3AsO_4 + 2H^+ + 2e \rightleftharpoons H_3AsO_3 + H_2O$	+0.56
I_2, I^-/Pt	$I_2 + 2e \rightleftharpoons 2I^-$	+0.53
$[Fe(CN)_6]^{3-}$, $[Fe(CN)_6]^{4-}$/Pt	$[Fe(CN)_6]^{3-} + e \rightleftharpoons [Fe(CN)_6]^{4-}$	+0.36
H^+, H_2/Pt	$H^+ + e \rightleftharpoons \frac{1}{2}H_2$	0.00
Ti^{4+}, Ti^{3+}/Pt	$Ti^{4+} + e \rightleftharpoons Ti^{3+}$	−0.06
S, S^{2-}/Pt	$D + 2e \rightleftharpoons S^{2-}$	−0.51

Then at temperature T,

$$E_T = E° + \frac{RT}{F}\cdot\log_e\frac{a_{Fe^{3+}}}{a_{Fe^{2+}}}$$

and when both ferrous and ferric ions are present in equal concentrations the second term disappears, and $E_T = E°$. That is, the e.m.f. of the cell is a direct measure of the standard oxidation-reduction potential when the concentration (= activity) of oxidised and reduced forms of the system are equal. In practice, accurate measurement of the standard O/R potential of the ferrous-ferric system must be carried out in the presence of hydrochloric acid to depress hydrolysis. This necessitates extrapolation of the results to zero acid concentration. The standard oxidation-reduction potentials of a number of systems in common use for volumetric analysis are given in Table 3.2.

The greater the value of the oxidation-reduction potential, the more powerful the oxidising agent and, conversely, the lower the value the more powerful the reducing agent. Thus, the position in the table of an oxidation-reduction system is an indication of its ability to oxidise or reduce other systems. Any system will oxidise any other system which occurs below it in the table and, similarly, will reduce any system situated above it in the table.

It is important to distinguish between the strength of an oxidising or reducing agent as expressed by its oxidation-reduction potential, and its actual capacity to oxidise or reduce. Thus, strong oxidising or reducing agents, as implied by the value of $E°$ for the system, may exhibit only limited ability actually to oxidise or reduce, due to buffering of the system. Such systems are said to be well poised.

The general equation for the calculation of oxidation-reduction potentials

In a single oxidation-reduction system represented by the equilibrium:

$$\underset{(\text{ox})}{\text{Oxidised form}} + ne \rightleftharpoons \underset{(\text{red})}{\text{Reduced form}}$$

where n is the number of electrons involved in the process, the electrode potential is given by the expression:

$$E_T = E° - \frac{RT}{nF}\cdot\log_e \frac{a_{red}}{a_{ox}}$$

$$= E° + \frac{RT}{nF}\cdot\log_e \frac{a_{ox}}{a_{red}}$$

where E_T is the observed potential at $T°$ absolute and $E°$ the standard-oxidation potential of the system.

Substituting in this expression for R (8.313 joules), F (96 500 coulombs), taking T at 298°K (25°), converting to common logarithms and expressing activities in terms of concentration then

$$E_{25} = E° + \frac{0.0592}{n}\cdot\log_{10}\frac{[\text{ox}]}{[\text{red}]}$$

The actual oxidation-reduction potential of a system can be calculated from a knowledge of the standard potential and the percentage of oxidised and reduced forms which are present, by substituting in the general equation:

$$E_T = E° + \frac{RT}{nF}\cdot\log_e\frac{[\text{ox}]}{[\text{red}]}$$

The results of such calculations can be expressed graphically, as shown in Fig. 3.2 which shows the relationship between the calculated oxidation-reduction potential of a number of systems and the percentage of oxidised forms present. The curves for systems such as MnO_4^-/Mn^{2+} which are dependent upon the hydrogen ion concentration are calculated on the basis that it is molar. An increase or decrease of hydrogen ion concentration could either raise or lower the position of the curve in the table.

Oxidation of iron(II) by permanganate

Ferrous iron is oxidized to the ferric state by potassium permanganate in acid solution according to the following equation:

$$MnO_4^- + 5Fe^{2+} + 8H^+ \rightleftharpoons Mn^{2+} + 5Fe^{3+} + 4H_2O$$

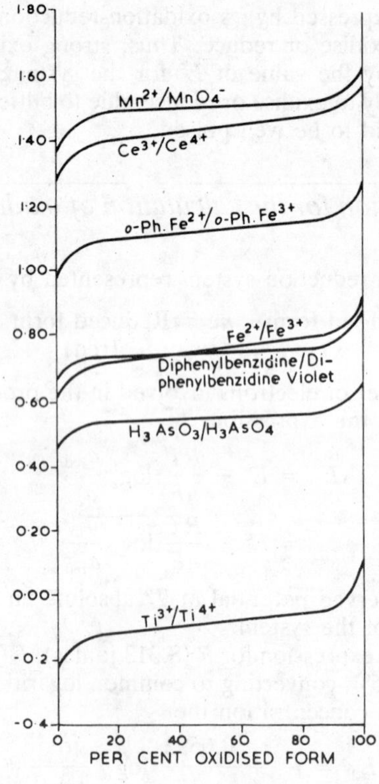

Fig. 3.2 Oxidation-reduction potentials

Since the reaction is carried out in dilute aqueous solution the water term of the equilibrium is constant and the equilibrium constant, K is given by the expression:

$$K = \frac{[Mn^{2+}]\cdot[Fe^{3+}]^5}{[MnO_4^-]\cdot[Fe^{2+}]^5\cdot[H^+]^8}$$

The reaction can now be considered as the summation of two separate oxidation-reduction systems as follows:

$$MnO_4^- + 8H^+ + 5e \rightleftharpoons Mn^{2+} + 4H_2O$$
$$5[Fe^{2+} \rightleftharpoons Fe^{3+} + e]$$

An equation for the oxidation-reduction potential of each system can be written:

$$E = 1.52 + \frac{0.0592}{5}\cdot\log\frac{[MnO_4^-]\cdot[H^+]^8}{[Mn^{2+}]}$$

and

$$E = 0.77 + \frac{0.0592}{5} \cdot \log \frac{[Fe^{3+}]^5}{[Fe^{2+}]^5}$$

$E°$ being 1.52 and 0.77 for the two systems, respectively.

Considering these two electrode systems as being combined together in a cell, the e.m.f. at equilibrium will be zero and the two equations can be equated,

$$\therefore \quad 1.5222 + \frac{0.0592}{5} \cdot \log \frac{[MnO_4^-] \cdot [H^+]^8}{[Mn^{2+}]} = 0.77 + \frac{0.0592}{5} \cdot \frac{[Fe^{3+}]^5}{[Fe^{2+}]^5}$$

$$\therefore \quad \log \frac{[Mn^{2+}] \cdot [Fe^{3+}]^5}{[MnO_4^-] \cdot [H^+]^8 \cdot [Fe^{2+}]^5} = \frac{5\,(1.52-0.77)}{0.0592}$$

$$\therefore \quad \log K = 63.34 \quad \text{and} \quad K = 2.19 \times 10^{63}$$

It is clear from the large value of K, the equilibrium constant, that the reaction will proceed to completion, the residual ferrous iron concentration being of negligible proportion.

Oxidation of iron(II) by cerium(IV) sulphate

Ferrous iron can also be oxidised to the ferric state by cerium(IV) sulphate according to the equation:

$$Ce^{4+} + Fe^{2+} \rightleftharpoons Ce^{3+} + Fe^{3+}$$

and

$$K = \frac{[Ce^{3+}] \cdot [Fe^{3+}]}{[Ce^{4+}] \cdot [Fe^{2+}]}$$

The two separate oxidation-reduction systems are the ferrous-ferric and the ceric-cerous electrode reactions, which may be expressed as follows:

$$E_{Fe^{3+}/Fe^{2+}} = 0.77 + 0.0592 \log \frac{[Fe^{3+}]}{[Fe^{2+}]}$$

and

$$E_{Ce^{4+}/Ce^{3+}} = 1.45 + 0.0592 \log \frac{[Ce^{4+}]}{[Ce^{3+}]}$$

Hence at equilibrium:

$$\log \frac{[Ce^{3+}] \cdot [Fe^{3+}]}{[Ce^{4+}] \cdot [Fe^{2+}]} = \frac{1.45 - 0.77}{0.0592}$$

$$\therefore \quad \log K = 11.62 \quad \text{and} \quad K = 4.17 \times 10^{11}$$

Again, the large value of K indicates that the reaction will proceed to completion.

Reduction of iron(III) by titanium(III) chloride

The reduction of ferric iron to the ferrous state by titanium(III) chloride (or sulphate) may be expressed by the following equation:

$$Fe^{3+} + Ti^{3+} \rightleftharpoons Fe^{2+} + Ti^{4+}$$

and

$$K = \frac{[Fe^{2+}]\cdot[Ti^{4+}]}{[Fe^{3+}]\cdot[Ti^{3+}]}$$

The separate oxidation-reduction systems may be expressed:

$$E_{Fe^{3+}/Fe^{2+}} = 0.77 + 0.0592 \log \frac{[Fe^{3+}]}{[Fe^{2+}]}$$

and

$$E_{Ti^{4+}/Ti^{3+}} = -0.06 + 0.0592 \log \frac{[Ti^{4+}]}{[Ti^{3+}]}$$

Hence, at equilibrium:

$$\log \frac{[Fe^{2+}]\cdot[Ti^{4+}]}{[Fe^{3+}]\cdot[Ti^{3+}]} = \frac{0.77 + 0.06}{0.0592}$$

$$\therefore \quad \log K = 14.03 \quad \text{and} \quad K = 1.07 \times 10^{14}$$

As in the above examples, the large value of K indicates that the reaction will proceed to completion.

Oxidation of arsenic trioxide by iodine

Consider the oxidation of arsenious acid ($As_2O_3 \equiv H_3AsO_3$) to arsenic acid ($As_2O_5 \equiv H_3AsO_4$) by iodine:

$$H_3AsO_3 + I_2 + H_2O \rightleftharpoons H_3AsO_4 + 2H^+ + 2I^-$$

Assuming the concentration of water to be constant, then

$$K = \frac{[H_3AsO_4]\cdot[I^-]^2\cdot[H^+]^2}{[H_3AsO_3]\cdot[I_2]}$$

The reaction can now be considered as the summation of the two separate oxidation-reduction systems:

$$H_3AsO_4 + 2H^+ + 2e \rightleftharpoons H_3AsO_3 + _2O$$

$$I_2 + 2e \rightleftharpoons 2I^-$$

Hence,

$$E = 0.56 + \frac{0.0592}{2}\cdot\log \frac{[H_3AsO_4]\cdot[H^+]^2}{[H_3AsO_3]}$$

and

$$E = 0.53 + \frac{0.0592}{2}\cdot\log \frac{[I_2]}{[I^-]^2}$$

$$\therefore \quad 0.56 + \frac{0.0592}{2}\cdot\log \frac{[H_3AsO_4]\cdot[H^+]^2}{[H_3AsO_3]} = 0.53 + \frac{0.0592}{2}\cdot\log \frac{[I_2]}{[I^-]^2}$$

$$\therefore \quad \log \frac{[H_2AsO_4]\cdot[H^+]^2\cdot[I^-]^2}{[H_3AsO_3]\cdot[I_2]} = \frac{2(0.53 - 0.56)}{0.0592}$$

$$\therefore \quad \log K = -\frac{0.06}{0.0592}$$
$$= -1.014$$

and

$$K = 9.68 \times 10^{-2}$$

The very small value of K indicates that the reaction does not proceed to completion and, in fact, the equilibrium is displaced from right to left. The reaction can be displaced completely from left to right if it is carried out in the presence of sodium bicarbonate to remove hydrogen ions.

Irreversible oxidations and reductions

Many oxidations and reductions are not based on reversible electrode reactions. This applies generally to the oxidation and reduction of covalent compounds, when the necessary loss or gain of electrons takes place very much less readily than with similar ionic processes. The loss of a volatile oxidation (or reduction) product, such as carbon dioxide in the oxidation of oxalic acid by potassium permanganate, similarly provides for an irreversible form of reaction:

$$5[(COOH)_2 \rightarrow 2CO_2 + 2H^+ + 2e]$$
$$2[MnO_4^- + 8H^+ + 5e \rightleftharpoons Mn^{2+} + 4H_2O]$$

Summation of these two equations gives:

$$2MnO_4^- + 5(COOH)_2 + 6H^+ \rightarrow 2Mn^{2+} + 10CO_2 + 8H_2O$$

This last equation shows that the oxalic acid alone is unable to supply sufficient hydrogen ions for the reaction and this explains the necessity for conducting the titration in the presence of dilute sulphuric acid.

Since the overall reaction is irreversible, the law of mass action does not apply and no equilibrium constant can be calculated.

Other examples of non-reversible oxidation and reduction used in quantitative analysis include the oxidation of iodides with potassium iodate, the reduction of iodine with sodium thiosulphate and the oxidation of sulphydryl (-SH) compounds with iron(II) or iodine.

The rate of oxidation-reduction reactions

It is possible to calculate the equilibrium constant of reversible oxidation-reduction reactions but this gives no indication of the speed at which equilibrium will be reached. In many cases the reaction is quite slow and can only be used as the basis for quantitative titration in the presence of a catalyst which increases the reaction velocity. Thus, the oxidation of oxalic acid by potassium permanganate is extremely slow at room temperature and in the absence of manganous ions. The reaction velocity is increased to a reasonable rate by heating to 60°. Even so, the reaction is slow to start, and potassium permanganate should be added cautiously at first until the necessary catalytic quantities of manganese(II) ion can be built up.

Similarly, in the standardization of ceric sulphate by arsenic trioxide, the reaction is slow at room temperature but can be catalysed by the addition of a trace of osmium tetroxide.

Oxidation-reduction indicators

Self-indicating. Potassium permanganate is a good example; its solutions are so intensely coloured that a single drop will impart a definite pink colour at the end point of a titration to a comparatively large volume of solution. Cerium(IV) sulphate, which is yellow, and iodine (brown), both reduce to colourless ions and are, themselves, sufficiently deeply coloured to provide good visual end points. The only objection to the use of self-indicating titrants is that the visual end point represents a slight over-titration.

External indicators. These are little used now. Some depend on the fact that excess titrant present immediately after the end point gives a visible reaction with some specific reagent. Others are based on some visible reaction of the titrated substance with a suitable reagent, so that the end point is marked by failure to elicit the reaction. An example of the latter type is the use of potassium hexacyanoferrate(III) as external indicator in the titration of iron(II) by potassium dichromate. Drops of the solution removed to a spotting tile during the titration will give a deep prussian blue colour with potassium hexacyanoferrate(III) because ferrous ions are still present. At the end point, iron(III) only is present and this does not give a colour with potassium hexacyanoferrate(III).

Internal indicators. Consider the oxidation of iron(II) by potassium permanganate in molar acid solution. As the titration proceeds, the changes in the oxidation potential of the solution will follow a curve (Fig. 3.2) which is given by the sum of the separate curves of the two systems. At the end point, when all the iron has been oxidised to the ferric state, there is a steep rise in potential and this can be observed either potentiometrically or by the use of oxidation-reduction indicators.

Oxidation-reduction indicators are substances which have different colours in their oxidised and reduced form. Ideally, the reactions should be reversible and should give a precise and easily observable colour change the end point. The two forms of the indicator comprise an oxidation-reduction system, the standard oxidation-reduction potential of which is intermediate between that of the titrated system and the titrant:

$$\mathrm{In_{ox}} + ne \rightleftharpoons \mathrm{In_{red}}$$

Most oxidation-reduction indicators are dyes, the reduced or leuco forms of which are colourless. Since dyes are intensely coloured, the indicator can be used at such a low concentration that there is no interference with the system under examination.

Consider an indicator at a potential of 0.06 volt above that of its standard oxidation potential (E°)

$$\therefore \quad E_T - E^\circ = 0.06 = 0.0592 \log_{10} \frac{[ox]}{[red]}$$

$$\therefore \quad \log_{10} \frac{[ox]}{[red]} = \frac{0.06}{0.0592}$$

$$= 1.014$$

$$\therefore \quad \frac{[ox]}{[red]} = \frac{10.33}{1}$$

The indicator is, therefore, over 90% oxidised and its colour will be indistinguishable from that of the oxidised form. Similarly, when the indicator is at a potential 0.06 volt lower than that of its E°, over 90% of the indicator will be reduced and its colour, for all practical purposes, will be that of the reduced form. Such an indicator will change colour over a potential range which is given by $E^\circ_{In} \pm 0.06$ volt. If the indicator is to be of practical use in a particular titration, this potential range must not overlap that of either oxidation-reduction equilibrium concerned in the reaction. This is ensured if the potential differences $(E^\circ_1 - E^\circ_{In})$ and $(E^\circ_{In} - E^\circ_2)$ are not less than 0.15 volt. The above calculation is based on a working temperature of 25°. Sharper changes over smaller potential ranges will be obtained at lower temperatures since the value of the denominator in the above expression will be smaller.

Diphenylamine; sodium diphenylaminesulphonate; *N,N'*-diphenylbenzidine. These three indicators, which are all used as solutions in concentrated sulphuric acid, are considered together since their mechanism of action is similar. The first step in the oxidation of diphenylamine (I) is its irreversible conversion into diphenylbenzidine (II):

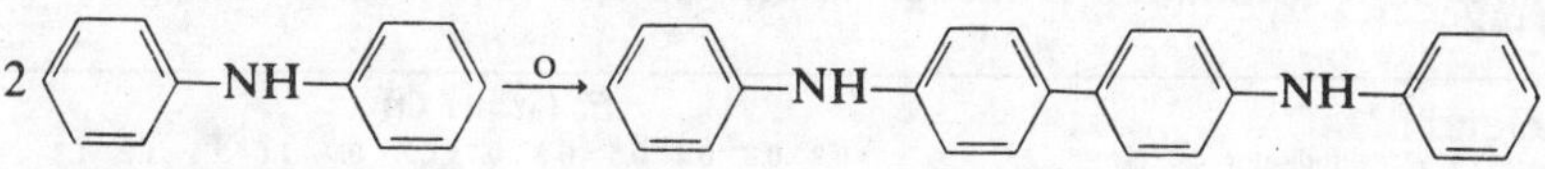

I II

Diphenylbenzidine (II) is colourless, but is reversibly oxidized into diphenylbenzidine violet (III):

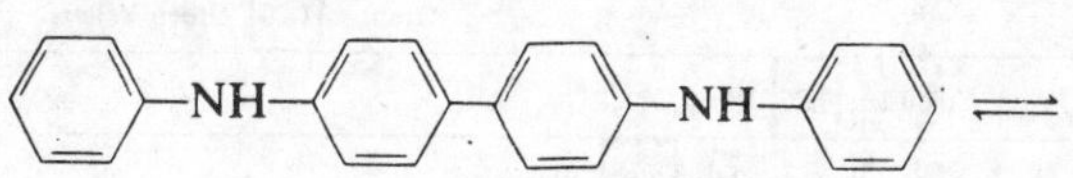

II

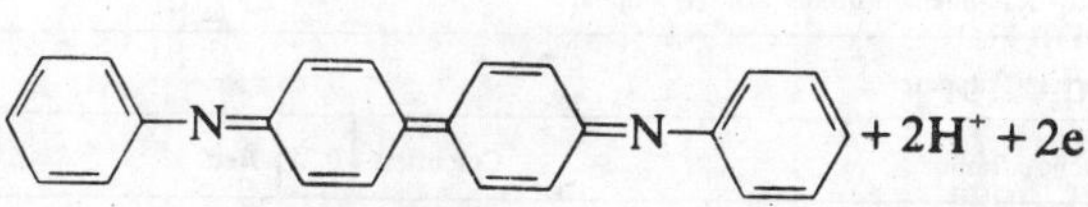

III

The oxidation potential of this system is 0.76 volt and it is used as an indicator in the titration of ferrous iron by potassium dichromate. The standard oxidation-reduction potentials $E^\circ_{Cr_2O_7^{2-}/Cr^{3+}}$ and $E^\circ_{Fe^{3+}/Fe^{2+}}$ are 1.36 and 0.77 volt, respectively, so that diphenylamine is only able to function as an indicator in this reaction when phosphoric acid is present in the solution. This decreases the concentration of ferric ions in the solution by complex formation and, hence, reduces the actual potential of the ferrous-ferric system to a level which is sufficiently low to permit the indicator to function.

Sodium diphenylaminesulphonate acts by a similar mechanism, its oxidation-reduction potential E° being 0.83 volt, but, unlike diphenylamine and diphenylbenzidine, it is readily soluble in water.

Ferroin Sulphate (Tris(1,10-phenanthroline)-iron(II)sulphate). Ferroin sulphate is a bright red complex formed by combination of the base, 1,10-phenanthroline with iron(II) sulphate:

3 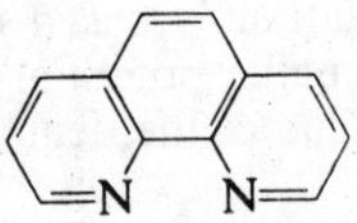 $+ Fe^{2+} \longrightarrow [(C_{12}H_8N_2)_3Fe]^{2+}$

This complex is readily oxidised reversibly to the corresponding *ortho*-phenanthroline–iron(III) complex, which is pale blue in colour:

$$[(C_{12}H_8N_2)_3Fe]^{2+} \rightleftharpoons [(C_{12}H_8N_2)_3Fe]^{3+} + e$$

Table 3.3. Oxidation-reduction indicators and their colour changes

Indicator		E° (at pH O) 0.2 0.3 0.4 0.5 0.6 0.7 0.8 0.9 1.0 1.1 1.2 1.3	
Sodium Dichloro 2,6-phenolindophenol	Colourless	0.26	Blue
Diphenylamine	Colourless	0.76	Blue
Sodium diphenylaminesulphonate	Colourless	0.85	Blue
Diphenylbenzidine	Colourless	0.76	Red violet
Erioglaucine	Orange	1.00	Green Yellow
Indigomonosulphonate	Colourless	0.26	Blue
Methylene Blue	Colourless	0.52	Blue
Nitro-1,10-phenanthroline iron(II) sulphate	Red	1.14	Magenta
Ferroin Sulphate	Red	1.14	Blue
Phenosafranine	Colourless	0.28	Red
N-Phenylanthranilic Acid	Colourless	1.08	Red Violet

The oxidation-reduction potential of this system is 1.14 volts. The complex is used as an indicator in the titration of iron(II) ($E^\circ_{Fe^{3+}/Fe^{2+}}$ = 0.77 V) by cerium(IV) sulphate ($E^\circ_{Ce^{4+}/Ce^{3+}}$ = 1.45 V), the complete colour range of the indicator lying well within the potential gap between the two systems.

Table 3.3 gives a list of oxidation-reduction indicators, showing the E° values at pH = 0 and the approximate potential range of visible colour change.

4
Technique of quantitative analysis

General information

Attention to detail is absolutely essential if success in quantitative analysis is to be achieved. A number of the more important general points which should be observed are enumerated below.

Cleanliness

The bench and apparatus must be kept scrupulously clean. A bench-cloth and a glass-cloth are required. All glassware should be rinsed with water (distilled or demineralised) before use if it has been standing in the cupboard. The outside of vessels should be wiped dry with the glass-cloth, but the latter should not be used on the inside of vessels. Glass vessels should be free from grease. In general, soap or detergent solution is satisfactory for cleaning glassware. The apparatus should then be *thoroughly* rinsed with tap water followed by *water* (distilled or dimineralized). On some occasions, however, a stronger cleaning agent may be required; this can be prepared by adding 15 g of powdered sodium dichromate to 500 ml of concentrated sulphuric acid. (*Note:* This cleaning agent is extremely corrosive.) A little of the 'cleaning mixture' is poured into the vessel to be freed from grease and allowed to come into contact with the whole of the interior surface. Several hours contact is desirable. Any surplus should then be drained into the stock 'cleaning mixture' bottle, and the vessel rinsed successively with tap water and then *water*. In the case of graduated apparatus, the first washing should be done with a large volume of cold water added quickly to prevent the apparatus becoming too hot as the solution and the water are mixed.

Tidiness

All reagent bottles must be returned to their *correct* positions on the reagent shelf *immediately* after use. Stoppers of reagent bottles must not be placed on the bench; they should be held in the left hand and replaced in the bottle after the reagent has been used. The apparatus should be arranged on the bench and in the cupboard in an orderly manner.

Labelling

All solutions, filtrates, precipitates etc. should be labelled systematically throughout the analytical procedure. If any liquid other than water is introduced into a wash-bottle, the bottle should be labelled appropriately and *immediately*.

Planning

Before commencing a determination, the directions for the assay to be performed must be read carefully. Care should be taken to ensure that all the details of the technique and the underlying basic principles are understood. The work should be planned so that there is no hold-up at any stage, for example, crucibles should be prepared and dried to constant weight before the particular solution and precipitate is ready for filtration. Two or more operations should be kept going at the same time e.g. while a precipitate is cooling in a desiccator, a weighing or a titration should be carried out.

Graduated apparatus

Graduated apparatus must not be heated and must not be used as containers for hot liquids, because the glass will expand and may not contract to its initial volume on cooling to room temperature.

Determinations in replicate

In general, determinations should be performed on at least two separate samples of the product under examination. Good agreement between the two results engenders confidence; lack of agreement provides evidence of incorrect work. Two results should not be averaged unless they are within 0.4%. Burette readings should agree within 0.05 ml for titration figures of between 20 and 30 ml when equal volumes of solution are being titrated. A further titration (or titrations) is necessary if this agreement is not obtained.

Records

All data should be recorded *directly* into a laboratory note-book and *not* on pieces of paper.

It is recommended, in a student's course in quantitative analysis, that the experimental observations, such as weighings and burette readings, should be recorded on the left-hand page, and the equations, description of the determination, calculations and conclusion should be recorded on the right-hand page of the laboratory note-book. Examples of typical laboratory records are given on p.142.

Balances

The analytical balance

The analytical balance possesses a high degree of sensitivity in order to give the true weight of samples (weight in this context being synonymous with mass). Factors affecting these requirements are:

(i) the length of the balance arms
(ii) co-planarity of the knife edges
(iii) the weight of the beam
(iv) the position of centre of gravity of the beam in relation to the central knife edge or pivot.

The influence of these factors is shown by considering the hypothetical balance beam ABC (Fig. 4.1) of weight W with knife edges at A, B and C and centre of gravity at G. Figure 4.1 shows the position of the beam when loads W_1 and W_2 have caused a small deflection θ.

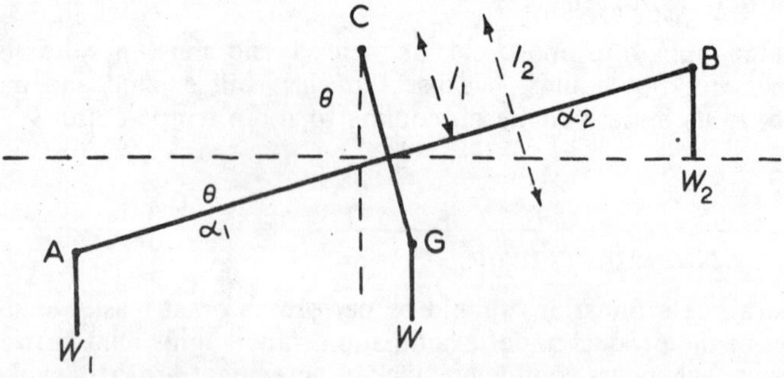

Fig.4.1 Hypothetical balance beam

At equilibrium, it may be shown that

$$\tan\theta = \frac{W_1a_1 - W_2a_2}{(W_1 + W_2)l_1 + Wl_2} \qquad (1)$$

When equal weights are on each arm, the balance beam should be horizontal, i.e. $\theta = 0$; this applies when the lengths (a_1 and a_2) of the arms are equal. If they are not, a true weight may be obtained by weighing the sample first on one pan to give an apparent weight W_3 and then on the other pan to give an apparent weight W_4. The true weight is calculated from the formula

$$\text{True Weight} = \sqrt{(W_3 \cdot W_4)}$$

The sensitivity of a balance can be defined as the deflection of a pointer caused by a standard difference of weight between the two sides, e.g. the deflection, in scale divisions, per mg weight difference. For small deflections, tan θ may be taken as θ expresed in radians and, incorporating the condition of equal balance arms in eq (1), the sensitivity is proportional to

$$\frac{\theta}{W_1 - W_2} = \frac{a}{(W_1 + W_2)l_1 + Wl_2} \tag{2}$$

where a represents the length of a balance arm. Increasing the length of each arm will increase the sensitivity of the balance only if it can be done without too great an increase in the weight W of the beam.

Sensitivity can be increased by reducing the magnitude of the denominator in (2) and manufacturers eliminate l_1 by making all three knife edges co-planar. Under this condition, the sensitivity is now proportional to a/Wl_2 and should remain constant with increase in load. If this co-planarity is lost due to excessive load and distortion of the beam, the sensitivity will decrease with increase in load. If, on the other hand, the central knife edge is below the other two, the sensitivity will increase with increase in load. The position of the centre of gravity can also be varied by means of a sensitivity bob. In this way G can be made to approach C and, hence, l_2 will be reduced, leading to increased sensitivity. In the limit, when G coincides with C, a horizontal position of equilibrium is possible, but the slightest additional weight on one side causes the beam to swing to an extreme position; the balance is now unstable.

The ordinary technique in weighing is to determine the weight of sample by difference and, hence, inequality of balance arms and loss of co-planarity of knife edges under load, with concomitant variations in sensitivity, may give rise to small errors. To avoid these errors, weighing by substitution has received considerable attention and is likely to achieve much wider use with the advent of synthetic sapphire for planes and knife edges. This hard material meets the objection of excessive wear on bearing surfaces at high loads when 'softer' materials are used. The technique involves both balance pans being fully loaded, and when a sample is placed upon the left-hand pan, weights must be *taken off* until equilibrium is attained. This is time-consuming on the conventional balance, but is convenient and rapid on certain types of single-pan balance (see below).

Use and care of the balance

It is presumed that the student is already well acquainted with the principles of the balance and its use in weighing. The following points are intended to draw attention to certain important practical details, using the simple **analytical balance**, which requires manual addition of weights and manipulation of a 'rider'.

(*a*) Material spilled on the balance pans or base must be cleaned up immediately.

(*b*) The balance should be tested to see that it swings freely and is in adjustment before each weighing is made.

(*c*) Sample must not be weighed directly upon the balance pans. A stoppered container must be used for weighing liquids and volatile, deliquescent or hygroscopic solids.

(*d*) Hot objects must be cooled to room temperature before introduction into the balance case.

(*e*) The balance door must be closed when the final weighing is made.

(*f*) The arrestment of the beam must be lowered *slowly* when setting the balance swinging.

(*g*) The beam must be raised and the pans arrested before an object or weight is added to or removed from the pans.

(*h*) The object to be weighed is usually placed on the left-hand pan and weights on the right-hand one.*

(*i*) All weights must be handled with forceps and *not* with the fingers. The forceps are manipulated with the right hand and the balance arrestment with the left hand. The heavy weight should be placed towards the centre of the balance pan. There are only two permissible places for weights to rest: on the balance pan or in the correct space in the box of weights.

(*j*) Before recording a weight, check the empty spaces in the box of weights as well as the weights on the balance pan.

(*k*) The balance must not be overloaded (maximum load usually 200 g).

(*l*) When a weighing has been completed, a check should be made to see that the beam is arrested, the weights are in their correct places in the box of weights, the 'rider' is on the sliding hook and the balance-case and box of weights are closed.

Use of a 'rider'

Most laboratories now use aperiodic or single-pan balances. Balances of this sort use a dial-operated system of weight addition and subtraction, with the smallest weights automatically visualised by means of a light pointer on an illuminated screen. The principle on which these modern balances operate, however, is fundamentally the same as that of the simple analytical balance, which requires the use of a beam rider for measurement of the smallest weights.

Weights smaller than 0.01 g are inconvenient to handle on the conventional analytical balance, therefore a 'rider' which is placed on a graduated scale along the top of the balance beam is used instead. The scale is divided into 20 or 10 equal parts (Fig. 4.2), each of the parts being subdivided.

If the beam is graduated as in Fig. 4.2A, a 10 mg rider, suitably shaped to hang on the beam, is used. If the rider is placed at 'a' (Fig. 4.2A) the effect is equal to that of a 0.01 g weight placed on the right-hand pan of the balance; if it is hung at 'b', the effective weight is equal to 0.006 g. The distance between two small divisions as shown in Fig. 4.2A represents a weight of 0.0002 g. The left-hand section of the scale (below zero) is seldom used; if the rider is placed on this scale it represents the effective weight of the rider applied to the left-hand pan. The rider is manipulated

*A left-handed person may use the reverse only if the balance beam is graduated as shown in Fig. 4.7 A.

from outside the balance case by means of a sliding hook which moves above the balance beam. When the balance is not in use the rider is kept suspended from the rider hook away from contact with the beam.

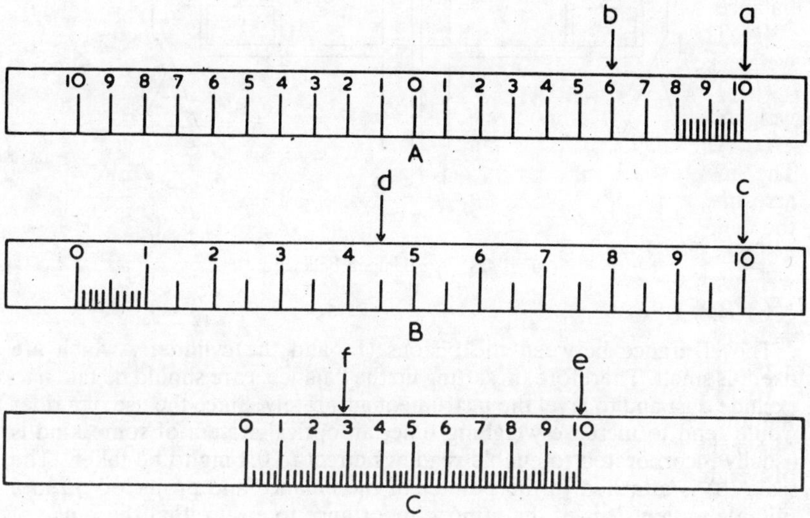

Fig. 4.2 Types of balance beam

In Fig. 4.2B another type of division of the beam is shown. A 5 mg rider is used for a balance with a beam of this type. The zero position of the rider is on the beam at division '0' and not on the hook. The effective weight of the rider is always on the right-hand pan. The rider at 'c' represents an effective weight of 0.01 g on the pan while at 'd' it represents an effective weight of 0.0045 g. The distance between two small divisions represents an effective weight of 0.0001 g.

A third type of beam division is shown in Fig. 4.2C. The graduations cover only half the effective length of the beam, and a 10 mg rider is used. The zero position of the rider is on the beam at division '0' and, as in 4.2C, the effective weight of the rider is always on the right-hand pan and at 'f' represents an effective weight of 0.0028 g. The scale is subdivided in fifths and each small division represents an effective weight of 0.0002 g.

Aperiodic balances

Aperiodic balances are those in which the swing of the beam is damped so that the beam comes to rest rapidly, usually in about 12 s or less. The damping is achieved by a piston arrangement on each arm of the balance as shown diagrammatically in Fig. 4.3.

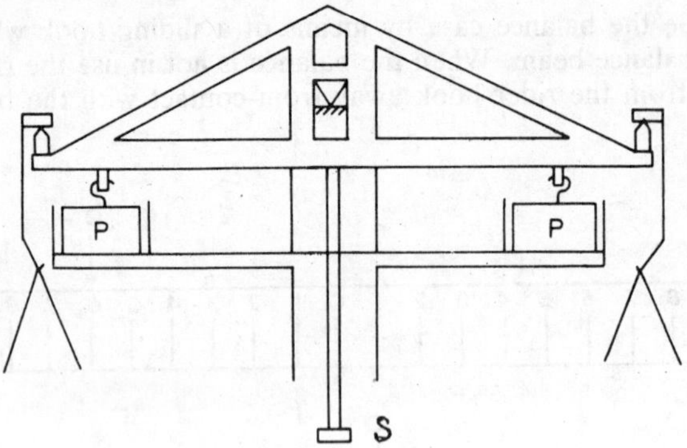

Fig. 4.3 Air-damped balance

The clearance between the pistons (P) and the cylinders, which are fixed, is small. Therefore, in setting up the balance, care should be taken to exclude dust and to level the instrument accurately. Since the use of a rider would tend to increase weighing time, an optical system of some kind is usually incorporated to enable readings direct to 0.1 mg to be taken. The scale (S) is attached to the pointer of the balance and projected on to a suitable screen. It is of the utmost importance to ensure that the range of the scale, e.g.. 0–10 (or 0–100) represents 10.0 (or 100.0) mg and this should be checked frequently at different loads.

Single-pan balances

If the right-hand pan is enclosed completely, the left-hand pan of the aperiodic balance now becomes the single pan of the conventional single-pan balance, i.e. one in which three knife edges are still retained. This is the variable-load balance which is operated in the same way as an ordinary balance, except that mechanical addition of weights to the 'right-hand' side is achieved by an array of dial controls, as ilustrated in Fig. 4.4.

When weighing by substitution is adopted, i.e. weighing under constant load, one of the outer knife edges may be replaced by a counterweight rigidly fixed to the balance beam. For maximum benefit from this design, the two arms of the balance are of unequal length, as illustrated in Table 10 column 3. The inequality in the arms leads to a reduction in the weight necessary for a counterweight and, hence, to a reduction in the weight of the beam. Damping is achieved by movement of a damping disc in a cylinder much in the same way as described for aperiodic balances.

A summary of data on the various types of balance is reproduced in Table 4.1.

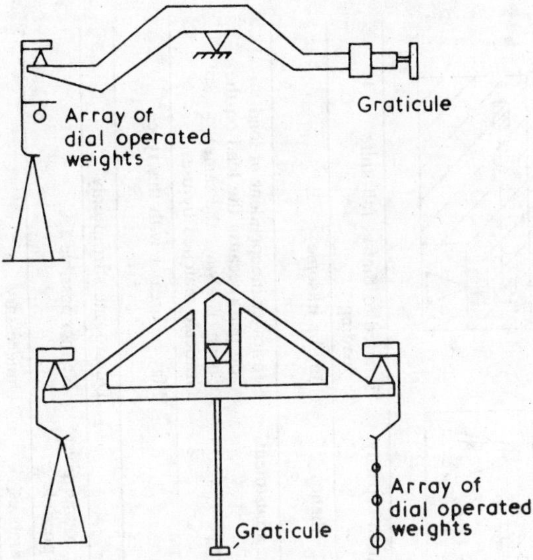

Fig. 4.4 Two types of single-pan balance (diagrammatical only)

Weighing the sample

An accurate weight of a sample may be obtained either by 'weighing by difference' or by 'weighing by addition'.

Weighing by difference

Method A Place sufficient sample for several analyses in a stoppered weighing bottle and weigh accurately. Take the bottle from the balance with the *tips* of a *clean* dry thumb and finger and remove the stopper (left hand), holding the bottle over the mouth of the vessel in which the sample is to be placed. Pour out the estimated correct amount of substance by *carefully* tilting the bottle from the horizontal and rotating it. While it is still over the receiving vessel, tilt the bottle towards the upright position and rotate the bottle about its vertical axis. Tap the bottle lightly with the finger so that sample remaining on the inside of the bottle lip slides back in, and that on the outside falls into the receiving flask. Insert the stopper slowly and carefully to prevent powder being blown out of the bottle. Reweigh the bottle and contents. The difference between the two weighings gives the weight of sample.

The disadvantage of this method lies in the difficulty of estimating the correct amount to be tipped out of the bottle. Since it is inadmissible to return any sample to the weighing bottle after it has been introduced into the receiving vessel, a number of additions of small portions are frequently required before the correct sample weight is obtained. The British Pharmacopoeia states that in assays 'the quantity actually used must not deviate by more than 10 per cent from that stated in the method'. The

Table 4.1 Comparative information on different types of modern, direct-reading, knife-edge balance

SUBJECT / TYPE	Normal three-knife, two-pan balance	Three-knife, single-pan balance	Two-knife, single-pan balance
	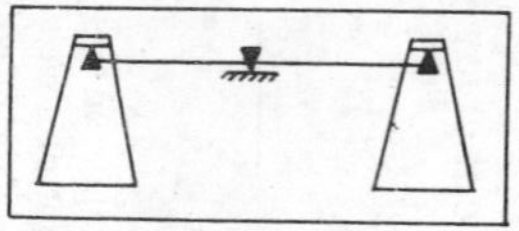	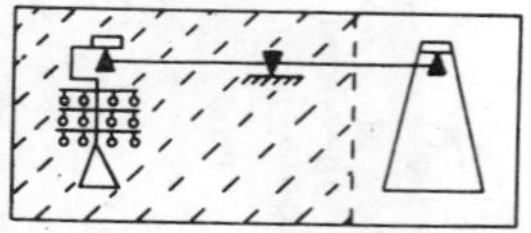	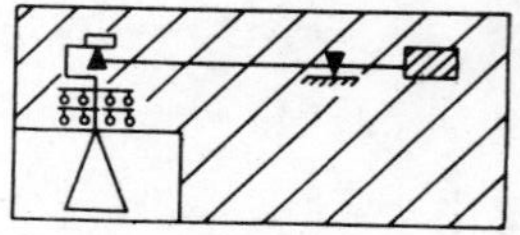
Accuracy	Usually best at loads less than maximum capacity		Limited to that at full knife loading
Variation of zero with air density	None	Slight	May be irksome
Sensitivity graticule	Only match for all loads on pan when adjustment of knives is ideal (involves professional skill)		Sensitivity independent of load on pan (because the load on the beam is always the same) Graticule matched by user
Variation of sensitivity with balance temperature	Slight		May be greater with this type
Stirrup error	Due to two stirrups		Due to one stirrup only
Weighings with counterpoise or tare	Both possible	Small tare or counterpoise may be possible	Neither possible
Adjustment of arm length	Equality required to very fine tolerances		Unnecessary
Other adjustments	All three knives to be adjusted parallel and co-planar		Only two knives to be made parallel
Wear of knives	Normal		Accelerated Knife loading always at maximum

following (Method B) is, therefore, recommended, despite the fact that it involves *four accurate* weighing to obtain *two* analytical sample portions instead of the *three accurate* weighings which would be required by Method A.

Method B. Weigh the weighing bottle approximately to the nearest 0.01 g. (The approximate weight of the students' own bottle should be known.) Introduce into the bottle slightly more than the specified amount of sample. Alternatively, if the material to be weighed is stable, weigh slightly more than the specified amount on a rough balance and then introduce into the weighing bottle. Weigh the bottle and contents accurately. Tip out the contents of the bottle into the receiving vessel, using the technique described under Method A. When most of the solid has been transferred (no attempt should be made to transfer the last few particles which adhere to the inside of the bottle), replace the stopper, taking the precautions mentioned above, and reweigh the bottle and remaining contents *accurately*.

General precautions for weighing the sample

(*a*) If the material is deliquescent or hygroscopic, exposure to the atmosphere should be reduced to a minimum and method B adopted.

(*b*) If the sample is being transferred to a conical flask or other vessel with a narrow aperture which is not narrow enough to require the use of a funnel, it is necessary to ensure that the mouth of the vessel is dry because of the possibility of contact with the weighing bottle. A funnel need not be used when transferring a solid from a weighing bottle to a conical flask in the 'weighing by difference' procedure.

(*c*) A record of the weighings must be made in the laboratory note-book *immediately*, in ink.

(*d*) Analytical samples of liquids should not be weighed by difference but by the method of 'weighing by addition', as outlined below.

Weighing by addition

The empty dry vessel is first weighed *accurately*. The vessel can be a clock-glass, a weighing bottle, a *small* beaker, a *small* conical flask, a crucible etc., depending upon the weight and nature of the material to be weighed. The sample is then introduced into the vessel and the container and sample weighed again *accurately*. It is essential that the vessel should not be too large in order that changes due to adsorption of moisture and electrification during weighing may be kept to a minimum. Solids require slightly different weighing procedures from those used for liquids and the general methods are outlined below.

Method A *Solids* (as in the preparation of standard solutions). Weigh the vessel (clock-glass, watch-glass or weighing bottle) *accurately*. Introduce the solid in portions until the correct weight has been added. Weigh the vessel and contents *accurately*. Hold the vessel over the receiver (beaker, or funnel in the mouth of the flask) and incline slightly so that the solid slides down slowly into the receiver. Wash the weighing vessel thoroughly with a jet of *water*, collecting *all* the washings in the receiver.

On the semi-micro scale, the solid can be weighed directly in the small conical flask or beaker in which the titration or other subsequent treatment will be performed. *Great care* must be taken in weighing small quantities to ensure that the vessel into which the sample is to be weighed directly has been allowed to remain on the balance pan for some time so that a constant weight is obtained. The material to be weighed should be introduced directly into the weighed container while it remains on the balance pan and then the container and its contents reweighed accurately.

Method B. *Water-miscible liquids*. Weigh accurately an empty, clean, stoppered weighing bottle. Remove the stopper, placing it on the balance pan. Add the liquid carefully from a teat pipette until the correct weight has been added. Insert the stopper into the weighing bottle and weigh the latter and its contents accurately. Liquid should not come into contact with the ground glass portion of the bottle in order that the stopper is not wetted. Take the bottle from the balance and, holding it at the base between the tips of the fingers and the thumb of the right hand, remove the stopper with the left hand, and pour the liquid gently down a guiding rod (also in the left hand) into the receiving vessel. A funnel need only be used if the receiver has a narrow neck. While the bottle is upturned and still held over the receiving vessel, transfer it to the left hand and wash down the bottle and guide rod with a jet of *water* from a wash-bottle held in the right hand.

Method C *Water-immiscible liquids*. If the titration (or other treatment) of the sample necessitates the use of a large flask or other container, the sample is usually weighed in a small vessel, e.g. a sample tube or other special container with a flattened end (see the Determination of Iodine Values, p.188).

Weigh the small vessel empty, introduce the sample and then accurately weigh the vessel and its contents. Introduce the weighing vessel and its contents into the larger vessel in which the determination is to be carried out.

If the determination can be carried out in a small vessel, and the liquid is non-volatile, the liquid may be weighed directly into the vessel.

Accuracy of weighings

It is not necessary to weigh a sample with a degree of accuracy which is appreciably greater than that which can be attained in the subsequent steps of the analysis.

In volumetric work, if the accuracy of the apparatus is ±0.2% and 1 g of sample is required for a particular assay, it would be unnecessary to weigh the sample to ±0.0001 g since this would constitute an error of only ±0.01% in the weighing. For samples of such size in volumetric work, it is not necessary to weigh closer than ±0.0005 g. In gravimetric work, however, all weighings should be within ±0.0002 g. The British and European Pharmacopoeias adopt a convention whereby the degree of precision required is indicated by the number of significant figures stated e.g. 1 g, 0.50 g, or 0.500 g.

Constant weight

In practice, it is unusual for two consecutive weighings of the same object to be completely identical. This is particularly true where the object has been submitted to some physical treatment, such as heating and cooling between weighings. In gravimetric analysis, an object or sample is said to be at constant weight when two consecutive weighings after heating and then cooling in the desiccator differ by not more than 0.0003 g. The British

Pharmacopoeia defines the term 'constant weight' used in relation to the determination of loss on drying or loss on ignition as meaning that two consecutive weighings do not differ by more than 0.5 mg per g of substance or residue for the determination, the second weighing being made after an additional hour of drying or after further ignition.

Technique of volumetric analysis

Volumetric analysis requires the accurate measurement of *volumes* of interacting solutions. The graduated apparatus most commonly used in volumetric work are measuring (graduated) flasks, burettes, pipettes and measuring cylinders.

Measuring (graduated) flasks

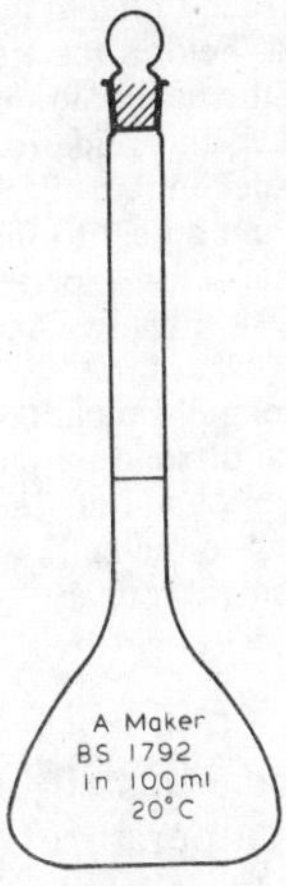

Fig. 4.5 Graduated flask

Measuring flasks are usually round or pear-shaped, flat-bottomed and with a long neck, which bears a single graduation mark extending right round the neck (Fig. 4.5). Such flasks, with one graduation, are made to *contain* a specified volume of liquid at 20°, when the level of the bottom of the meniscus (top, if convex liquid surfaces, as with mercury) coincides with the mark. The long, narrow neck makes for accurate adjustment, the height of the liquid being sensitive to small changes of volume.

The litre is the standard unit of volume for all volumetric glassware. It is equal to 1 cubic decimetre (dm^3), 1000 cubic centimetres or 1000 millilitres. Thus, the millilitre and cubic centimetre are the same. Volumetric apparatus is usually standardised in millilitres.

Volumetric apparatus is manufactured to two sets of tolerances (Grade A and Grade B, respectively) which have been laid down by the National Physical Laboratory and the British Standards Institution (BS 1792). Volumetric flasks are required to be marked with the nominal capacity expressed in ml, the temperature at which standardised (20°C), the letters 'In' to indicate that the flask is graduated to contain, the letter 'A' or 'B' to indicate the class of accuracy to which the flask has been graduated, the marker's name and the BS standard number (Fig. 4.5). Permitted tolerances for graduated flasks in common use are shown in Table 4.2. The British Pharmacopoeia requires apparatus of Grade A standard to be used and, unless otherwise directed, requires analytical operations to be conducted at 20°C. Glassware used in the United States is also standardized at 20°, despite the fact that analytical operations are conducted at 25°.

Table 4.2 Tolerances in the capacity of measuring flasks

Capacity (ml)	Tolerance Class A(±ml)	Tolerance Class B (±ml)
5	0.02	0.04
10	0.02	0.04
25	0.03	0.06
50	0.05	0.10
100	0.08	0.15
200	0.15	0.30
250	0.15	0.30
500	0.25	0.50
1000	0.40	0.80
2000	0.60	1.20

The preparation of solutions of definite concentration
Transfer the appropriate weight of solid (weighed by addition, see p.000) quantitatively to a beaker and dissolve in *water* (or other specified solvent).* Transfer the solution quantitatively by means of a glass rod, a funnel and a jet of *water* from a wash-bottle, as follows. Hold the beaker in the right hand and pour the liquid gently down the guiding rod, held in the left hand, into the funnel placed in the mouth of the graduated flask. Transfer the beaker to the left hand while still upturned and still held over the funnel. Wash the beaker and guide rod with a jet of *water* from the wash-bottle held in the right hand. Wash the funnel until the flask is two-thirds full, remove the funnel, swirl the flask to mix the contents and adjust to the mark. The final adjustment should be made by adding *water* dropwise from a teat pipette; time must be allowed for water to drain down the inside of the neck of the flask. Finally shake the flask thoroughly to mix the contents.

If the solid is easily soluble in water, it may be added directly to a dry funnel in the mouth of the flask. The solid should slide easily down the funnel and the final trace of solid on the weighing vessel and funnel can be washed into the flask until the latter is about half full. The flask is then swirled gently until solution is effected and the volume made up to the mark as described above. The dry stopper is then inserted and the solution mixed thoroughly by inverting and rotating the flask.

For precise work, the temperature of the solution should be adjusted to 20° before making up to the mark (see later for effect of temperature).

Standard solutions, if they are to be stored, are usually transferred to stock bottles. If such a solution is to be transferred, the receiving vessel should be rinsed with two or three successive small quantities of the solution before the main bulk is added. When a standard solution is used some time after preparation, the container and its contents should be shaken well before any of the solution is withdrawn. This shaking mixes the condensed water droplets, which have collected on the inside of the container above the solution with the bulk of the solutions.

*Purified water (distilled or demineralized) is the solvent described in this account. Other solvents are also used consequently when *water* is mentioned in this account it is intended that the appropriate solvent should be used as required.

Accurate dilutions of standard solutions

Accurate dilutions can be prepared by pipetting the standard solution into a graduated flask and diluting to the mark with *water*. For example, 100 ml of 0.2M solution can be prepared from M solution by accurately pipetting 20 ml of the latter into a 100 ml graduated flask and diluting to the mark with *water*. The factor for the dilution is the same as for the original standard solution.

Pipettes

Pipettes are of two kinds:

(a) *transfer* pipettes which have one mark and are used to *deliver* a specified volume of liquid under certain specified conditions

(b) *graduated* (or *measuring*) pipettes which have graduated stems and are employed to deliver various small volumes as required; they are not employed for measuring very exact volumes of liquids.

Permitted tolerances and delivery times as laid down by the British Standards Institution (BS 1583:1961) for bulb transfer pipettes in common use are shown in Table 4.3.

Table 4.3 Tolerances and delivery times for one-mark pipettes

Capacity (ml)	Tolerance (±ml)		Delivery times (s)		
	Class A	Class B	Min Class A	Min Class B	Max Classes A and B
1	0.007	0.015	7	5	15
2	0.01	0.02	7	5	15
3	0.015	0.03	10	7	20
4	0.015	0.03	10	7	20
5	0.015	0.03	15	10	25
10	0.02	0.04	15	10	25
15	0.025	0.05	20	15	30
20	0.03	0.06	25	20	40
25	0.03	0.06	25	20	40
50	0.04	0.08	30	20	50
100	0.06	0.12	40	30	60
200	0.08	0.16	50	40	70

All transfer pipettes have a single graduation mark. The capacity and the temperature at which it was graduated (Ex) and a reference to delivery time in seconds is stated on the bulb. In addition, Class A pipettes also state the time of outflow (25 in Fig. 4.6). The method of calibration obviates the need to specify drainage time, though a waiting time of 3 s after apparent cessation of outflow is still essential. The stated times apply only for water and aqueous solutions.

Use of the transfer pipette Rinse the pipette with *water* before use and allow to drain. Remove the drop of water remaining in the tip by touching against filter paper and wipe the outside of the pipette to prevent dilution of the solution to be pipetted. Rinse the pipette with two or three small portions of the solution, each portion being used to wet the whole inside surface before being allowed to run out. Suck the liquid up into the pipette until just above the single graduation mark; close the upper end of the pipette with the tip of the index finger. Remove the pipette from the bulk solution and gently wipe the outside of the stem free from adhering liquid. Then, holding the pipette vertical and with the graduation mark at eye level, gently relax the pressure on the finger, so that the level of the liquid slowly falls. When the bottom of the meniscus coincides with the mark* increase pressure with the finger, to prevent further escape of liquid, and remove the drop adhering to the tip by gently touching against a porcelain tile. Introduce the pipette into the receiving vessel and allow the liquid to run out with the tip of the pipette *touching* the inside of the vessel at an angle of 60°, but *not* dipping into the delivered liquid. When all the liquid has run in, allow the pipette to drain in this position for 3 s (*waiting time*) and then remove the pipette. (*Note*: the small drop of liquid which remains in the tip of the emptied pipette is taken into account in the calibration and must *not* be added to be delivered liquid by blowing down the pipette.) A draining time is necessary in the case of liquids which are more viscous or have a much larger surface tension than water, e.g. strong solution of iodine.

After use, rinse the pipette with *water* and allow to drain.

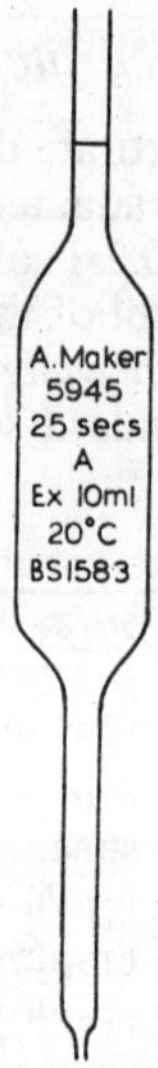

Fig. 4.6
Transfer pipette

Automatic pipettes

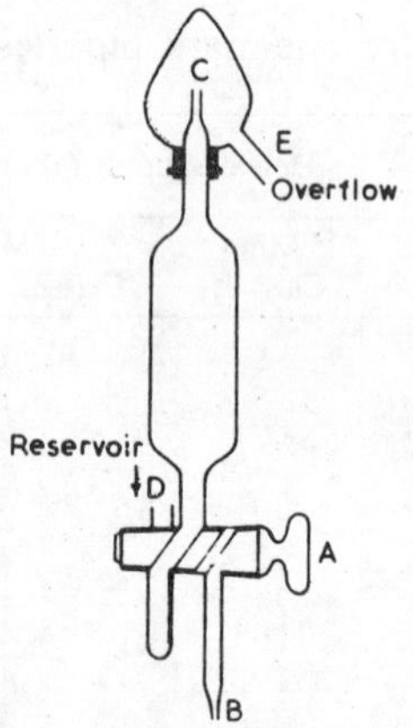

Fig. 4.7
Automatic pipette

These can be used to advantage with certain corrosive or toxic liquids. One type, shown in Fig. 4.7, is frequently used for iodine monochloride solution required in the determination of iodine values of oils. This solution is prepared with glacial acetic acid. The pipette delivers its stated volume of liquid when filled with the liquid in question from tip (B) to tip (C) and allowed to run out and drain in the normal manner. D is connected to an aspirator which is fixed above the pipette so that the solution feeds in under gravity. E is a waste overflow.

*The National Physical Laboratory describes a method of reading the meniscus in graduated glassware, viz. a dark horizontal line on a white background is placed 1 mm below the meniscus. A slight adjustment of the position of the dark line causes the meniscus to stand out sharply against the white background.

Manipulation of the automatic pipette Turn the two-way tap clockwise to open so that solution commences to flow into the pipette. When about 5 ml has run into the pipette, turn A clockwise through 180°, so that solution now flows from the pipette to fill the delivery tube B. As soon as B is full to the tip, again turn A clockwise through 180°, so that the body of the pipette is filled completely to the tip, closing the tap A by turning clockwise through 90° when solution commences to overflow at C. The pipette is now full from tip to tip and ready for use. Remove the drop of solution from tip B, run out and drain for 15 s in the usual way.

Burettes

A burette is a graduated vessel (Fig. 4.8) which is used for the accurate delivery of variable volumes of liquids. Burettes are made of varying overall capacity from 1 ml up to 100 ml; 50 ml burettes are of a size which is convenient for most volumetric operations. As with other forms of graduated glassware for volumetric work, they are produced to either Class A or Class B specifications (BS 846:1962). All Class A and some of Class B burettes have graduations which extend right round the barrel of the burette to reduce parallax errors in reading the burette. Class B burettes are usually graduated on one side only. Permitted tolerances on capacity for burettes in common use are shown in Table 4.4. The tolerance represents the maximum error allowed at any point and also the maximum difference allowed between the errors at any two points. Thus, a tolerance of ±0.05 ml implies that the burette may be in error at any point by ±0.05 ml, provided that the difference between the errors at any two points is not more than 0.05 ml.

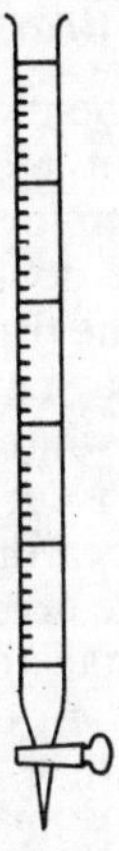

Fig. 4.8 Burette

Table 4.4 Tolerances on capacity for burettes

Nominal capacity (ml)	Scale subdivision (ml)	Tolerance on capacity (±ml)	
		Class A	Class B
1	0.01	0.006	0.01
2	0.02	0.01	0.02
5	0.02	0.01	0.02
5	0.05	0.02	0.04
10	0.02	0.01	0.02
10	0.1	—	0.05
25	0.05	0.03	0.05
25	0.1	0.05	0.1
50	0.1	0.05	0.1
100	0.2	0.1	0.2

Use of the burette See that the burette tap is lubricated with a thin film of lubricant. Before use, rinse the burette twice with small portions of the solution (about 5 ml), allowing the burette to drain out well between the addition of each portion. Fill the burette with solution until the latter is slightly above the zero mark. Open the burette tap to fill the tip and expel all air bubbles and, with the zero at eye level, carefully and slowly run the liquid out until the bottom of the meniscus is level with or just below the zero mark. Remove the drop on the tip of the burette by momentarily touching it against a porcelain tile or flask. Read the burette accurately after the film of solution adhering to the wall of the burette above the surface of the liquid has been allowed to drain for about 15 s. The burette must be read at eye level in order to avoid errors due to parallax; this is particularly important when using Class B apparatus. A piece of white paper or a white tile held behind the burette and just touching it at the appropriate level will assist observation of the meniscus and make for easy and accurate reading. Burettes (50 ml) are graduated in millilitres and tenths of a millilitre, but, with a little experience, eye estimation is possible to a fifth of a division. Readings can be recorded to the nearest 0.02 ml. At the completion of a titration, 15 s must be allowed to elapse before the final reading is made, to allow for drainage.

Titration technique

The most suitable vessels for general work are conical flasks. Beakers are not recommended. If the latter are used, a stirring rod must also be employed to mix solutions during the titration. With a conical flask this can be done safely and easily by gently swirling the flask during the titration. The titration vessel should be kept well polished in order that the end point may be seen clearly. The solution being titrated is generally viewed against a white background (white tile).

Near the end point, it is advisable to split the drops of titrant. Partially open the tap so that a fraction of a drop flows out and remains attached to the burette tip. Touch the liquid against the inside of the flask and wash the small volume of titrant into the bulk of the liquid with a few drops of *water*. In any event the upper internal portion of the flask should be washed down with a little *water* just before the end point.

When it is considered that the end point has been reached, note the burette reading and then, add a further drop of titrant, when a further distinct colour or other change should occur, unless the indicator change is to colourless. (*Note*: this cannot be done if the solution is required for subsequent titration when excess titrant would interfere.)

If the colour change at the end point is gradual, it is useful to have a comparison solution available. For example, if methyl orange is being used as indicator and the end point is gradual, two flasks containing the same volume of solution of approximately the same composition as the liquid being titrated can be prepared, one slightly acidic (red solution) and the other slightly alkaline (yellow solution). These comparison solutions will assist in deciding the colour change which indicates the end point.

Titrations should be carried out in duplicate and results should agree to within 0.05 ml (based on a 20 ml titration). If such agreement is not obtained, further titrations are necessary.

Any solution remaining in the burette after a series of titrations should be rejected and must not be returned to the stock solution. The burette is then washed out with distilled water after use and allowed to drain.

Measuring (graduated) cylinders

These are vessels (Fig. 4.9) unstoppered (A) or stoppered (B), of capacities varying from 5 ml up to about 2 l. The smaller cylinders up to 100 ml are usually graduated in millilitres or fractions of a millilitre, whilst larger cylinders are graduated in units of 2, 5, 10, 20 or 50 ml, according to size. Cylinders are used for measuring out volumes of solution when only approximate volumes are required.

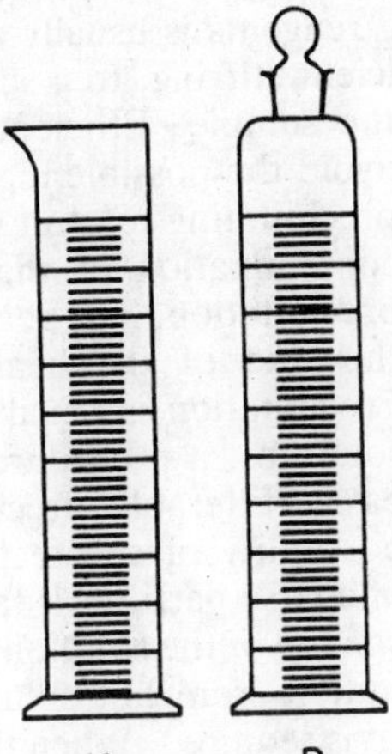

Fig. 4.9
Measuring cylinders

Effect of temperature upon volumetric measurement

The effect of normal temperature changes on the volume of glass apparatus is negligible. However, graduated glassware must be neither heated nor filled with hot liquids because the glass will expand and may not contract to its initial volume on cooling to room temperature.

The effect of temperature variation on aqueous solutions is somewhat greater [cubical expansions of water and dilute aqueous solutions can be taken as approximately 0.00025 (0.025%) per centigrade degree in the region 20° to 30°]. For general student work, corrections for the slight changes in temperatures in the laboratory can be ignored when aqueous solutions are used. However, when titrations in non-aqueous solvents (p.000) are carried out, temperature corrections are important because the organic solvents used in volumetric work have much higher coefficients of expansion than that of water.

Technique of gravimetric analysis

Much of the technique which has already been described is common to both gravimetric and volumetric analysis. In gravimetric work, certain additional techniques are necessary, namely, *precipitation, filtration, washing of the precipitate* and *drying* (or *ignition*) *of the precipitate*; these operations are considered in detail in the succeeding sections.

Precipitation

Gravimetric precipitations are usually made in beakers. Except during the actual precipitation, the beaker is covered with a clock-glass. A thin stirring rod, rounded at each end, is also required (see Fig. 4.10). The solution of the precipitating reagents is usually added slowly from a teat pipette or burette, with efficient stirring, to a suitably diluted solution of the sample. Efficient stirring is necessary to avoid the possible high local concentrations of precipitating reagent which would tend to give contamination of the precipitate due to co-precipitation. The reagent is introduced down the side of the beaker to avoid splashing. Precipitation is usually made from hot dilute solutions, a procedure which tends to give an easily filterable precipitate and reduces the possibility of co-precipitation. The formation of coarse particles is favoured by *slow* addition of the precipitant, vigorous *stirring* of the solution during precipitation and by carrying out the precipitation from hot solutions. Only a moderate excess of precipitating reagent is required. When the precipitate has settled somewhat, a few more drops of precipitant should be added to test for the completeness of precipitation; if further precipitation occurs, then the process of addition of precipitant, stirring and allowing the precipitate to settle and testing again must be carried out until there is no doubt about completeness of precipitation. The stirring rod should not come into contact with the sides or bottom of the beaker during stirring in order to avoid scratching particles of glass from the surfaces. Also, a scratched surface is difficult to wash clean. Reagents should be examined for the clarity before use and filtered if necessary.

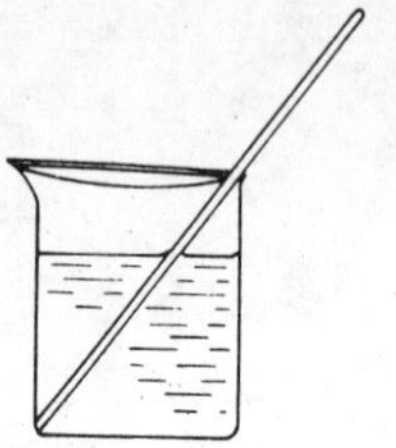

Fig. 4.10
Beaker, stirring rod and watch-glass

Before filtration, precipitates may need to be digested by allowing them to stand overnight, or by heating the precipitate and its supernatant liquid nearly to boiling point for some time. Digestion in this manner increases the degree of coarseness of the precipitate. A further check for completeness of precipitation should be carried out when the supernatant liquid becomes clear during the period of digestion. No further precipitate should be produced when a few drops of the precipitant are added. However, should further precipitation occur, more precipitant must be added as described above.

If the precipitate is much more soluble in hot water than in cold, the solution is allowed to cool to room temperature before filtration. Otherwise, solutions are filtered hot to speed up the filtration.

Filtration

Apparatus used in filtration

Various types of filtering apparatus are used in gravimetric analysis.

(*a*) Gooch crucible of either porcelain or silica in which ready-made, disposable, glass filter mats may only be used for precipitates at temperatures below 200°.

(*b*) Sintered glass or sintered silica (Vitreosil) crucibles.

Gooch or sintered crucibles have now almost completely superseded the use of filter paper, largely because filtration and washing of the precipitates can be carried out so much more rapidly and efficiently with their aid. The ignition of precipitates in crucibles also avoids the necessity of first drying and then burning off the filter paper. Filtration through paper may, however, occasionally be advantageous with gelatinous precipitates.

Gooch crucibles. A Gooch crucible should be supported by a soft rubber collar in a glass adapter which passes through a rubber bung fitting into a Buchner flask (Fig. 4.11). The rubber collar should not project below the bottom or above the top of the crucible. The tip of the adapter should project below the side arm of the Buchner flask to avoid the risk of the filtrate being sucked out of the flask. A trap is required to prevent any backflow from the pump contaminating the filtrate. Furthermore, it reduces the effects of changes in water pressure which may cause the filter pad (see later) to lift from the base of the crucible.

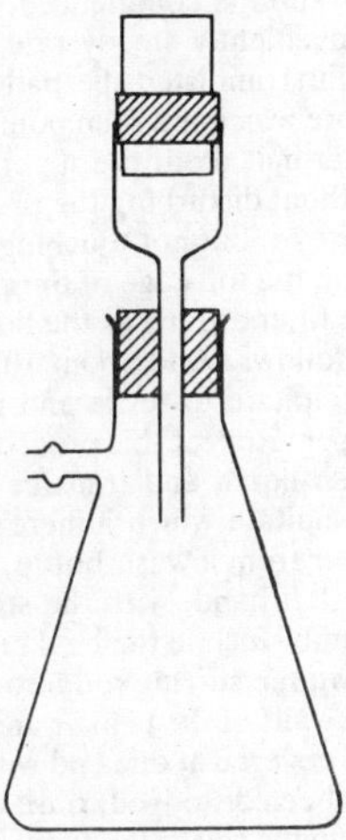

Fig. 4.11 Gooch crucible assembly

Disposable glass fibre filter discs. Standard glass filter discs are available (Whatman; 2.1 cm diameter) for use as filtration beds in Gooch crucibles. These filter discs are cheap, easy to handle, stable up to temperatures of at least 180° and, like the asbestos filter beds formerly used, are readily removed for cleansing and disposal at the end of the analysis.

Preparation of the Gooch crucible for filtration Place the crucible in the suction filtration apparatus (Fig 4.11). Sit a glass fibre filter disc gently on top of the porous base of the crucible. Gently wash the crucible and filter disc with *water* to remove any loose fibres, and apply gentle suction to drain off excess water. Suck the pad as dry as possible, covering the mouth of the crucible during the suction to prevent dust particles being drawn in. Dry (or ignite) the prepared Gooch crucible under the conditions specified for the drying (or ignition) of the precipitate, cool in a desiccator and weigh. Reheat, cool, and weigh; repeat until constant weight is attained.

Sintered glass or sintered silica crucibles. The filtering beds are fused-in sintered discs of the same material as the crucible. The sintered glass type of crucible is most frequently used by students. They are available in various degrees of porosity numbered 1, 2, 3 and 4, indicating decreasing pore diameter with increasing number. A number 3 crucible is suitable for precipitates of medium particle size, such as silver chloride, while a number 4 is necessary for fine precipitates, such as barium sulphate. Sintered glass crucibles should not be heated at temperatures above 400°. If the precipitate has to be ignited, or requires drying at temperatures above 400°, sintered silica crucibles (Vitreosil) should be used.

Sintered crucibles are prepared for use in filtration by washing and passing *water* through under suction and then drying (or igniting) to constant weight under the conditions which are specified for the drying (or ignition) of the filtered precipitate.

Technique of filtration and washing the precipitate

Using a Gooch crucible Care must be taken to avoid dislodging the filter disc. Suction must, therefore, be applied before pouring liquid into the crucible and should be adjusted, before filtration is commenced, to a low level just sufficient to draw liquid through the filter at a conveniently steady rate.

First, moisten the pad with a few drops of *water* from a teat pipette. Apply suction, add more *water* and then pour the liquid gently down a stirring rod on to the central portion of the filter mat. Pour the supernatant liquid above the precipitate down the glass rod into the filter without disturbing the precipitate more than is necessary. Keep the lower end of the glass rod close to, but not touching, the filter pad. Do not allow the liquid to rise above a level of 1 cm from the top edge of the crucible. Keep the beaker inclined and a stream of liquid passing into the filter as long as the liquid filters freely. Wash the precipitate in the beaker by decantation as follows. Add about 10 ml of the wash liquid to the precipitate, stir the mixture, allow the precipitate to settle and pour off the supernatant liquid down the glass rod on to the filter paper. Repeat this procedure three or four times, then stir the precipitate with about 20 ml of wash liquid and transfer as much of the solid as possible to the filter. Wash the traces of precipitate which adhere to the sides and bottom of the beaker into the filter with a jet of *water* from a wash-bottle, as shown. Hold the wash-bottle in the right hand and the beaker in the left hand, with the stirring rod pressed firmly against the lip of the beaker with the left thumb. Incline the beaker and direct a stream of wash liquid against the precipitate to wash it down the stirring rod into the filter. Small amounts of precipitate which adhere tenaciously to the walls of the beaker and the rod should be dislodged by rubbing with a 'policeman' (a glass rod covered at one end with a small piece of rubber or plastic tubing). When all the precipitate has been dislodged, rinse the 'policeman' with wash liquid and transfer the trace of precipitate down the rod to the filter. Finally, hold the beaker and stirring rod up to the light and examine carefully for traces of precipitate; repeat the treatment if necessary. (*Note*: a 'policeman' should not be used during precipitation or the early stages of filtration and should be removed from the beaker immediately after use.)

When the precipitate has been transferred quantitatively to the filter, it should be washed immediately with wash liquid from a wash-bottle. Unwashed precipitate should not be allowed to stand for any length of time, because it will dry out and the mass will crack and, in this form, it cannot be washed properly. The jet of liquid should be directed on to the side of the crucible. The filter should be allowed to drain completely between each washing. From time to time, test for the completeness of washing by collecting small samples of the washings in a test-tube as they pass through the funnel, applying appropriate qualitative tests. When negative tests are obtained, the precipitate and crucible are ready for drying.

In some determinations, it is advantageous to wash the precipitate finally with ethanol to remove most of the water and thus reduce the time required for subsequent drying of the precipitate.

Using sintered glass or sintered silica crucibles The technique is similar to that described for Gooch crucibles except that a filter mat is not used.

Drying and ignition of the precipitate When the precipitate has been washed free from excess of the precipitant and other soluble impurities, it is converted to a substance of constant composition before it is weighed. The details of the actual temperature of drying or the temperature of ignition are given in the description of each determination. General techniques are outlined below.

Phase separating filtration

Special filter papers (Whatman No. 1PS), impregnated with a water-repellent, are available for effecting rapid, small-scale separations of immiscible mixtures of aqueous and organic liquids. The papers, which are used in a conventional filter funnel, are hydrophobic. Organic liquids readily pass through the paper, whilst aqueous liquids (and solids) are retained, so that rapid separation of two such liquids is readily attained.

Variables in quantitative analysis

Variations in the results obtained for replicate assays may arise from a number of different factors, of which the following are the most significant:

(1) inhomogeneity of the medicament
(2) sampling errors
(3) the precision of the assay method
(4) random errors, including those due to the operator.

It will be apparent, therefore, that, if the analysis is to provide a reasonable assessment of the quality of the substance and to be capable of revealing inhomogeneity, suitable sampling procedures must be applied, the precision of the assay method must be known and the extent of random error established.

Sampling for analysis

Any form of analysis, however carefully devised and however carefully conducted, is circumscribed by the limitations of the sampling procedure and the purpose for which it is applied. Quality control procedures designed to check products at the time of manufacture for conformity with a manufacturing specification must clearly be representative of the batch as a whole. Otherwise, there is a risk of either accepting a bad batch or rejecting a good one. In practice, this means taking such a number of samples as may be determined by the size of the batch, or, if the process is a continuous one, ensuring that samples are withdrawn for analysis at regular time intervals. The number or frequency of sampling will also be determined by the risk of inhomogeneity in the product. Clearly, this will be much lower in solutions than in suspensions, emulsions, semisolids (ointments and creams) or solids (powders and granules).

Samples drawn from solutions may generally be assumed to be representative of the whole. Aqueous solutions which have been stored undisturbed for any period of time should, nonetheless, be stirred or shaken thoroughly before sampling to ensure that any solvent condensed on the cooler parts of the vessel above the solution is completely remixed into solution. Bulk heterogeneous systems, such as suspensions, emulsions, ointments, creams and powders, which are liable to undergo a degree of separation on standing, should likewise be thoroughly remixed before samples are withdrawn for analysis.

Sampling as part of 'in process' control procedures should, similarly, be representative of the process as a whole. Procedures should be designed to ensure that adequate checks are made, not only at times when the process is running normally, but also when the product is at greatest risk of being outside specification. This applies both to batch processes and continuous processes. Thus, a proportion of samples taken should relate to critical times whenever these may arise, as, for example, when continuous processes are restarted after shut-down, or in packaging operations by sampling at the beginning, in the middle and the end of the operation.

Precision of assays and sampling

The number of samples required is determined not only by the nature of the product and the variables of the process, such as batch size, but also by the precision with which the analytical procedure can be applied. The precision of an assay is inherent in the analytical method, but may be modified by the sensitivity of any instrumentation used and by the skill of the analyst. It is determined by the number of replicates tested and is assessed by the standard deviation of the method, or the coefficient of variation, which, for high precision, should be small.

The **standard deviation** (*s*) is a measure of the scatter of the results about the arithmetic **mean**. This may be calculated from a set of results as follows:

$$s = \sqrt{\frac{\Sigma\ (X - \overline{X}^2)}{n - 1}}$$

where X = observed result

$\overline{X}$ = mean

n = No. of determinations

Standard deviation provides the basis for the calculation of **limits of error** or **confidence limits**, which give the range within which the true value of the result lies. Their calculation is based on the theory of the normal distribution of results about the mean. The limits, however, cannot be stated exactly, but only in terms of probability levels, for example, as follows:

Limits	Probability that results will lie within the stated limits
mean ± standard deviation	0.683
mean ± (1.96 × standard deviation)	0.95
mean ± (2.58 × standard deviation)	0.99

Thus, where the probability is 0.95, nineteen out of twenty results may be expected to lie within the limits given by the mean ± (1.96 × standard deviation). The actual limits depend upon the number of tests (n), and can be calculated using the factor *t*. This is obtained from a table of *t* for the appropriate confidence level at the point for (n − 1), which is equivalent to the number of degrees of freedom (ϕ), i.e. if n = 10, ϕ = 9. The measure of reliability of the results is then given by the expression:

$$\text{mean} \pm \frac{ts}{\sqrt{n}}$$

Thus,

(a) for a single determination ($\phi = 0$), the 95% confidence limits are given by:

$$\text{mean} \pm 1.96s$$

(b) for duplicates ($\phi = 1$; $t = 12.7$) the 95% confidence limits are given by:

$$\text{mean} \pm \frac{12.7s}{\sqrt{2}}$$

(c) for ten determinations ($\phi = 9$; $t = 2.26$) the 95% confidence limits are given by:

$$\text{mean} \pm \frac{2.26s}{\sqrt{10}}$$

Analysis of variance

Where it is necessary to assess the contribution of variables due to both the precision of the assay itself and sampling errors, use is made of the square of the standard deviation, *variance* (V) as these values are additive. Thus,

$$\text{Variance (total)} = \frac{\text{Variance (sampling)}}{\text{No. of samples}} + \frac{\text{Variance (assay)}}{\text{No. of samples} \times \text{No. of assays}}$$

Suppose that variance of sampling is 0.06 and variance of assay is 0.04, then, if only one sample is taken and assayed four times,

$$V_{\text{total}} = \frac{0.06}{1} + \frac{0.04}{1 \times 4} = 0.07$$

Likewise, if four samples are taken and each is assayed only once,

$$V_{\text{total}} = \frac{0.06}{4} + \frac{0.04}{4 \times 1} = 0.025$$

and, similarly, if four samples are taken and each is assayed twice, then

$$V_{\text{total}} = \frac{0.06}{4} + \frac{0.04}{4 \times 2} = 0.020$$

It is evident, therefore, that much greater overall precision is attained by increasing the number of samples taken from the batch than by increasing the number of assays on each sample. Such methods form the basis for the construction of sampling plans.

Analysis of variance can also be used to compare the reliability of analytical methods in the hands of different operators in the same laboratory or between laboratories. They may also be extended to test the reliability of one or more different analytical methods for their suitability when applied in a particular assay. The variance ratio test is used. In this, the **variance ratio, F**, obtained by dividing the larger by the smaller variance obtained in two comparable sets of operations, is compared with the theoretical value of F for the same acceptance limits and number of degrees of freedom. The difference is significant if the calculated value of F is greater than the theoretical value.

The **accuracy** of an assay method is the measure of its ability to give the true result. This can only be established by concurrence of results obtained in two or more independent analytical methods. Where the accuracy of one method is established, the accuracy of any new method can only be ascertained by means of a **t-test**. This determines whether or not the mean value found by the new method is significantly different from that obtained by the established method.

Pharmacopoeial specifications

In contrast to release specifications in the factory, which are capable of full statistical control of sampling and analytical procedures, pharmacopoeial specifications lay down the minimum standard acceptable for the product at the time it reaches the patient. Samples may, therefore, be restricted in size and wholly unrepresentative of the original production batch from which they are drawn. Thus, any one batch of product may well become distributed in small packs over a wide geographical area, stored under widely varying conditions, albeit within prescribed limits, and even subjected to breaking open of the original containers with exposure to atmospheric effects in unsealed, poorly sealed, or partially filled containers. The size of such samples may also be determined by the number of unit doses ordered in the prescription being sampled. In these circumstances, the analyst has little control over sampling factors and his results are circumscribed by this lack of control. For these reasons, pharmacopoeial specifications generally, though not always, set lower standards than manufacturing specifications, even where little deterioration of the product is anticipated.

Additional reading and statistical tables

L. Saunders and R. Fleming, *Mathematics and statistics for use in the biological and pharmaceutical sciences*, Pharmaceutical Press, London, 1971.

5
Acidimetry and alkalimetry

Standard volumetric solutions

Expression of concentration

Traditionally, the concentrations of standard volumetric solutions have been expressed in terms of **normality**, which is derived as shown below from the equivalent weights of the reacting substances. The normality of a solution is, therefore, determined by the reaction involved and may have different values when used for different purposes. This is considered by many analysts to be a fundamental disadvantage. In consequence, most national and international drug standards compendia have changed the method of expressing concentrations from normality to **molarity** in accordance with modern analytical practice. The definition of these expressions is elaborated below.

Normality

A **normal solution** (designated N) is a solution which contains one gram equivalent of substance per litre of solution. Solutions of strengths other than normal are designated appropriately, e.g.

Twice nomal	2 N		
Half normal	0.5 N	or	N/2
Deci-normal	0.1 N	or	N/10
Centi-normal	0.01 N	or	N/100

Equivalent weight

The equivalent weight of a substance is determined by the actual reaction under consideration. The equivalent weight of an acid or base is that weight of it which contains 1.0079 g of replaceable hydrogen (or its equivalent, such as the hydroxyl group). Thus, the equivalent weights of hydrochloric acid (HCl) and sodium hydroxide (NaOH) are equal to their molecular weights, whilst that of sulphuric acid (H_2SO_4) or of oxalic acid $(COOH)_2$ which contain two replaceable hydrogens is equal to half the respective molecular weights.

In oxidation-reduction titrations, the equivalent weight is that weight which yields or combines with 1.0079 g of 'available' hydrogen or 7.9997 g of 'available' oxygen, 'available' being defined as available for use in the oxidation or reduction reaction under consideration.

The equivalent weight in precipitation reactions, as in argentimetry, is defined as that weight of substance which contains or combines with 1 g atom of a univalent metal.

The values taken for atomic and equivalent weights were originally based on the atomic weight $^{16}O = 16$. For the convenience of spectroscopists, atomic weights are now based on $^{12}C = 12$ but this involves only an insignificant change in the values hitherto accepted for the purposes of quantitative analytical chemistry.

Calculation of equivalent weight

Hydrochloric acid

$$HCl \equiv H$$
$$\therefore \quad 36.47\text{ g } HCl \equiv 1000\text{ ml N}$$

Sodium Carbonate

$$Na_2CO_3 + 2HCl \rightarrow 2NaCl + H_2O + CO_2$$
$$\therefore \quad Na_2CO_3 \equiv 2HCl \equiv 2H$$
$$\therefore \quad 106\text{ g } Na_2CO_3 \equiv 2000\text{ ml N}$$
$$\therefore \quad 53\text{ g } Na_2CO_3 \equiv 1000\text{ ml N } HCl$$

Sodium Acid Oxalate

The equivalent varies with the reaction under consideration:

(i) *as an acid*

$$NaHC_2O_4 + NaOH \rightarrow Na_2C_2O_4 + H_2O$$
$$\therefore \quad NaHC_2O_4 \equiv NaOH \equiv H$$
$$\therefore \quad 112\text{ g } NaHC_2O_4 \equiv 1000\text{ ml N } NaOH$$

(ii) *as a reducing agent*

$$NaHC_2O_4 + \bar{O} \rightarrow NaOH + 2CO_2$$
$$\therefore \quad NaHC_2O_4 \equiv \bar{O} \equiv 2H$$
$$\therefore \quad 112\text{ g } NaHC_2O_4 \equiv 2000\text{ ml N}$$
$$\therefore \quad 56\text{ g } NaHC_2O_4 \equiv 1000\text{ ml N } NaOH$$

Molarity

A molar solution (designated M) is one which contains the gram-molecular weight of the substance in one litre of solution. The strengths of molar solutions are independent of the reaction under consideration. The relationship between molarity and normality is a function of the equivalence of the acid or base. Thus, for **monobasic acids** and **monoacidic bases**, Molar (M) and Normal (N) solutions are identical in concentration, since

$$100\text{ ml M } HCl \equiv HCl \equiv H \equiv 1000\text{ ml N } HCl$$

and

$$HCl \equiv NaOH$$

In contrast, for **dibasic acids** and **diacidic bases**, Molar (M) and 2 Normal

(2N) are identical in concentration. Likewise, 0.5M and N solutions and all others in those same proportions are identical in concentration, since

$$1000 \text{ ml M } H_2SO_4 \equiv H_2SO_4 \equiv 2H \equiv 2000 \text{ ml N } H_2SO_4$$

and

$$1000 \text{ ml M } Na_2CO_3 \equiv Na_2CO_3 \equiv 2HCl \equiv 2H \equiv 1000 \text{ ml 2N } Na_2CO_3$$

The equivalence of standard solutions

If an acid A is titrated to neutrality against a base B and the volume of acid and base used are v_A and v_B respectively, then, since the two solutions contain the same number of gram-equivalents, or gram molecules (moles),

$$v_A \times \text{normality}_A = v_B \times \text{normality}_B$$

and

$$v_A \times \text{molarity}_A = v_B \times \text{molarity}_B$$

Thus, if the molarity (or normality) of one solution is known, and the equivalent volumes of the two solutions are determined by titration, the molarity (or normality) of the second solution can be found.

Preparation of standard solutions

If the chemical is available in a pure state, weigh out an exact quantity, dissolve it in water and make up to volume. Substances which are not usually obtainable in a pure stage, e.g. mineral acids and caustic alkalis, are prepared as approximate solutions and standardised against a known pure solid standard, e.g. A.R. potassium hydrogen phthalate for alkalis and anhydrous sodium carbonate for acids. Solutions must be standardised under the same conditions as those to be used in any subsequent determination. If the unstandardised solution has to be used in the burette in the determinations, then it must be placed in the burette for standardisation and not in the titration vessel. The indicator used in standardisation must, in general, be that which has to be used in the subsequent determination.

Factors

In the preparation of standard solutions, it is often convenient to make solutions which, although of known strengths, are only approximately M, 0.5M or 0.1M etc. The relationship of the exact strength of such a solution to the nominal normality of the solution is then indicated by a factor. The factor is defined as the number of millimetres of exactly M, 0.5M etc. solution which are equivalent to 1 ml of the solution of the same nominal molarity. It is the number by which the actual volume of the solution of approximate strength must be multiplied to obtain the equivalent volume of a standard solution of exact molarity.

Calculation of the factor in the standardisation of 0.5M (approx.) H_2SO_4 *by* 0.5M Na_2CO_3.

Suppose 20 ml 0.5M Na_2CO_3 was neutralised by 19.40 ml of 0.5M (approx.) H_2SO_4. Then,

$$19.40 \text{ ml } 0.5\text{M (approx.) } H_2SO_4 \equiv 20 \text{ ml } 0.5\text{M } Na_2CO_3$$

$$\therefore \quad 1 \text{ ml } 0.5\text{M (approx.) } H_2SO_4 \equiv \frac{20 \times 0.5}{19.40}$$

$$\equiv 0.515 \text{ ml M } Na_2CO_3$$

Therefore, the strength of the H_2SO_4 solution is 0.515M. The **factor** of the solution is 0.515 and all volumes of this solution when multiplied by 0.515 will give the equivalent volume of molar solution.

It should be noted that, in equating the two solutions, the one of *unknown* strength is placed on the *left-hand side* of the equation, i.e.

$$19.40 \text{ ml of } 0.5\text{M (approx.)} \equiv 20 \text{ ml } 0.5\text{M } Na_2CO_3$$

Adherence to this rule will ensure correct calculation of the factor of an unknown solution and avoidance of the pitfall of reciprocal factors.

A tenfold quantitive dilution of a M solution gives a 0.1M solution with the same factor as the original M solution, e.g. *n* ml 0.515M H_2SO_4 diluted with water to 10*n* ml gives a solution which is 0.0515M H_2SO_4.

Method of recording quantitive analysis in the practical note-book

All measurements such as weighings and burette readings should be recorded on the left-hand page of the note-book.

A statement of the determination to be carried out, the equations, the description of the method, the calculations and the final statement of the result should be recorded on the right-hand page of the note-book.

On p.142 is a specimen pair of pages from a practical note-book.

Standardisation of M H_2SO_4

One of the methods of standardisation depends upon the use of exsiccated sodium carbonate:

$$H_2SO_4 + Na_2CO_3 \rightarrow Na_2SO_4 + CO_2 + H_2O$$

Method Heat sufficient A.R. (AnalaR) exsiccated sodium carbonate in a nickel crucible at 260°–270° for half an hour to remove any trace of moisture. Cool in a desiccator and weigh an appropriate portion into a beaker containing *water* (to avoid caking which otherwise occurs when it is wetted in the confines of a funnel). Dissolve with the aid of heat if necessary, cool and transfer to an appropriate graduated flask. Dilute to volume and mix thoroughly.

Pipette 20 ml of this solution into a conical flask, add methyl orange indicator (1 drop) and titrate with the sulphuric acid solution until the first shade of orange colour is obtained. Repeat with another 20 ml portion. The two titrations should agree to within 0.05 ml, otherwise repeat until agreement is obtained.

The method can also be applied to the standardisation of other strengths of acid solutions.

Standard solutions of hydrochloric acid, sulphuric acid and sodium carbonate remain stable if properly stored.

Standardisation of M NaOH

Solutions of sodium hydroxide are always standardised by titration with standard acids of equivalent normality. They always contain a small but variable amount of carbon dioxide fixed as sodium carbonate so that the reactions expressed by equations (1–3) must be considered when standardising sodium (or potassium) hydroxide solutions:

$$NaOH + HCl \rightarrow NaCl + H_2O \quad (1)$$

$$Na_2CO_3 + HCl \rightarrow NaHCO_3 + NaCl \quad (2)$$

$$NaHCO_3 + HCl \rightarrow NaCl + H_2CO_3 \quad (3)$$

$$H_2CO_3 \downarrow CO_2 + H_2O$$

As the hydrochloric acid is run into the sodium hydroxide solution the reactions expressed by equations (1), (2) and (3), occur in that order. The several end points of these reactions can be determined by suitable choice of indicators. Phenolphthalein, for example, is sensitive to carbon dioxide, i.e. a solution which is sufficiently alkaline to be *just pink* in colour will be changed to colourless by carbon dioxide. Immediately the reaction expressed by equation (2) is completed, the next drop of HCl will liberate CO_2 (equation 3), and the indicator will change from pink to colourless. Thus the phenolphthalein end point represents the point at which all hydroxide and half the carbonate have been neutralised. Methyl orange, on the other hand, is insensitive to carbon dioxide so that the methyl orange end point is obtained when all three reactions are complete. The methyl orange factor is, therefore, a measure of total alkali (hydroxide + carbonate). Therefore, the methyl orange and phenolphthalein factors are different.

Method Pipette 20 ml of approximately M NaOH into a conical flask, add 2 drops of phenolphthalein solution and titrate with M HCl until the solution is just colourless. Note the burette reading for calculation of the phenolphthalein factor, add 1 drop of methyl orange solution and continue the titration until the first orange colour is obtained. Record the burette reading and calculate the methyl orange factor. The standardisation should be carried out in duplicate.

Note on the preparation of carbonate-free sodium hydroxide solution

Place a small amount of paraffin wax in a glass-stoppered bottle and warm the bottle in a water bath to melt the wax. Shake to coat the whole of the inside surface of the bottle with paraffin wax and cool rapidly to form a solid film. Prepare a 50% w/w solution of sodium hydroxide in water and store in the bottle for 24 h to allow sodium carbonate to crystallise out. Decant the clear supernatant liquid and dilute with CO_2-free water to the required strength.

Alternatively, carbonate-free sodium hydroxide can be prepared by ion exchange techniques (Part 2).

SPECIMEN PAGES FROM A PRACTICAL NOTEBOOK

Determination of the percentage of Na_2CO_3 in the sample of exsiccated Sodium Carbonate

Results

	1	2
1st weight of weighing bottle + sample =	21.2312 g	21.1901 g
2nd weight of weighing bottle + sample =	20.2472 g	20.2380 g
Weight of sample =	0.9840 g	0.9521 g
Burette readings: Final	17.02	16.45
Initial	0.00	0.00
Volume of 1.065м HCl used	17.02 ml	16.45 ml

This determination depends upon the reactions expressed by the following equations:

$$Na_2CO_3 + 2HCl \rightarrow 2NaCl + CO_2 + H_2O$$
$$\therefore\ Na_2CO_3 \equiv 2HCl$$
$$\therefore\ 106.0\text{ g } Na_2CO_3 \equiv 2000\text{ ml м solution}$$
$$\therefore\ 53.00\text{ g } Na_2CO_3 \equiv 1000\text{ ml м solution}$$
$$\therefore\ 0.05300\text{ g } Na_2CO_3 \equiv 1\text{ ml м solution}$$

∴ 1 g of sample should give a burette reading of about 20 ml of м solution.

Method Accurately weigh about 1 g sample into a conical flask containing a little water and titrated with м HCl solution.

Indicator Methyl orange—1 drop

End point The orange colour nearest to yellow

Calculations I

$$\text{Weight of sample} = 0.9840\text{ g}$$
$$\text{Burette reading} = 17.02\text{ ml } 1.065\text{м HCl}$$
$$\therefore\ 0.9840\text{ g sample} = 17.02 \times 1.065\text{ ml м}$$
$$\therefore\ 1\text{ g sample} \equiv \frac{17.02 \times 1.065}{0.9840}\text{ ml of м}$$
$$\equiv 18.42\text{ ml м HCl}$$

II

$$\text{Weight of sample} = 0.9521\text{ g}$$
$$\text{Burette reading} = 16.45\text{ ml } 1.065\text{м HCl}$$
$$\therefore\ 0.9521\text{ g sample} \equiv 16.45 \times 1.065\text{ ml м}$$
$$\therefore\ 1\text{ g sample} \equiv \frac{16.45 \times 1.065}{0.9521}\text{ ml м}$$
$$\equiv 18.40\text{ ml м HCl}$$

$$\text{Average of I and II, 1 g sample} \equiv 18.41\text{ ml м}$$
$$\equiv 18.41\text{ ml} \times 0.05300\text{ g } Na_2CO_3$$
$$\equiv 0.9757\text{ g } Na_2CO_3$$

Conclusion: The sample of exsiccated Sodium Carbonate was found to contain 97.57% Na_2CO_3.

Direct titration of strong acids

Titration of strong acids with sodium hydroxide solution gives salts which are not hydrolysed in aqueous solution. The solutions are therefore neutral at the equivalence point. The pH changes extremely rapidly in the region of the equivalence point and an indicator which changes colour anywhere within the range between pH 4 and pH 10 is suitable. Methyl orange is often used for such titrations.

Hydrochloric acid *Determination of the percentage w/w of* HCl

$$HCl + NaOH \rightarrow NaCl + H_2O$$

$$\therefore \quad 36.46\text{ g HCl} \equiv 1000\text{ ml M NaOH}$$

$$\therefore \quad 0.03646\text{ g HCl} \equiv 1\text{ ml M NaOH}$$

Method Weigh out accurately 2 ml of sample (see page 122 for technique of weighing water-miscible liquids) using a stoppered weighing bottle to prevent loss of hydrochloric acid vapour (this would also attack the metal parts of the balance and weights). Transfer the acid into a stoppered flask by washing out with about 30 ml of *water*. Titrate with M NaOH using methyl orange as indicator, taking the first definite orange colour as the end point.

COGNATE DETERMINATIONS

Perchloric Acid

Sulphuric Acid

Thiamine Hydrochloride. Determination of Cl *present as hydrochloride.* This is a direct titration with 0.1M NaOH using bromothymol blue solution as indicator. The end point is given by the bluish-green colour of the indicator (pH 7.0).

Determination of aldehydes and ketones in essential oils

The determination of aldehydes depends upon the reaction with hydroxylamine hydrochloride expressed by the following equation:

$$R\cdot CHO + NH_2OH,\ HCl \rightarrow R\cdot CH{:}N\cdot OH + H_2O + HCl$$

The liberated hydrochloric acid can be titrated with standard alkali. The reaction, which is reversible, goes to completion if the hydrochloric acid is removed by titration with alkali. As the end point is flat and somewhat difficult to detect, the titration is continued to the full yellow colour of the methyl orange indicator. Hydroxylamine hydrochloride is hydrolysed in solution and is acidic to methyl orange. To compensate for the acid present due to hydrolysis of the reagent, the latter is first neutralised to the full yellow colour of methyl orange. Therefore, the reagent consists mainly of hydroxylamine hydrochloride but a small proportion of the free hydroxylamine also is present; both base and salt may react with aldehydes. Thus the acid titrated during the actual assay is somewhat less than should theoretically have been formed and an empirical factor (×1.008) is introduced into the calculation to allow for this error.

0.5M *Potassium Hydroxide in Ethanol* (60%)

Preparation Dissolve sufficient potassium hydroxide in ethanol (60%) to produce a solution containing 28.05 g of KOH per litre. NB Potassium hydroxide contains only *c.* 80% of KOH.

Standardisation Pipette 0.5M HCl (20 ml) into a conical flask; add phenol red solution and titrate with approx. 0.5M KOH in ethanol (60%).

Hydroxylamine Hydrochloride Reagent in Ethanol (60%)

Preparation Dissolve hydroxylamine hydrochloride (34.75 g) in ethanol (950 ml of 60%) and add methyl orange [5 ml; 0.2 w/v solution in ethanol (60%)]. Add 0.5M KOH in ethanol (60%) dropwise until the full yellow colour of the indicator is obtained. Adjust the volume to 100 ml with ethanol (60%). Check the pH of the solution by means of the following test: to 10 ml add one drop of 0.5M KOH in ethanol (60%)—no change in colour is produced. To a further 10 ml add one drop of 0.5M HCl—the colour changes slightly towards orange.

Cinnamon Oil *Determination of the percentage w/w of cinnamic aldehyde* C_9H_8O

Method By means of a wire loop, suspend a glass-stoppered tube (about 25 mm in diameter and 150 mm long) from the stirrup of a balance. Allow 5 min for temperature equilibration to occur and then weigh the tube. By means of a teat pipette, introduce the sample (about 1 g) and reweigh. Add from a measuring cylinder toluene (5 ml) and hydroxylamine hydrochloride reagent in 60% ethanol (15 ml), shake vigorously and titrate with 0.5M KOH in ethanol (60%) until the red colour changes to full yellow. Continue the titration with vigorous shaking until the lower layer retains the full yellow of the indicator after shaking vigorously for 2 min and allowing to separate. The reaction is slow and a titration indicates the approximate amount of aldehyde in the sample. Add a further 0.5 ml titrant to the above mixture to ensure that the full yellow colour of the indicator has been obtained, and use this as a standard colour for matching the end point of two further accurate determinations.

The reaction is not quantitive unless an adequate excess of hydroxylamine hydrochloride is used. To ensure that this is so, the volume of hydroxylamine hydrochloride reagent in ethanol (60%) added must exceed the burette reading of 0.5M KOH in alcohol (60%) by between 1 and 2 ml.

1 ml 0.5M KOH in ethanol (60%)
$\equiv$ 0.06661 (0.06608 × 1.008) g of cinnamic aldehyde C_9H_8O

Lemon Oil *Determination of the percentage w/w of aldehydes calculated as citral* $C_{10}H_{16}O$

It can be determined by the above method using 10 g of sample and 7 ml of hydroxylamine hydrochloride reagent in ethanol (60%). Since Lemon Oil contains less aldehydes and more terpene than Cinnamon Oil a correspondingly larger weight of sample must be used in the assay. Addition of toluene is unnecessary since the proportion of terpene is sufficiently large to cause the titration liquid to separate into two layers. 1 ml 0.5M KOH in ethanol (60%) $\equiv$ 0.07673 (0.07611 × 1.008) g of citral $C_{10}H_{16}O$.

COGNATE DETERMINATIONS

Orange Oil
Terpeneless Lemon Oil
Terpeneless Orange Oil

Dill Oil *Determination of the percentage w/w of carvone* $C_{10}H_{14}O$

Carvone, a ketone, is determined similarly to aldehydes in volatile oils. The reagents are prepared in 90% ethanol to effect complete solution of the oil. This facilitates the reaction, which is very much slower than with the aldehydes. For similar reasons, the titration is carried out at an elevated temperature by immersing the tube in a water-bath maintained between 75° and 80°. The stopper of the tube should be loosened before heating as considerable pressures will otherwise build up rapidly inside the sealed tube, with the result that the glass stopper is liable to be ejected. The indicator, dimethyl yellow, gives a sharper colour change than methyl orange in strong ethanol solutions.

Each ml of M KOH ≡ 0.1514 (0.1502 × 1.008) g of carvone

Industrial alcohol must *not* be used in any of the above assays since it is liable to contain small quantities of carbonyl compounds which would react with the hydroxylamine reagent and so vitiate the results.

COGNATE DETERMINATIONS

Caraway Oil
Peppermint Oil

Direct titration of weak acids

Titration of weak acids with sodium hydroxide solution gives salts which will be hydrolysed in aqueous solution to a greater or lesser extent depending upon the dissociation constant of the acid. The pH of the solution at the equivalence point will be above pH 7 and the indicator most frequently used in such titrations is phenolphthalein.

Acetic Acid (33%) *Determination of the percentage w/w of* CH_3COOH

$$CH_3COOH + NaOH \rightarrow CH_3COONa + H_2O$$
$$\therefore \quad 60.05 \text{ g } CH_3COOH \equiv 100 \text{ ml M NaOH}$$
$$\therefore \quad 0.6005 \text{ g } C_2H_4O_2 \equiv 1 \text{ ml M NaOH}$$

∴ 5 g of sample (≡ *c*, 33%) is approximately equivalent to 27 ml M.

Method Weigh out accurately 5 g of sample (p.000 for technique of weighing water-miscible liquids) using a stoppered weighing bottle. Wash out the acid into a stoppered flask using about 50 ml of *water*. Titrate with M NaOH, using phenolphthalein as indicator, and take the first shade of pink as the end point.

Dilute Phosphoric Acid *Determination of the percentage w/w of* H_3PO_4

This determination depends upon the partial neutralisation of the phosphoric acid to give disodium hydrogen phosphate. The latter tends to hydrolyse and this is suppressed in the assay by the addition of sodium chloride and limitation of the volume of water.

$$H_3PO_4 + 2NaOH \rightarrow Na_2HPO_4 + 2H_2O$$
$$\therefore \quad H_3PO_4 \equiv 2NaOH \equiv 2000 \text{ ml M NaOH}$$
$$\therefore \quad 98.00 \text{ g } H_3PO_4 \equiv 2000 \text{ ml M NaOH}$$

$$\therefore \quad 0.0490\,g\ H_3PO_4 \equiv 1\,ml\ \text{M}\ NaOH$$

$\therefore$ 10 g of 10% solution is equivalent to approximately 20 ml M NaOH.

Method Weigh a clean dry weighing bottle, add the sample (about 10 g) and reweigh. Transfer the contents of the weighing bottle to a conical flask by means of a glass rod and small funnel. Wash the weighing bottle, rod and funnel with *water* and collect the washings in a conical flask. *Not more than* 30 *ml of water should be used.* Add sodium chloride (10 g; weighed on a rough balance) and titrate with M NaOH using 2 drops of solution of phenolphthalein as indicator.

COGNATE DETERMINATIONS

Benzoic Acid Prepare 30 ml (approx.) of neutralised ethanol (96%) by adding 10 drops of phenol red and titrating with 0.5M sodium hydroxide until a full red colour is obtained. Use 15 ml of this neutralised ethanol to dissolve about 2.5 g of sample, accurately weighed, add 20 ml of water to precipitate the acid in a finely divided state and titrate with 0.5M sodium hydroxide until a full red colour is obtained.

Citric Acid

Chlorambucil Titration in aqueous acetone.

Mustine Injection Ethanol is added to keep mustine base in solution during the titration.

Nicotinic Acid Tablets Nicotinic acid is extracted from the powdered tablets with hot 96% ethanol (neutralised to phenolphthalein), the extract diluted with water and titrated.

Potassium Hydrogen Phthalate

Salicylic Acid Determined as for *Benzoic Acid.*

Salicylic Acid Ointment

Undecanoic Acid

Special modifications in the direct titration of weak acids

Acid Value of fixed oils

Acid value is defined as the number of milligrams of potassium hydroxide required to neutralise the free acid in 1 g of substance. The method of determination is specified and is outlined below. The amount of free acid present in most of the fixed oils is generally low and the acid value is used to eliminate low-grade and rancid oils, which tend to have higher acid values.

Method Weigh accurately a 250 ml conical flask. Introduce the sample (about 10 g) and reweigh. Mix ether (25 ml), 96% ethanol (25 ml) and 1 ml of solution of phenolphthalein (1.0% in 96% ethanol) and, if necessary, neutralise by titration with a few drops of 0.1M aqueous KOH. Add this solution to the flask and shake to dissolve the fatty acids which are present in the oil. The oil itself is usually insoluble. Heat is unnecessary except for solid fats and waxes, for which it is essential to ensure complete extraction of the free fatty acids. Cool to room temperature before commencing the titration and continue the titration until the pink colour of the indicator persists for 15 s. Do not wash down with water during this titration. Strict adherence to this time factor is essential as the glycerides of the oil become hydrolysed in the presence of excess potassium hydroxide.

$$R{\cdot}COOH + KOH \rightarrow R{\cdot}COOK + H_2O$$

$$\therefore \quad KOH \equiv H$$

$$\therefore \quad 5.61\text{ g KOH} \equiv 10\,000\text{ ml } 0.1\text{M}$$

$$\therefore \quad 5.61\text{ mg KOH} \equiv 1\text{ ml } 0.1\text{M}$$

$$\therefore \quad \text{Acid value} = \frac{\text{burette reading } (\times\ 0.1 \text{ molarity of alkali}) \times 5.61}{\text{weight of sample (in g)}}$$

COGNATE DETERMINATIONS

Acid value of Oleic Acid This is determined as above, but using 1 g only. In this example, the acid value determination is really a determination of equivalent weight.

White Beeswax Xylene is added to increase solubility.

Aspirin *Determination of the percentage of* $C_9H_8O_4$

The determination of aspirin is based on the following reactions:

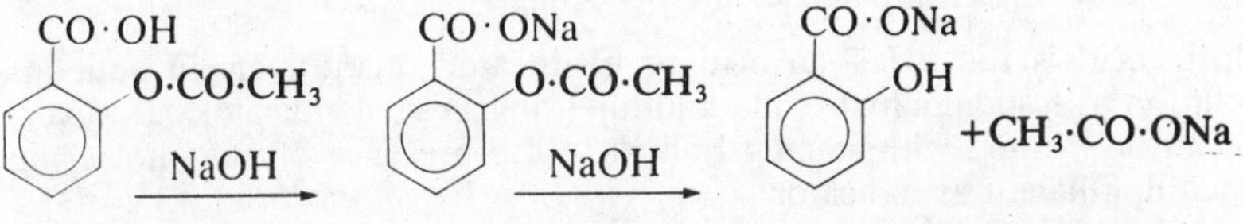

Aspirin Sodium salicylate

Aspirin is liable to be contaminated with salicylic acid which may be present as a result of incomplete conversion to aspirin during manufacture or through hydrolysis on storage. Hydrolysis also releases acetic acid, which can often be detected in samples of aspirin by odour. The quality of aspirin is, therefore, controlled by a combination of a double assay procedure and a separate colorimetric limit test for salicylic acid based on the reaction of its free phenolic group with iron(III) chloride. In the assay, the sample is dissolved in ethanol (96%), because of its low solubility in water (1 in 300), and titrated directly with sodium hydroxide using phenolphthalein as indicator. Aspirin and contaminating acetic and salicylic acids are titrated in this procedure. However, for calculation purposes in determining compliance with the standard, it is assumed that all material titrated is aspirin.

$$CH_3{\cdot}CO{\cdot}O{\cdot}C_6H_4{\cdot}COOH + NaOH \rightarrow CH_3{\cdot}CO{\cdot}O{\cdot}C_6H_4{\cdot}CO{\cdot}ONa + H_2O$$

$$\therefore \quad C_9H_8O_4 \equiv NaOH \equiv 1000\text{ ml M}$$

$$\therefore \quad 180.2\text{ g } C_9H_8O_4 \equiv 1000\text{ ml M}$$

$$\therefore \quad 0.01802\text{ g } C_9H_8O_4 \equiv 1\text{ ml } 0.1\text{M}$$

Aspirin alone is determined by refluxing the neutralised sample solution with excess standard sodium hydroxide solution to hydrolyse the acetoxy group, cooling and back-titrating the excess acid with standard acid. The difference between the volumes of standard sodium hydroxide solution consumed in the first and second titrations is a measure of the salicylic acid and acetic acid contamination of the product and must not be more than 0.4 ml for each 0.5 g of sample, equivalent to *c.* 1.45%.

Method Accurately weigh approximately 0.5 g of sample. Dissolve in ethanol (96%; 10 ml) and titrate with 0.1M NaOH using phenolphthalein as indicator. Add 50 ml 0.1M NaOH from a pipette. Immerse the flask in a boiling water bath for 15 min. Cool, taking precautions to exclude carbon dioxide, by covering the mouth of the flask with a small inverted beaker and allowing a stream of cold water to flow over it. Back-titrate the excess of alkali with 0.1M HCl using phenolphthalein as indicator. The pink colour of the indicator may gradually fade in contact with alkali. If this occurs, add a few more drops of indicator.

Ammonium Chloride

Addition of formaldehyde leads to its decomposition to methylene imine with release of hydrochloric acid which can then be titrated directly.

$$NH_4Cl + HCHO \rightarrow CH_2{=}NH + H_2O + HCl$$

Boric Acid *Determination of the percentage of* H_3BO_3

Boric acid is too weak an acid to be titrated quantitively in aqueous solution with sodium hydroxide solution using a visual indicator. However, it can be titrated with standard alkali in the presence of mannitol using phenolphthalein as indicator.

$$H_3BO_3 + NaOH \xrightarrow{\text{mannitol}} NaBO_2 + 2H_2O$$

$$\therefore \quad H_3BO_3 \equiv NaOH \equiv 1000 \text{ ml M NaOH}$$

$$\therefore \quad 61.84 \text{ g } H_3BO_3 \equiv 1000 \text{ ml M NaOH}$$

$$\therefore \quad 0.0618 \text{ g } H_3BO_3 \equiv 1 \text{ ml M NaOH}$$

Boric acid is esterified in the presence of polyhydric alcohols, such as glycerol or mannitol, forming a monobasic acid which is strong enough to give a satisfactory end point. It is stated that if glycerol is added a glycerylboric acid is formed. At least 30% glycerol or mannitol is needed to prevent hydrolysis of the titratable acid.

Method Accurately weigh the sample (about 1 g), add *water* (100 ml) and mannitol (15 g) and titrate with M NaOH using phenolphthalein as indicator.

Borax *Determination of the percentage of* $Na_2B_4O_7,10H_2O$

Borax behaves in aqueous solution as a mixture of sodium metaborate and boric acid. The latter is too weak an acid to be titrated directly, but in the presence of glycerol or mannitol it forms complexes of considerably greater acidic strength. In consequence, borax can be titrated directly with standard NaOH in the presence of mannitol, using phenolphthalein as indicator.

$$Na_2B_4O_7,10H_2O \xrightarrow{\text{mannitol}} 2NaBO_2 + 2H_3BO_3 + 7H_2O$$

$$2H_3BO_3 + 2NaOH \xrightarrow{\text{mannitol}} 2NaBO_2 + 4H_2O$$

$$\therefore \quad 381.4 \text{ g } Na_2B_4O_7,10H_2O \equiv 2000 \text{ ml M NaOH}$$

$$\therefore \quad 0.1907 \text{ g } Na_2B_4O_7,10H_2O \equiv 1 \text{ ml M NaOH}$$

Method Prepare a solution of mannitol (20 g) in water (100 ml) and neutralise to phenolphthalein with 0.1M NaOH. Add the sample (aprox. 3 g, accurately weighed) heating, if necessary, to dissolve. Cool rapidly and titrate the solution with 0.1M NaOH using phenolphthalein as indicator.

Busulphan

This substance is hydrolysed by refluxing with water and the liberated methanesulphonic acid is titrated with standard alkali using phenolphthalein as indicator.

$$CH_3 \cdot SO_2 \cdot O(CH_2)_4O \cdot SO_2 \cdot CH_3 + 2H_2O \rightarrow HO(CH_2)_4OH + 2CH_3 \cdot SO_2 \cdot OH$$

Phenylbutazone

Because phenylbutazone is almost insoluble in water, aqueous acetone is used as the solvent. This reduces the apparent pK_a of the acid; bromothymol blue is, therefore, used as indicator. It is essential to carry out a blank titration on the same volume of solvent solution, the blank titration being subtracted from the first titration before calculating the result.

COGNATE DETERMINATIONS

Cycloserine Propan-2-ol is used as the titration solvent and carbonate-free 0.1M NaOH as titrant.

Nicoumalone

Sulphinpyrazone

Direct titration of strong bases

Titration of strong bases with strong acids gives salts which are not hydrolysed in aqueous solution and the solution is, therefore, neutral. The pH changes in the region of the equivalence point are sufficiently large to permit a wide choice of indicator. Methyl orange is usually used.

Sodium Carbonate Decahydrate *Determination of the percentage of* Na_2CO_3

This determination depends upon the reaction expressed by the following equation:

$$Na_2CO_3 + 2HCl \rightarrow 2NaCl + H_2O + CO_2$$

$$\therefore \quad Na_2CO_3 \equiv 2HCl \equiv 2000 \text{ ml M HCl}$$

$$\therefore \quad 106 \text{ g } Na_2CO_3 \equiv 2000 \text{ ml M HCl}$$

$$\therefore \quad 0.05300 \text{ g } Na_2CO_3 \equiv 1 \text{ ml M HCl}$$

Method Accurately weigh the sample (approx. 3 g). Dissolve in water (25 ml) and titrate with M HCl using methyl orange indicator.

COGNATE DETERMINATIONS

Sodium Bicarbonate
Compound Sodium Bicarbonate Tablets
Sodium Bicarbonate Intravenous Infusion
Sodium Carbonate Monohydrate

Borax *Determination of* $Na_2B_4O_7,10H_2O$

Borax can be titrated quantitatively with standard acid:

$$Na_2B_4O_7, 10H_2O + 2HCl \rightarrow 4H_3BO_3 + 2NaCl + 5H_2O$$

$$\therefore\ Na_2B_4O_7, 10H_2O \equiv 2HCl \equiv 2000\ \text{ml M HCl}$$

$$\therefore\ 381.4\ \text{g}\ Na_2B_4O_7,10H_2O \equiv 2000\ \text{ml M HCl}$$

$$\therefore\ 0.1907\ \text{g}\ Na_2B_4O_7,10H_2O \equiv 1\ \text{ml M HCl}$$

To obtain the correct end point, the indicator must not be affected by the weak acid, H_3BO_3. This condition is met by the use of methyl red.

Method Dissolve the sample (about 2.3 g accurately weighed) in *water* (100 ml) with the aid of heat. Cool and titrate with 0.5M HCl using methyl red as indicator.

Contamination of Borax with either boric acid or sodium carbonate

Since borax is capable of being titrated with either acid or alkali (p.147), mixtures of borax with either sodium carbonate or boric acid can be assayed by a double titration procedure.

The official assay serves also to detect admixture of either boric acid or alkali with the sample, as explained below.

Method Accurately weigh the sample (3.0 g approx.), dissolve it in *water* (60 ml) and titrate with 0.5M hydrochloric acid using methyl red solution as indicator. Record the volume of 0.5M hydrochloric acid required. Reserve the titration liquid and repeat this portion of the assay. Calculate the volume of 0.5M HCl required to neutralise exactly 1 g of sample. These calculated volumes should *not* differ from one another by more than 0.05 ml. Boil the titration liquid to expel carbon dioxide. Cool the solution, add mannitol (20 g) and titrate with M sodium hydroxide using phenolphthalein solution as indicator.

Special modifications in the direct titration of strong bases

Calcium Hydroxide Solution

Although this is a fairly strong alkali and is titrated with hydrochloric acid solution, methyl orange is not used as indicator. The solution absorbs carbon dioxide from the atmosphere to form carbonate but since the free $Ca(OH)_2$ and not the total alkali content is required, phenolphthalein is used as indicator.

Calcium Hydroxide *Determination of the percentage of* $Ca(OH)_2$

This substance absorbs CO_2 from the atmosphere and some $CaCO_3$ is always present. The object of the determination is to find the percentage of $Ca(OH)_2$; this substance is separated from the $CaCO_3$ by means of the solubility of the former but not the latter in sucrose solution.

Method Shake gently in a litre bottle to give a fine suspension about 3 g (accurately weighed) with 10 ml ethanol (96%) previously neutralised to phenolphthalein. Add 10% solution of sucrose (490 ml) previously neutralised to phenolphthalein, shake the mixture vigorously for about 5 min and then at frequent intervals for 4 h. Filter an aliquot portion (250 ml) and titrate the M HCl using phenolphthalein as indicator. The burette reading is equivalent to the $Ca(OH)_2$ in half the weight of sample taken.

Sodium Hydroxide *Determination of the total alkali calculated as* NaOH, *and carbonate calculated as* Na_2CO_3

This determination depends upon the reactions expressed by the following equations:

$$Na_2CO_3 + BaCl_2 \rightarrow 2NaCl + BaCO_3$$

$$NaOH + HCl \rightarrow NaCl + H_2O$$

$$BaCO_3 + 2HCl \rightarrow BaCl_2 + H_2O + CO_2$$

$$\therefore \quad NaOH \equiv HCl$$

$$\therefore \quad 40.001\text{ g } NaOH \equiv 1000\text{ ml M}$$

$$\therefore \quad 0.0400\text{ g } NaOH \equiv 1\text{ ml M } HCl$$

and

$$BaCO_3 \equiv Na_2CO_3$$

$$\therefore \quad 105.98\text{ g } Na_2CO_3 \equiv 2000\text{ ml M}$$

$$\therefore \quad 0.05299\text{ g } Na_2CO_3 \equiv 1\text{ ml M } HCl$$

Barium chloride solution is added to precipitate the carbonate as $BaCO_3$. The free alkali can then be titrated with standard acid using phenolphthalein as indicator. Finally, the carbonate (now as barium carbonate) can be titrated with standard acid using bromophenol blue as indicator. The end point of the first titration is equivalent to the conversion of the NaOH into sodium chloride. The carbonate does not react since it has been converted into barium carbonate, which is insoluble as long as the solution is not acidic.

Method Weigh accurately about 2 g of sample. Add *water* (about 25 ml) and barium chloride solution (5 ml) and titrate with M HCl using phenolphthalein as indicator. Titrate *slowly* with *continuous shaking* to prevent the acid attacking the carbonate. Take the burette reading at the end point, i.e. when the solution becomes just colourless. Now add bromophenol blue solution (15 drops) (the solution will now be blue) and continue the titration with M HCl until a full green colour is obtained. The first full green colour will disappear on shaking as the acid attacks the undissolved carbonate. Therefore, titrate until the full green colour remains permanent on shaking.

Calculate the total alkali from the total burette reading and the carbonate from the difference in the readings at the two end points using phenolphthalein and bromophenol blue as indicators.

COGNATE DETERMINATION

Potassium Hydroxide

Sodium Benzoate *Determination of the percentage of* $C_6H_5 \cdot CO \cdot ONa$

This determination depends upon the reaction expressed by the following equation:

$$C_6H_5 \cdot COONa + HCl \rightarrow C_6H_5 \cdot COOH + NaCl$$

Any usual indicator would be affected by the benzoic acid $C_6H_5 \cdot COOH$ liberated in the titration, i.e. benzoic acid is a sufficiently strong acid to give a pH which will be 'acid' to the usual indicators. Therefore, a modification must be introduced in order to remove this acid.

Use is made of th- fact that benzoic acid is more soluble in ether than it is in water. When a substance is shaken with two immiscible solvents in which it is soluble, the substance distributes itself so that the concentration in each liquid is in the ratio of the solubilities of the substance in each of the solvents separately, e.g.

Benzoic acid is soluble in water 1 in 350 } Ratio of solubilities
Benzoic acid is soluble in ether 1 in 3 } is *c.* 1:120

Therefore, very approximately (since ether is miscible with water to some extent), when benzoic acid is shaken with ether and water, the concentration in the ether layer will be 120 times that in the water layer. Using equal volumes of ether and water and 1 g of acid, only 0.085 g ($\frac{1}{120}$) of the acid will remain in the aqueous layer. This fact is utilised in the determination.

$$C_6H_5 \cdot COONa + HCl \rightarrow C_6H_5 \cdot COOH + NaCl$$

$$\therefore \quad 144.1 \text{ g } C_6H_5 \cdot COONa \equiv 1000 \text{ ml M } HCl$$

$$\therefore \quad .07205 \text{ g } C_6H_5 \cdot COONa \equiv 1 \text{ ml } 0.5\text{M } HCl$$

Method Dissolve the sample, about 3 g accurately weighed in a little *water* in a beaker and transfer quantitively to a separator. Alternatively, introduce the sample directly into the separator via a wide-mouthed funnel. About 50 ml of *water* in all should be used to effect solution and transference. Add ether (50 ml) and bromophenol blue indicator (20 drops) (yellow in strong acid, green in slightly acid, blue in alkali). Titrate with 0.5M HCl until a full green colour is obtained, In viewing the colour of the solution, place white paper half way round the separator and *examine in daylight.* The separating funnel must be stoppered and shaken at intervals during the titration. It does not matter if the end point is slightly over-run at this stage in the determination, since there is still sufficient benzoic acid in the aqueous layer to give a pH which is acid to the indicator. This acid must be further reduced in concentration if the correct end point is to be obtained. Run off the aqueous layer into a clean separating funnel and rinse the stem with about 2 ml of *water*. Wash the ether layer remaining with two portions each of 5 ml of *water* and add to the aqueous layer in the second separator. Again rinse the stem of the first separator. To this aqueous solution add 20 ml of ether and shake well. The benzoic acid still left in the aqueous layer becomes redistributed between the ether and aqueous layer and so, after shaking, very little benzoic acid remains in the aqueous solution; the aqueous layer, therefore, becomes blue again. Continue the titration, shaking well after each addition, until a full green colour is obtained. A bluish-green colour indicates under-titration and a yellowish-green colour indicates over-titration.

Thiopentone Sodium *Determination of the percentage of* Na

Thiopentone sodium is a mixture of the monosodium derivative of 5-ethyl-5-(1-methybutyl)-2-thiobarbituric acid and anhydrous sodium carbonate. It is determined by direct titration with 0.1M HCl, using methyl red as indicator, until the pink colour of the indicator is obtained. The solution is then boiled gently for 2 min to expel carbon dioxide, cooled and the titration continued until the pink colour is restored.

COGNATE DETERMINATIONS

Methohexitone Injection
Thiopentone Injection

Direct titration of weak bases

Titration of weak bases with standard mineral acid solutions gives salts which will be hydrolysed in aqueous solution to a greater or lesser extent depending upon the dissociation constant of the base. The pH of the solution at the equivalence point will be below pH 7 and, therefore, methyl red is the indicator used most frequently in such titrations.

Aminophylline *Determination of the percentage of ethylenediamine*

The sample is dissolved in water and titrated with 0.1M H_2SO_4 to the green colour of bromocresol green.

Back-titrations

Determinations involving back-titrations consist in the addition of excess of a standard volumetric solution (from a pipette) to a weighed amount of sample, and determination of the excess not required by the sample. Hence the amount of volumetric solution used by the substance is determined.

In general this method is used for:

(1) volatile substances, e.g. ammonia, some of which would be lost during the titration,

(2) insoluble substances, e.g. *Calcium Carbonate*, which requires excess volumetric solution to effect a quantitive reaction,

(3) substances for which a quantitive reaction proceeds rapidly only in presence of excess of the reagent, e.g. *Lactic Acid*,

(4) substances which require heating with a volumetric reagent during the determination in which decomposition or loss of the reactants or products would occur in the process, e.g. *Formaldehyde* solution.

Lactic Acid *Determination of the percentage w/w of lactic acid and its condensation products together calculated as* $CH_3 \cdot CHOH \cdot COOH$

Lactic acid consists of a mixture of lactic acid (2-hydroxypropionic acid), lactoyl-lactic acid and various other condensation polymers derived from it. The object of the official assay is to determine both lactic acid itself and its condensation products. Hydrolysis of lactoyl-lactic acid and the condensation polymers with excess sodium hydroxide produces lactic acid which, together with the free lactic acid originally present, is neutralised by the sodium hydroxide solution. The excess of sodium hydroxide is back-titrated with standard hydrochloric acid.

$$\begin{array}{l} CH_3{\cdot}CHOH{\cdot}CO{\cdot}\underset{\displaystyle CH_3{\cdot}CH{\cdot}COOH}{\underset{|}{O}} + H_2O + 2NaOH \rightarrow 2CH_3{\cdot}CHOH{\cdot}COONa \end{array}$$

Lactoyl-lactic acid

$$CH_3{\cdot}CHOH{\cdot}COOH + NaOH \rightarrow CH_3{\cdot}CHOH{\cdot}COONa + H_2O$$

$$\therefore \quad C_3H_6O_3 \equiv NaOH \equiv 1000\text{ ml M } NaOH$$

$$\therefore \quad 90.08\text{ g } C_3H_6O_3 \equiv 1000\text{ ml M } NaOH$$

$$\therefore \quad 0.009008\text{ g } C_3H_6O_3 \equiv 1\text{ ml } 0.1\text{N } NaOH$$

Method Weigh the sample (*c.* 0.2 g) in a weighing bottle. Wash out the sample via a funnel into a conical flask, using about 20 ml of *water*. Lactic acid is viscous and so the weighing bottle must be washed out very carefully. Add NaOH (50 ml) from a pipette. Allow to stand for 30 min and back-titrate the excess of alkali with 0.1M HCl using phenolphthalein as indicator. The pink colour of phenolphthalein gradually fades in contact with alkali. If this occurs, add a few more drops of indicator.

Note on the use of phenolphthalein in this determination It would appear that methyl orange might be a suitable indicator in this determination since sodium hydroxide solution is being titrated with hydrochloric acid. But the choice of indicator is always governed by the weakest acid (or base) present in the system. In this case lactic acid is liberated at the end point and, therefore, phenolphthalein is used as indicator.

Method of writing up

Weight of sample	0.2135 g
Volume of NaOH available	50 ml (0.1005M)
Back-titration	48.12 ml (0.0995M)

Then,

Volume of NaOH solution available for sample	= 50.25 ml 0.1M
Volume of NaOH solution not used by sample	= 47.87 ml 0.1M
Volume of NaOH required by sample	= 2.38 ml 0.1M

$$\therefore \; 1\text{ g sample} \equiv \frac{2.38}{0.2135}\text{ ml } 0.1\text{M}$$

$$\equiv 11.15\text{ ml } 0.1\text{M } NaOH$$

Chalk *Determination of the percentage of* $CaCO_3$ *calculated with reference to the dried substance*

There are two types of standard in the British Pharmacopoeia.

(1) A limit of the amount of pure chemical that the official substance shall contain. This may be a *minimum* only, e.g. Boric Acid contains not less than 99.5% H_3BO_3 or a *minimum and maximum*, e.g. Citric Acid contains not less than 99.5% and not more than the equivalent of 101% $C_6H_5O_7$, calculated with reference to the anhydrous substance. A

maximum is included when, for example, a substance containing water of crystallisation can effloresce under certain conditions. Where no maximum is stated, however, and the standard is expressed in terms of the chemical formula for the substance of the monograph, an upper limit of 100.5% is implied (British Pharmacopoeia general notice).

(2) A limit of the amount of pure chemical in the sample when *calculated with reference* to the dried or ignited substance, e.g. $CaCO_3$.

The dried substance is not titrated but the determination is carried out on the ordinary undried material; meanwhile, a quantity of the substance is dried at the required temperature and the percentage loss on drying ascertained. The percentage purity of the sample, calculated with reference to the substance dried at the specified temperature, can then be calculated.

This method of determination is better than one involving the assay on the dried material because:

(a) it is quicker since both parts of the determination can be carried out simultaneously and
(b) dried substances are difficult to weigh, since they absorb atmospheric moisture very quickly.

Standards of this type necessitate the inclusion of a limit of moisture under tests for purity in the Official Monograph, as otherwise a sample of the substance containing large percentages of water would still comply with the standard.

Loss on drying. Determine the loss on drying as described for sodium chloride (p.27).

Assay. Chalk is insoluble in water and the carbon dioxide released on titration with acid would interfere with the indicator unless removed. The sample, in water, is treated with excess M hydrochloric acid, the solution boiled to remove carbon dioxide, cooled and the excess acid titrated with M sodium hydroxide using methyl orange as indicator.

$$CaCO_3 + 2HCl \rightarrow CaCl_2 + H_2O + CO_2$$
$$\therefore \quad CaCO_3 \equiv 2HCl \equiv 2000 \text{ ml M HCl}$$
$$\therefore \quad 100.1 \text{ g } CaCO_3 \equiv 2000 \text{ ml M HCl}$$
$$\therefore \quad 0.05004 \text{ g } CaCO_3 \equiv 1 \text{ ml M HCl}$$

Method Weigh the sample (about 1.5 g) into a conical flask. Add M HCl (50 ml) from a pipette, heat gently until all the material has dissolved, and then boil for 1–2 min to remove CO_2. Cool and back-titrate the excess hydrochloric acid using methyl orange as indicator.

Calculation

A sample of chalk contained 98.90% $CaCO_3$ (from titration) and on drying lost 0.7620% of its weight.

$$\therefore \quad 100 \text{ g sample (undried)} \equiv 98.90 \text{ g } CaCO_3$$
$$\therefore \quad (100 - 0.762) = 99.238 \text{ g dried sample} \equiv 98.90 \text{ g } CaCO_3$$

$$\therefore \quad 100 \text{ g dried sample} \equiv \frac{98.90 \times 100}{99.238}$$
$$\equiv 99.66 \text{ g } CaCO_3$$

∴ *The sample of chalk contains 99.66%* $CaCO_3$ *calculated with reference to the substance dried to constant weight.*

COGNATE DETERMINATION

Lithium Carbonate

Alkali metal salts of aliphatic acids

When these substances are heated strongly the organic portion of the molecule is destroyed and the metals (Na or K) are converted quantitively into Na_2CO_3 or K_2CO_3; these carbonates can be determined by titration.

Sodium Acid Citrate *Determination of the percentage of* $C_6H_6Na_2O_7, 1\frac{1}{2}H_2O$

$$C_6H_6Na_2O_7, 1\tfrac{1}{2}H_2O \xrightarrow{\text{heat}} Na_2CO_3$$
$$\therefore \quad 263.1 \text{ g } C_6H_6Na_2O_7, 1\tfrac{1}{2}H_2O \equiv 2000 \text{ ml M}$$
$$\therefore \quad 0.06578 \text{ g } C_6H_6Na_2O_7, 1\tfrac{1}{2}H_2O \equiv 1 \text{ ml } 0.5\text{M HCl}$$
$$\therefore \quad 2 \text{ g} \equiv \text{approx. } 30 \text{ ml } 0.5\text{M}$$

Method Weigh the sample (about 1.9–2.1 g) accurately into a clean, dry crucible. The weight must be kept within the limits stated, because if too much sample is weighed and 50 ml of 0.5M acid is used there will be no back-titration. If insufficient sample is weighed, difficulty will be experienced in washing the filter paper free from excess of acid.

Place this crucible containing the sample inside another crucible to serve as a jacket crucible. The initial heating should be gently carried out over a small Bunsen flame, since the molten mass tends to split and bubble over the side of the crucible. Constant attention is needed at this stage, until the mass no longer swells up and charring commences. Do not allow the emitted gases to ignite as this may cause mechanical expulsion of particles of residue. When the residue is black, continue to heat for 10 min at a moderate temperature. Strong heat should be avoided to prevent fusion of the sodium carbonate with the silica of the crucible. Remove the jacket crucible and burn off any tar adhering to the inside of the inner crucible using a moderate Bunsen flame. *It is essential to burn off all the tarry material.* Allow the crucible to cool. Place the crucible with its contents on its side in a 250 ml beaker containing *water* (about 10 ml). To the beaker partially covered with a clock-glass add 0.5M HCl (25 ml) from a pipette and boil gently, breaking up the mass in the crucible with a glass rod, to ensure that all the carbonate reacts with the acid. Filter and collect the filtrate in a conical flask. Wash the beaker and contents with 5 ml portions of boiling water. Pass the washings through the filter and collect them in the conical flask. (If the filtrate is yellow or brown, due to incomplete removal of tarry matter, it is useless to continue.) Allow the filter to drain completely between each washing. When washing the filter paper pay special attention to the top of the paper and the fold. *It is essential to wash efficiently and yet the volume of water used must be kept to the minimum.* After washing, test the beaker, crucible and filter for freedom from acidity with *one drop* of *diluted* methyl orange indicator.

Cool the flask and back-titrate the excess of 0.5M acid with 0.5M NaOH. The end point is the last shade of orange before yellow, i.e. the point where the next drop of alkali will turn the solution yellow (it is useful to have a comparison flask containing indicator and about the same amount of water made alkaline with one drop of alkali).

COGNATE DETERMINATIONS

Sodium Citrate Tablets
Sodium Lactate Intravenous Infusion
Compound Sodium Lactate Intravenous Infusion

Back-titrations with blank determinations

In general, blank determinations are used if the volumetric solution is unstable or if it alters in strength during the assay. It is necessary to perform blank determinations in assays which involve heating a liquid containing excess of standard alkali, cooling and back-titrating the excess. Heating and cooling an alkaline liquid results in an apparent change in strength if certain indicators are used. This may be due to interaction of the reagent with the glass or to the absorption of atmospheric CO_2. The amount of the change will be dependent upon the conditions used. In effect, the alkali must be standardised under the conditions to be used in the determination; this is called a blank determination. Examples of this method of determination are given below.

Aspirin Tablets *Determination of the percentage of* $C_9H_8O_4$

This determination depends upon the reactions expressed by the following equations:

$$CH_3 \cdot CO \cdot O \cdot C_6H_4 \cdot CO \cdot OH + 2NaOH \rightarrow CH_3 \cdot CO \cdot ONa + C_6H_4(OH) \cdot CO \cdot ONa$$

$$\therefore \quad CH_3 \cdot CO \cdot O \cdot C_6H_4 \cdot CO \cdot OH \equiv 2NaOH \equiv 200 \text{ ml M}$$

$$\therefore \quad 180.2 \text{ g } C_9H_8O_4 \equiv 200 \text{ ml M}$$

$$\therefore \quad 0.04504 \text{ g } C_9H_8O_4 \equiv 1 \text{ ml } 0.5\text{M}$$

Method Weigh and powder twenty tablets. Accurately weigh about 0.5 g of the sample into a titration flask. Add 0.5M NaOH (50 ml) from a pipette. Immerse the flask in a boiling water bath for 15 min. Cool, taking precautions to exclude carbon dioxide, by covering the mouth of the flask with a small inverted beaker and allowing a stream of cold water to flow over it. Back-titrate the excess alkali with 0.5M hydrochloric acid using phenol red solution as indicator. Carry out a blank determination by heating, cooling and titrating a further 50 ml 0.5M NaOH *without the addition of the aspirin sample* under the same conditions.

Formaldehyde Solution *Determination of the percentage w/w of* CH_2O

Formaldehyde can be oxidized by means of H_2O_2:

$$HCHO + \overline{O} \xrightarrow{H_2O_2} H \cdot COOH$$

Formic acid is volatile but loss may be prevented by performing the oxidation in the presence of excess standard alkali which is subsequently back-titrated with standard acid.

$$\therefore \quad HCHO \equiv HCOOH \equiv 1000 \text{ ml M NaOH}$$

$$\therefore \quad 30.03 \text{ g HCHO} \equiv 1000 \text{ ml M NaOH}$$

$$\therefore \quad 0.03003 \text{ g HCHO} \equiv 1 \text{ ml M NaOH}$$

Method Add the sample (about 3 g), accurately weighed, to a mixture of hydrogen peroxide (25 ml from a pipette) and M sodium hydroxide (50 ml) in a conical flask and warm on a water-bath until effervescence ceases (usually about 30 min). Cool (p.000) and titrate the excess of alkali with M hydrochloric acid, using phenolphthalein solution as indicator.

A blank determination is required because alkali is heated and then cooled. Since a trace of mineral acid may be present in hydrogen peroxide this reagent is added from a pipette.

Saponification Value of fixed oils

Saponification Value is defined as the number of mg of potassium hydroxide required to neutralise the fatty acids resulting from the complete hydrolysis of 1 g of the substance when determined by the undermentioned method. The Acid Value (p.146) of most edible oils is small relative to the Saponification Value and is included in the latter. The Saponification Value is a measure of both free and combined acids.

Method Accurately weigh the sample (approximately 2 g) in a small glass sample tube. Introduce the tube and sample into a 250 ml conical flask. Pipette ethanolic KOH (M approx. 25 ml) into the flask. Fit a reflux condenser and heat with the flask immersed in a boiling water-bath for 1 h. Remove the water-bath and wash down the condenser with not more than 5 ml of neutral alcohol and add solution of phenolphthalein (1 ml). (More indicator than usual is required because phenolphthalein does not function well as an indicator in strong ethanolic solution.) Remove the condenser, place a beaker over the neck of the flask and cool under a tap. When quite cold, titrate the excess potassium hydroxide with 0.5M HCl. Perform a blank determination at the same time. When performing the exercise in duplicate, it is preferable to carry out the determination and blank determination simultaneously.

$$R{\cdot}COOR' + KOH \equiv R{\cdot}COOK + R'OH$$

$$\therefore \quad 56.1 \text{ g KOH} \equiv 2000 \text{ ml } 0.5\text{M}$$

$$\therefore \quad 28.05 \text{ mg KOH} \equiv 1 \text{ ml } 0.5\text{M}$$

Calculation

b = burette reading for blank
a = burette reading for sample

$$\text{then Saponification Value} = \frac{(b - a) \times (\text{molarity}) \times 28.05}{\text{weight of sample (in g)}}$$

The Saponification Value of the following oils is determined in this way:

Almond Oil
Arachis Oil
Castor Oil
Cocoanut Oil
Cod-liver Oil
Cottonseed Oil
Halibut-liver Oil
Sesame Oil
Theobroma Oil

All edible oils have Saponification Values lying between 188 and 196 and hence, this test alone is of little value for identification purposes. The Saponification Value is a measure of the equivalent weight of the acids present and is, therefore, useful as an indication of purity. Adulteration with mineral oils would be shown by low Saponification Values, whereas rancidity, which leads to the formation of low molecular weight acids, would be indicated by an abnormally high Saponification Value.

COGNATE DETERMINATIONS

Cetostearyl Alcohol
Emulsifying Wax
Wool Alcohols; Wool Fat; Hydrous Wool Fat

These substances contain steroidal esters which are difficult to hydrolyse and the time of refluxing is extended to 4 h. Ideally, these substances should be free from acids and esters, so the Saponification Value acts as a limit test for both.

Determination of esters

The determination of esters is performed by hydrolysing the substance to an alcohol and an acid using excess of standard ethanolic KOH solution, and then back-titrating the excess alkali. A blank determination is performed.

$$R\cdot COOR' + KOH \rightarrow R\cdot COOK + R'OH$$

$$\therefore \quad R\cdot COOR' \equiv KOH \equiv 2000\ \text{ml}\ 0.5\text{M}$$

Method Accurately weigh the sample (approx. 2 g) in a glass sample tube. Transfer to a 250 ml flask, and add ethanol (5 ml; 96%, previously boiled, cooled and neutralised to phenolphthalein). Neutralise the free acid in the solution by titrating with 0.1M ethanolic potassium hydroxide until just pink to phenolphthalein. Add 0.5M ethanolic KOH (20 ml) and reflux with the flask immersed in a boiling water bath for 1 h. Cool, add *water* (20 ml) and back-titrate the excess alkali with 0.5M HCl, adding a further 0.2 ml of phenolphthalein solution. Carry out a blank determination, heating and cooling under the same conditions. The difference in readings gives the amount of alkali required to saponify the ester.

The following esters are determined in this way:

Benzyl Benzoate 1 ml 0.5M KOH ≡ 0.1061 g $C_{14}H_{12}O_2$
Benzyl Benzoate Application
Dimethyl Phthalate 1 ml 0.5M KOH ≡ 0.04854 g $C_{10}H_{10}O_4$
Ethyl Oleate 1 ml 0.5M KOH ≡ 0.1553 g $C_{20}H_{38}O_2$
Methyl Salicylate 1 ml 0.1M KOH ≡ 0.01521 g $C_8H_8O_3$
Peppermint Oil 1 ml 0.5M KOH ≡ 0.099 g $C_{12}H_{22}O_2$ (Menthyl Acetate)

Determination of Ester Value

Ester Value is defined as the number of mg of potassium hydroxide required to neutralise the acids resulting from the complete hydrolysis of 1 g of material. Ester Value is a measure of the combined acids present in the substance and is determined by the method described above:

$$\text{Ester Value} = \frac{m \times (\text{molarity of acid}) \times 28.05}{w}$$

where w = weight (in g) of the substance taken and m = difference in burette readings for sample and blank determinations.

Beeswax. The Ester Value is determined by subtraction of the Acid Value from the Saponification value.

The Ester Value here determines the quality of the beeswax, which contains cerotic acid (approx. 20%) and melissyl pamitate (approx. 80%).

Ratio Number.

$$\text{Ratio Number} = \frac{\text{Ester Value}}{\text{Acid Value}}$$

Ratio number is applied only to beeswax (3.3 to 4.3) and is a further check on the relative proportions of the acid and ester constituents.

Benzyl Alcohol *Determination of the percentage w/w of* $C_6H_5CH_2OH$

This substance is determined by heating with acetic anhydride in presence of pyridine to produce an ester, benzyl acetate:

$$C_6H_5CH_2OH + (CH_3CO)_2O \rightarrow C_6H_5CH_2OCOCH_3 + CH_3COOH$$

Thus, each molecule of acetic anhydride reacts with the alcohol to produce one molecule of acetic acid, which is then titrated with standard alkali using phenolphthalein as indicator. The alkali also reacts with the excess of acetic anhydride:

$$(CH_3CO)_2 + 2NaOH \rightarrow 2CH_3COONa + H_2O$$

The pyridine does not interfere with the titration, because it is a weak base and phenolphthalein is used as indicator. A blank determination is performed.

$$\therefore \quad C_6H_5CH_2OH \equiv (CH_3CO)_2O \equiv CH_3COOH \equiv NaOH$$

and

$$(CH_3CO)_2O \equiv 2NaOH$$

$$\therefore \quad 108.1\text{ g } C_6H_5CH_2OH \equiv 1000\text{ ml M NaOH}$$

$$\therefore \quad 0.1081\text{ g } C_6H_5CH_2OH \equiv 1\text{ ml M NaOH}$$

Method Reflux 1.5 g (accurately weighed) with a mixture of acetic anhydride (1 part by volume) and pyridine (7 parts by volume) (25 ml) on a water-bath for 30 min. Cool, dilute with *water* (25 ml) and titrate the excess acetic anhydride and acetic acid (formed in the reaction) with M sodium hydroxide to phenolphthalein. Perform a blank determination with the reagent (25 ml). The difference in titre between the blank and the test determinations is equivalent to the benzyl alcohol.

COGNATE DETERMINATIONS

Dienoestrol Determined as for Benzyl Alcohol but with the following modifications:

(a) reflux time—2 h
(b) after refluxing, cool in ice water, and remove the dienoestrol acetate by filtration through a No. 4 sintered glass crucible
(c) titrate with 0.5M sodium hydroxide.

Peppermint Oil Determination of free alcohol.

Hydroxyl Value

This is determined on higher molecular weight alcohols in essentially the same way as for Benzyl Alcohol, but normally using stearic anhydride in place of acetic anhydride. Hydroxyl Value is calculated as the number of g of potassium hydroxide required to neutralize the stearic acid liberated by hydrolysis of the stearate ester derived from 1 g of the alcohol containing substance.

$$\text{Hydroxyl Value} = \frac{(b - a \times 56.11)}{w}$$

where b = blank titration
a = test titration
w = weight in g of substance.

The following substances are determined in this way:

Castor Oil
Cetomacrogol 1000
Cetomacrogol Emulsifying Wax
Cetostearyl alcohol

Determination of organically combined nitrogen

Hydrolysis with distillation of liberated volatile bases

This method is applied to the salts of volatile organic bases (amphetamine) and compounds (Meprobamate and Neostigmine Methylbromide) which decompose quantitively when heated with acid or alkali to yield volatile bases. The liberated base is determined either by absorption in excess standard mineral acid and back-titration with standard alkali, or by absorption into a solution of boric acid followed by direct titration with standard acid.

Meprobamate *Determination of the percentage of* $C_9H_{18}N_2O_4$

The determination depends on the decomposition:

$$MeCPr^n\cdot(CH_2O\cdot CO\cdot NH_2)_2 + H_2SO_4 + 2H_2O \rightarrow MeCPr^n(CH_2OH) + 2CO_2 + (NH_4)_2SO_4$$

$$(NH_4)_2SO_4 + 2NaOH \rightarrow Na_2SO_4 + 2H_2O + 2NH_3$$

$$\therefore\ MeCPr^n(CH_2\cdot O\cdot CO\cdot NH_2)_2 \equiv 2NH_3$$

$$\therefore\ 218.3\text{ g } C_9H_{18}N_2O_4 \equiv 2000\text{ ml M } H_2SO_4$$

$$\therefore\ 0.01091\text{ g } C_9H_{18}N_2O_4 \equiv 1\text{ ml } 0.05\text{M } H_2SO_4$$

Method Accurately weigh the sample (about 0.3 g) into a suitable flask. Add 25% v/v sulphuric acid (25 ml), and boil under reflux for 3 h. Cool, transfer to an ammonia distillation apparatus, add an excess of 5M sodium hydroxide and distil the ammonia into 0.05M H_2SO_4 (50 ml). Back-titrate the excess acid with 0.1M NaOH using methyl red as indicator.

COGNATE DETERMINATIONS

Amphetamine Sulphate
Dexamphetamine Sulphate
Nikethamide Injection
Neostigmine Methylsulphate
Neostigmine Tablets
Pyrazinamide

Reduction with distillation of liberated ammonia

Potassium Nitrate *Determination of the percentage of* KNO_3

If KNO_3 is treated with Devarda's alloy and alkali the NO_3^- ion is reduced to ammonia which is distilled into excess of 0.1M hydrochloric acid.

$$HNO_3 + 8H \equiv NH_3 + 3H_2O$$

$$\therefore \quad KNO_3 \equiv N \equiv NH_3 \equiv 1000 \text{ ml M HCl}$$

$$\therefore \quad 101.1 \text{ g } KNO_3 \equiv 1000 \text{ ml M HCl}$$

$$\therefore \quad 0.01011 \text{ g } KNO_3 \equiv 1 \text{ ml } 0.1\text{M HCl}$$

Method Dissolve the sample (about 0.3 g), accurately weighed, in *water* (300 ml) in an ammonia distillation apparatus, add Devarda's alloy (3 g) and sodium hydroxide solution (10 ml) and distil. Collect the distillate in 0.1M hydrochloric acid (50 ml) and titrate the excess of acid with 0.1M sodium hydroxide, using methyl red solution as indicator. Repeat the operation without the potassium nitrate. The difference between the titres is the acid required to neutralise the ammonia formed from the potassium nitrate.

Kjeldahl type determinations

In the following substances, the nitrogen is more firmly combined than in the foregoing examples. A much more severe treatment is required before the nitrogen is obtained quantitively as ammonia. The organic portion of the molecule is destroyed, and the carbon oxidised until a clear liquid is obtained. The method consists in digestion with concentrated sulphuric acid using either sodium or potassium sulphate to raise the boiling point of the acid and catalysts, such as mercury, selenium and copper, to hasten the reaction. The method is unsatisfactory for nitrogen present as nitro, azo, cyano or hydrazo groupings.

The weight of sample will vary from compound to compound, depending upon the percentage of nitrogen present. Similarly, some sulphuric acid is consumed in the process, being reduced to sulphur dioxide, consequent upon the oxidation of carbon and hydrogen in the molecule being determined. The volume of sulphuric acid, therefore, varies from one determination to the next to compensate for this loss, and to ensure that the same volume of sulphuric acid is always present at the completion of the digestion period. The constancy of this volume, and of the weight of anhydrous sodium sulphate, ensures a constant digestion temperature.

Primidone *Determination of the percentage of* $C_{12}H_{14}N_2O_2$

$$C_{12}H_{14}N_2O_2 \equiv N \equiv NH_3$$

$$\therefore \quad 218.3\text{ g } C_{12}H_{14}N_2O_2 \equiv 1000\text{ ml M}$$

$$\therefore \quad 0.01091\text{ g } C_{12}H_{14}N_2O_2 \equiv 1\text{ ml } 0.05\text{M } H_2SO_4$$

Method Accurately weigh the sample (0.2 g) into a long-necked flask (Kjeldahl flask) and add 3 g of mercury sulphate mixture (usually available in tablets, each containing 1 g anhydrous sodium sulphate and the equivalent of 0.1 g mercury) and nitrogen-free sulphuric acid (8.5 ml). Ensure that all solids are weighed into the bottom of the flask. Shake gently to mix the contents. Support the flask on a stand in a fume cupboard, inclining the neck of the flask at about 60°; place a small funnel in the neck of the flask. Heat the flask with a small flame until all frothing ceases and then increase the size of the flame so that the liquid refluxes in the neck of the flask. Continue the heating until a clear, colourless liquid is obtained (this may take several hours in some Kjeldahl determinations) and then boil gently for a further 2 h. Cool the solution and, when cold, dilute to 75–85 ml with *water*, add a piece of granulated zinc (this prevents bumping during distillation) and a solution of sodium hydroxide (15 g) and sodium thiosulphate (2 g) in *water* (25 ml). Immediately connect the flask to a distillation apparatus, mix the contents and distil the liberated ammonia into 0.05M H_2SO_4 (50 ml). Titrate with 0.1M NaOH using methyl red as indicator. A blank determination is performed and the difference between the two titres is equivalent to the ammonia liberated.

COGNATE DETERMINATION

Pentamidine Isethionate Approximate weight of sample 0.4 g; nitrogen-free sulphuric acid 8 ml.

Blood products

Blood products are standardised on their protein content. This is determined by a Kjeldahl determination of nitrogen but, since other nitrogenous substances, such as urea and amino acids, are present, some preliminary fractionation is essential. This is achieved by precipitation of the protein from solution as molybdic acid complex, and then submitting this fraction to a modified Kjeldahl determination of nitrogen.

Dried Human Plasma *Determination of the percentage w/v of protein in an aqueous solution of the substance, equal in volume to the volume of plasma from which it was obtained*

Method Dissolve the substance in a volume of *water* equal to the volume of *Water for Injection* stated on the label. Transfer 2 ml (pipette) of the solution to a round-bottomed centrifuge tube (capacity 15 ml). Add sodium molybdate solution (7.5% w/v; 2 ml) and a mixture of nitrogen-free sulphuric acid and purified water (1:30; 2 ml), shake and centrifuge for 5 min. Carefully decant the supernatant liquid and drain the tube by inverting on a filter paper. When thoroughly drained, add nitrogen-free sulphuric acid to the residue in the tube, mix and transfer to a Kjeldahl flask using, in all, 10 ml of acid. Add 1 g of mixture of anhydrous sodium sulphate (10 parts) and copper(II) sulphate (1 part). Gently add hydrogen peroxide solution (100 vol; 1 ml) and heat until the solution becomes clear, if necessary repeating the addition of hydrogen peroxide and heating. Transfer the solution to an ammonia distillation apparatus, make alkaline with sodium hydroxide (10M; 50 ml) and steam distil the liquid into standard hydrochloric acid (0.1M; 25 ml). Cool and titrate the excess acid with 0.1M NaOH using methyl red–methylene blue as indicator. Repeat using D-glucose (0.0250 g) as a blank, taking the difference between test and blank titrations as equivalent to the sample.

$$1 \text{ ml } 0.1\text{M HCl} \equiv 1.401 \text{ mg of nitrogen (N)}$$

6
Titration in non-aqueous solvents

Theory

Substances which are either too weakly basic or too weakly acidic to give sharp end points in aqueous solution can often be titrated in non-aqueous solvents. The reactions which occur during many non-aqueous titrations can be explained by means of the concepts of the Lowry Brönsted Theory. According to this theory an *acid* is a proton donor, i.e. a substance which tends to dissociate to yield a proton, and a *base* is a proton acceptor, i.e. a substance which tends to combine with a proton. When an acid HB dissociates it yields a proton together with the conjugate base B of the acid.

$$\underset{\text{acid}}{HB} \rightleftharpoons \underset{\text{proton}}{H^+} + \underset{\text{base}}{B^-}$$

Alternatively, the base B will combine with a proton to yield the conjugate acid HB of the base B, for every base has its conjugate acid and *vice versa*. It follows from these definitions that an acid may be either an electrically neutral molecule, such as HCl, or a positive charged cation such as $C_6H_5NH_3^+$. A base may be either an electrically neutral molecule, such as $C_6H_5NH_2$, or a negatively charged anion, such as Cl^-.

Some examples of acids and bases are set out below:

Acids Bases

$$HCl \rightleftharpoons H^+ + Cl^-$$
$$C_6H_5NH_3^+ \rightleftharpoons H^+ + C_6H_5NH_2$$
$$HSO_4^- \rightleftharpoons H^+ + SO_4^{2-}$$

Substances which are potentially acidic can function as acids only in the presence of a base to which they can donate a proton. Conversely, basic properties do not become apparent unless an acid also is present.

Solvents

Aprotic solvents are neutral, chemically inert substances such as toluene and chloroform. They have a low dielectric constant, do not react with either acids or bases and, therefore, do not favour ionisation. The fact that picric acid gives a colourless solution in toluene, which becomes yellow on adding aniline shows that picric acid is not dissociated in toluene solution. It also shows that in the presence of the base aniline it functions as an acid, the development of yellow colour being due to the formation of the picrate ion.

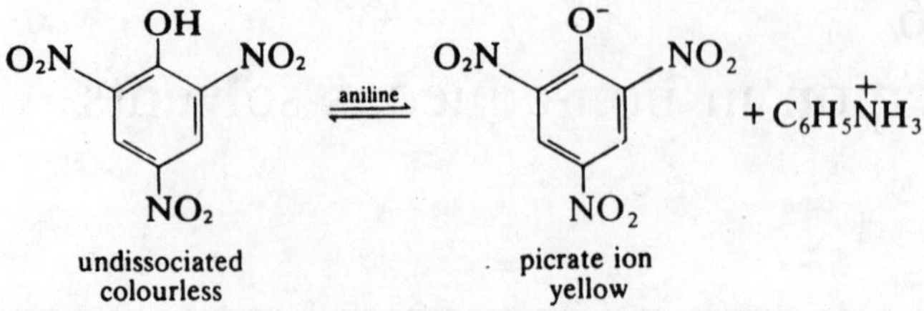

Since dissociation is not an essential preliminary to neutralisation, aprotic solvents are often added to 'ionising' solvents to depress solvolysis (which is comparable with hydrolysis) of the neutralisation product and so sharpen the end point.

Protophilic solvents are basic in character and react with acids to form solvated protons.

$$\underset{\text{acid}}{HB} + \underset{\substack{\text{basic}\\\text{solvent}}}{Sol.} \rightleftharpoons \underset{\substack{\text{solvated}\\\text{proton}}}{Sol.H^+} + \underset{\substack{\text{conjugate base}\\\text{of acid}}}{B^-}$$

A weakly basic solvent has less tendency than a strongly basic one to accept a proton. Similarly, a weak acid has less tendency to donate protons than a strong acid. As a result, a strong acid, such as perchloric acid, exhibits more strongly acidic properties than a weak acid, such as acetic acid, when dissolved in a weakly basic solvent. On the other hand, all acids tend to become indistinguishable in strength when dissolved in strongly basic solvents owing to the greater affinity of strong bases for protons. This is called the levelling effect. Strong bases are levelling solvents for acids; weak bases are differentiating solvents for acids.

Protogenic solvents are acidic substances, for example sulphuric acid. They exert a levelling effect on bases.

Amphiprotic solvents have both protophilic and protogenic properties. Examples are water, acetic acid and the alcohols. They are dissociated to a slight extent. The dissociation of acetic acid, which is frequently used as a solvent for titration of basic substance, is shown in the equation below.

$$CH_3COOH \rightleftharpoons H^+ + CH_3COO^-$$

Here the acetic acid is functioning as an acid. If a very strong acid such as perchloric acid is dissolved in acetic acid, the latter can function as a base and combine with protons donated by the perchloric acid to form an 'onium' ion.

$$HClO_4 \rightleftharpoons H^+ + ClO_4^-$$

$$CH_3COOH + H^+ \rightleftharpoons \underset{\text{onium ion}}{CH_3COOH_2^+}$$

Since the $CH_3COOH_2^+$ ion readily donates its proton to a base, a solution of perchloric acid in glacial acetic acid functions as a strongly acidic solution.

When a weak base, such as pyridine, is dissolved in acetic acid, the acetic acid exerts its levelling effect and enhances the basic properties of the pyridine. It is possible, therefore, to titrate a solution of a weak base in acetic acid with perchloric acid on acetic acid and obtain a sharp end point when attempts to carry out the titration in aqueous solution are unsuccessful.

$$
\begin{array}{llcl}
 & HClO_4 + CH_3COOH & \rightleftharpoons & CH_3COOH_2^+ + ClO_4^- \\
 & C_5H_5N + CH_3COOH & \rightleftharpoons & C_5H_5NH^+ + CH_3COO^- \\
 & CH_3COOH_2 + CH_3COO^- & \rightleftharpoons & 2CH_3COOH \\
\hline
\text{Adding} & HClO_4 + C_5H_5N & \rightleftharpoons & C_5H_5NH^+ + ClO_4 \\
\hline
\end{array}
$$

Temperature effects

Non-aqueous solvents, in general, have greater coefficients of expansion than water, so that small temperature differences can cause significant errors unless suitable correction factors are used. Standardisation and titration should be carried out as far as possible at the same temperature. If this is not possible however, the volume of titrant may be corrected by applying the following formula:

$$V_c = V[1 + 0.0011(t_1 - t_2)]$$

where V_c = corrected volume of titrant
V = volume of titrant measured
t_1 = temperature at which titrant was standardised
t_2 = temperature at which titration was carried out

Titration of alkali metal salts of organic acids

Preparation of 0.1M perchloric acid

Method Slowly add perchloric acid (72%; 8.5 ml) to glacial acetic acid (900 ml) with continuous and efficient mixing. Similarly add acetic anhydride (30 ml); adjust the volume to 1 l with glacial acetic acid and allow the solution to stand for 24 h before use.

The acetic anhydride reacts with the water in the perchloric acid and acetic acid and renders the mixture virtually anhydrous. Although excess acetic anhydride is not always disadvantageous, care must be taken to avoid an excess when primary and secondary amines (which acetylate readily to give non-basic products) are to be titrated. The perchloric acid must be well diluted with acetic acid before adding the acetic anhydride. Failure to observe this precaution leads to formation of the *explosive* acetyl-perchlorate.

Standardisation of 0.1M perchloric acid

Alkali and alkaline earth salts of organic acids function as bases in acetic acid solution.

$$RCOOM \rightleftharpoons RCOO^- + M^+$$
$$CH_3COOH_2 + RCOO^- \rightleftharpoons RCOOH + CH_3COOH$$

Potassium hydrogen phthalate may be used as a standardising agent for acetous perchloric acid. The reaction is expressed by the following equation:

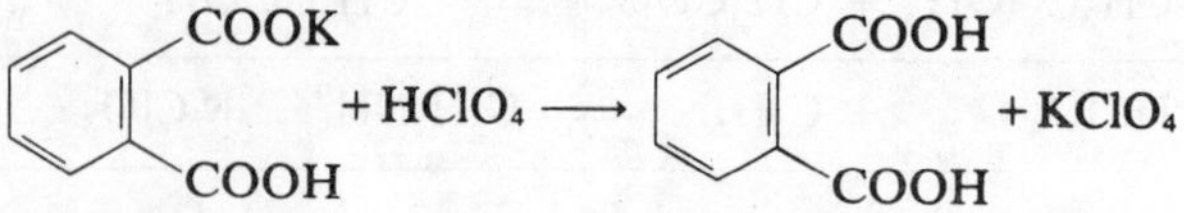

$$\therefore \quad 204.14\text{ g } C_8H_5O_4K \equiv HClO_4 \equiv 1000\text{ ml M}$$
$$\therefore \quad 0.02041\text{ g } C_8H_5O_4K \equiv 1\text{ ml } 0.1\text{M } HClO_4$$

Method Accurately weigh the potassium hydrogen phthalate (0.5 g approx.) into a 100 ml conical flask. Attach a reflux condenser fitted with a silica gel drying tube and add glacial acetic acid (25 ml). Warm until the salt has dissolved. Cool and titrate with 0.1M perchloric acid.

Indicator Use 2 drops of either 0.5% w/v acetous crystal violet (end point blue to blue green) or 0.5% w/v acetous oracet blue B (end point blue to pink).

Indicators

The following indicators are in common use:

Indicator	*Colour change*		
	basic	*neutral*	*acidic*
Crystal Violet (0.5% in glacial acetic acid)	violet	blue-green	yellowish-green
1-Naphtholbenzein (0.2% in glacial acetic acid)	blue or blue-green	orange	dark-green
Nile Blue A (1% in glacial acetic acid)	blue		blue-green
Oracet Blue B (0.5% in glacial acetic acid)	blue	purple	pink
Quinaldine Red (0.1% in methanol)	magenta		almost colourless

The same indicator must be used throughout for standardisation, titration and neutralisation of mercuric acetate solution, if used (p.169).

Potentiometric titration (see Part 2)

In cases where the end point is difficult to determine with visual indicators, the end point may be determined by titrating potentiometrically and plotting dE/dV against V.

Sodium Acetate *Determination of the percentage of* $CH_3\cdot COONa,3H_2O$

$$CH_3COONa + HClO_4 \rightarrow CH_3COOH + NaClO_4$$
$$\therefore \quad 136.1\text{ g } CH_3COONa, 3H_2O \equiv HClO_4 \equiv 1000\text{ ml M}$$
$$\therefore \quad 0.01361\text{ g } CH_3COONa, 3H_2O \equiv 1\text{ ml } 0.1\text{M}$$

Method Accurately weigh about 0.25 g of sample and dissolve in glacial acetic acid (50 ml). Add acetic anhydride (5 ml) and allow to stand for 30 min to remove water by conversion to acetic acid. Titrate with 0.1M $HClO_4$ using 1-naphtholbenzein as indicator.

COGNATE DETERMINATIONS

Potassium Acetate

Potassium Citrate

Sodium Cromoglycate A mixture of propan-2-ol and 1,4-dioxan is used as solvent for the sample and titrant solution.

Titration of amines and amine salts of organic acids

A wide range of primary, secondary and tertiary amines can be assayed by titration with perchloric acid in non-aqueous media, including, for example, such compounds as *Adrenaline, Chlordiazepoxide, Codeine, Diazepam, Ethionamide, Metronidazole* and *Pyrimethamine*. It is also equally applicable to amine and other organic base salts with organic acids (*Adrenaline Acid Tartrate*; *Chlorhexidine Acetate*), to some of their solid dose preparations (*Bisacodyl Suppositories; Ethionamide Tablets; Nitrazepam Tablets*) and to amino acids (*Glycine; Aminocaproic Acid*).

Nitrazepam *Determination of the percentage of* $C_{15}H_{11}N_3O_3$

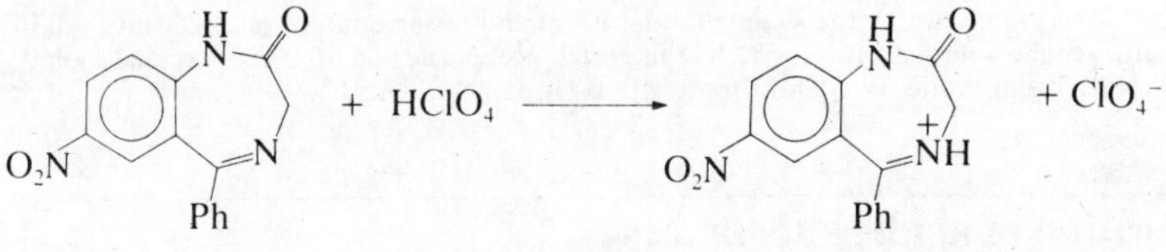

$$C_{15}H_{11}N_3O_3 \equiv HClO_4 \equiv 1000 \text{ ml M}$$

$$\therefore \quad 281.3 \text{ g } C_{15}H_{11}N_3O_3 \equiv 1000 \text{ ml M}$$

$$\therefore \quad 0.0281 \text{ g } C_{15}H_{11}N_3O_3 \equiv 1 \text{ ml } 0.1\text{M}$$

Method Dissolve approximately 0.5 g, accurately weighed, in acetic anhydride (50 ml) and titrate with 0.1M perchloric acid using nile blue A as indicator.

Titration of halogen acid salts of bases

The halide ions chloride, bromide and iodide are too weakly basic to react quantitively with acetous perchloric acid. Addition of mercuric acetate (which is undissociated in acetic acid solution) to a halide salt replaces the halide ion by an equivalent quantity of acetate ion, which is a strong base in acetic acid.

$$2R \cdot NH_2 \cdot HCl \rightleftharpoons 2RNH_3^+ + 2Cl^-$$

$$(CH_3COO)_2Hg \text{ (undissociated)} + 2Cl^- \rightarrow HgCl_2 \text{ (undissociated)} + 2CH_3COO^-$$

$$2CH_3COOH_2 + 2CH_3COOO^- \rightleftharpoons 4CH_3COOOH$$

Typical amine salts determined in this way include *Amitriptyline Hydrochloride*, *Chlordiazepoxide Hydrochloride*, *Chlorpromazine Hydrochloride*, *Ephedrine Hydrochloride*, *Hyoscine Hydrobromide*, *Lignocaine* (Lidocaine) *Hydrochloride*, *Pethidine* (meperidine) *Hydrochloride* and *Propanolol Hydrochloride*. The method is also applicable to halide salts of other organic bases such as biguanides (*Proguanil Hydrochloride*) and quarternary ammonium salts (*Gallamine Triethiodide*; *Neostigmine Bromide*; *Pancuronium Bromide*; *Suxamethonium Chloride* and *Bromide*).

Chlorpromazine Hydrochloride *Determination of the percentage of* $C_{17}H_{19}ClN_2S$, HCl

The determination depends upon the reactions expressed by the following equations:

$$C_{17}H_{19}ClN_2S,HCl \rightleftharpoons 2C_{20}H_{31}ON,H^+ + 2Cl^-$$
$$(CH_3COO)_2Hg + 2Cl^- \rightarrow HgCl_2 + 2CH_3 \cdot COO^-$$
$$2CH_3COOH_2{}^+ + 2CH_3COOO^- \rightleftharpoons 4CH_3COOH$$
$$C_{17}H_{19}ClN_2S,HCl \equiv Cl^- \equiv CH_3COOO^- \equiv HClO_4 \equiv 1000 \text{ ml M}$$
$$\therefore \quad 355.3 \text{ g } C_{17}H_{19}ClN_2S,HCl \equiv 1000 \text{ ml M } HClO_4$$
$$\therefore \quad 0.03553 \text{ g } C_{17}H_{19}ClN_2S,HCl \equiv 1 \text{ ml } 0.1\text{M } HClO_4$$

Method Accurately weigh the sample (about 0.4 g) and dissolve in acetone (200 ml). Add mercuric acetate solution (10 ml; 5% w/v in glacial acetic acid) and 0.2 ml of crystal violet solution (0.5% in acetic acid) and titrate with 0.1M perchloric acid.

Titration of acidic substances

A number of weakly acidic substances can be determined by non-aqueous titration with potassium, sodium or lithium methoxide in toluene-methanol or, alternatively, with tetrabutylammonium hydroxide in methanol. The former reagents are now little used, but details of their preparation and use may be found in the third edition of this book. Titration with tetrabutylammonium hydroxide is used in the assay of cyclic imides (*Ethosuximide*), thiazides (*Bendrofluazide; Hydrochlorthiazide*), sulphonamides (*Sulphafuruzole*) and phenols (*Danthron*; *Dichlorophen*; *Oxyclozanide*).

Preparation of 0.1M tetrabutylammonium hydroxide in toluene-methanol

Method Dissolve tetrabutylammonium iodide (40 g) in absolute methanol (90 ml), add finely powdered purified silver oxide (20 g) and shake vigorously for 1 h. Centrifuge a few ml of the mixture and test the supernatant liquid for iodide. If a positive reaction is obtained, add an additional 2 g of silver oxide and shake for a further 30 min. Repeat this procedure until the liquid is free from iodide, filter the mixture through a fine sintered glass filter and rinse the reaction vessel with 3 portions, each of 50 ml, of dry toluene. Add the washings to the filtrate and dilute to 1 l with dry toluene. Flush the solution with carbon dioxide-free nitrogen for 5 min and protect from carbon dioxide and moisture during storage.

Standardisation of 0.1M tetrabutylammonium hydroxide

Method Accurately weigh about 60 mg of benzoic acid into dimethylformamide (10 ml), previously neutralized to the full blue colour of thymol blue (0.3% w/v in methanol; 3 drops) by titration with 0.1M tetrabutylammonium hydroxide. Allow the benzoic acid to dissolve and titrate in an atmosphere of carbon dioxide-free nitrogen with 0.1M tetrabutylammonium hydroxide.

$$C_6H_5COOH \equiv 100 \text{ ml M}$$
$$\therefore \quad 0.01221 \text{ g } C_7H_6O_2 \equiv 1 \text{ ml } 0.1\text{M}$$

Ethosuximide *Determination of the percentage of* $C_7H_{11}NO_2$

The determination depends upon the reaction expressed by the following equation:

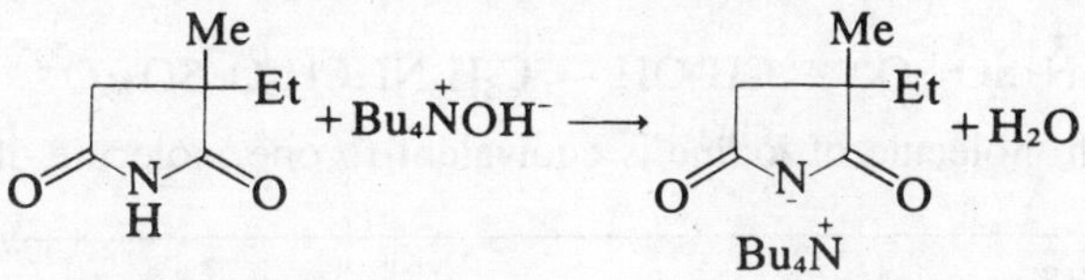

$$\therefore \quad C_7H_{11}NO_2 \equiv Bu_4\overset{+}{N}OH \equiv 1000 \text{ ml M}$$
$$\therefore \quad 141.2 \text{ g } C_7H_{11}NO_2 \equiv 1000 \text{ ml M}$$
$$\therefore \quad 0.01412 \text{ g } C_7H_{11}NO_2 \equiv 1 \text{ ml } 0.1\text{M}$$

The titration is carried out in dimethylformamide using magneson solution as indicator.

COGNATE DETERMINATIONS

Fluorouracil
Mercaptopurine

Phenobarbitone *Determination of the percentage of* $C_{12}H_{12}N_2O_3$

Phenobarbitone functions as a weak acid in aqueous media, but it is relatively insoluble. It is much more soluble in pyridine, from which it is precipitated quantitively as its silver salt on addition of silver nitrate. The supernatant pyridine, which contains the displaced HNO_3, can be titrated with NaOH solution.

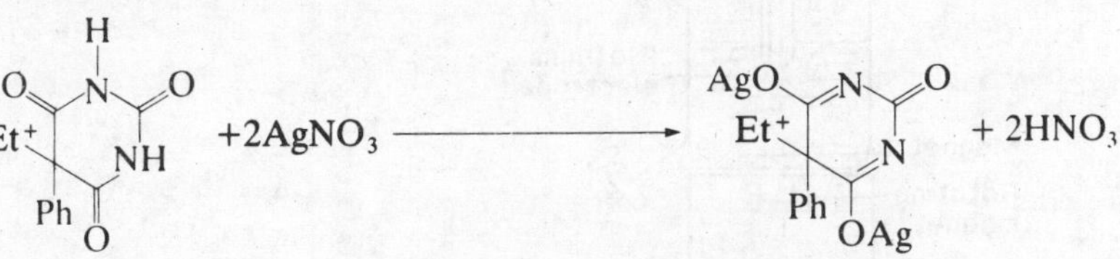

$$\therefore \quad 232.2 \text{ g } C_{12}H_{12}O_3N_2 \equiv 2HNO_3 \equiv 2000 \text{ ml M}$$
$$\therefore \quad 0.01161 \text{ g } C_{12}H_{12}O_3N_2 \equiv 1 \text{ ml } 0.1\text{M } AgNO_3$$

Method Accurately weigh approximately 0.1 g of sample and dissolve in pyridine (5 ml). Add a solution of silver nitrate (8.5% w/v) in pyridine (10 ml); add thymolphthalein solution (0.25 ml) and titrate with 0.1M ethanolic NaOH.

Aquametry

Water, in small quantities, is generally controlled by a loss on drying under specified conditions. When present in appreciable amount, the water content may often be readily determined by titration with **Karl Fischer Reagent** (KFR), which contains iodine, sulphur dioxide, anhydrous methanol and anhydrous pyridine. The reaction of KFR is stated to proceed in two stages:

$$3C_5H_5N + SO_2 + I_2 + H_2O \rightarrow 2C_5H_5N,HI + C_5H_5\overset{+}{N}\cdot SO_2\cdot O^-$$

$$C_5H_5\overset{+}{N}\cdot SO_2\cdot O^- + CH_3OH \rightarrow C_5H_5\overset{+}{N}HCH_3O\cdot SO_2\cdot O^-$$

Thus, each molecule of iodine is equivalent to one molecule of water.

Apparatus

Many forms of apparatus for the determination of water by the Karl Fischer method are available commercially. Only one design will be considered here and the apparatus described is suitable for materials having water contents of about 0.1%.

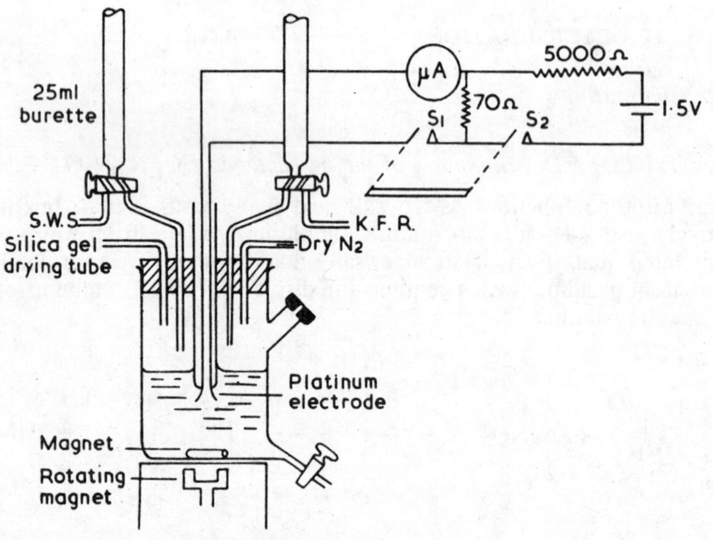

Fig. 6.1 Karl Fischer titration apparatus. SWS is standard solution of water in methanol. The dry N_2 inlet and outlet may be omitted

Figure 6.1 illustrates the main features of the apparatus. The reaction vessel should have a working capacity of at least 60 ml, the side tube being for the insertion of liquid samples. The galvanometer (100 μA full scale deflection) should have a resistance of less than 100 Ω. Other apparatus required includes 10 and 20 ml pipettes, a weighing pipette, all of which should be fitted with hypodermic needles (1 mm bore), volumetric flasks (100 ml) closed with latex vaccine caps, two or three drying bottles (containing freshly activated silica gel) and a final wash-bottle containing KFR for drying the nitrogen supply.

Note 1 Precautions must be taken to prevent atmospheric moisture from entering the apparatus, by using suitable guard tubes filled with freshly activated silica gel.

Note 2 Lubricate the stop-cocks with silicone grease and not petroleum jelly which reacts with KFR.

Note 3 (a) When titrating with KFR only the cathode is polarised, the anode is depolarised by iodide ions. At the end point free iodine depolarises the cathode and, therefore, the current rises rapidly, giving a deflection which persists for not less than 30 s.

(b) When titrating with the water-methanol reagent the electrodes are both depolarised, hence the current is large. At the end point (when all the iodine is reduced) the iodide ions polarize the cathode and, therefore, the current decreases rapidly almost to zero. The zero of the meter scale is reached only when an excess of the water-methanol reagent is added.

Preparation of Karl Fischer reagent

Method Into a glass-stoppered flask (about 750 ml capacity) place anhydrous methanol (400 ml, containing $\not>$0.03% water) and pure dry pyridine (80 g). Immerse the flask in a freezing mixture and slowly pass sulphur dioxide into the cold solution, with continuous agitation, until the increase in weight is 20 g. Care must be taken at all times to prevent access of moisture. Finally add iodine (45 g), shake to dissolve and allow the mixture to stand for 24 h before use. The reagent prepared in this manner will have a water equivalent of 3.5 mg/ml approximately.

Standardisation

Karl Fischer reagent is not stable and must be rigorously protected from both light and moisture. It must be standardised immediately before use by titration against a known amount of water, either as a standard water-methanol reagent or, preferably, as solid crystalline sodium tartrate dihydrate. Although the latter is almost insoluble in methanol, it yields its water of hydration to the solvent, and its precise water content can be determined by drying at 150° to constant weight.

Standardisation with water-methanol reagent

Method (1) Prepare a standard solution containing 0.15% w/v *water* in anhydrous methanol in a thoroughly dried volumetric flask (100 ml) fitted with a vaccine cap and filled with dry nitrogen. Partly fill the flask with anhydrous methanol using the hypodermic needle, allowing the displaced nitrogen to escape through a second needle protected by a guard tube. Add about 0.15 g of *water*, accurately weighed (*w* g) from a weighing pipette, and adjust to volume with anhydrous methanol. Remove the needles and shake thoroughly.

(2) Run into the reaction vessel sufficient KFR to cover the electrodes. Pass dry nitrogen through the apparatus, switch on the magnetic stirrer and leave for about 15 min, to remove traces of moisture. Close switch S_1/S_2 and run in sufficient water-methanol reagent to reduce the meter reading to about 10 μA and leave for a further 15 min. The galvanometer reading should remain constant throughout this period, thus indicating that the apparatus is dry and free from air leaks. Add more water-methanol reagent to give a meter reading of about 2 μA. Partially drain the reaction vessel, if necessary, and run in exactly 10 ml of KFR and titrate with the water-methanol reagent until the meter again reads 2 μA. Repeat until successive titrations agree to within 0.10 ml and open switch S_1/S_2. Calculate the strength of the water-methanol reagent in terms of the KFR, expressing the result as follows:

$$1 \text{ ml water-methanol reagent} = M \text{ ml KFR}$$

(3) Allow most of the liquid in the reaction vessel to drain and transfer to it, by means of a dry pipette, 10 ml of anhydrous methanol. Pass dry nitrogen through the apparatus, switch on the magnetic stirrer and close switch S_1/S_2. Add the KFR until a definite brown colour persists for at least 20 s when the meter should read about 80 μA (more or less, depending on the size and distance apart of the electrodes). Record accurately the volume of KFR added (v_1 ml). Titrate with the water-methanol reagent until the meter gives a small positive reading of about 2 μA. Record the volume (v_2). Repeat until successive titrations do not differ by more than 0.10 ml.

Repeat the titration using 10 ml of the *standard solution of water in methanol*. Record accurately the volume of KFR (v_3) and water-methanol reagent (v_4) added. Open switch S_1/S_2.

Calculate the water equivalent (F), in mg water per ml, of the KFR from the formula:

$$F = \frac{100w}{(v_3 - v_1) + M(v_2 - v_4)}$$

The water content of the water-methanol reagent is MF mg/ml and this value may be used for subsequent restandardisation of the KFR.

Standardisation with Sodium Tartrate Dihydrate

Method Transfer, by means of a dry pipette, 40 ml of anhydrous methanol to the titration vessel and titrate with the reagent until the galvanometer shows a steady deflection which persists for not less than 30 s. Rapidly add approximately 0.3 g of sodium(+)-tartrate of known water content, accurately weighed, store for 1 min and continue the titration with the reagent to the same end point. Similarly titrate a further 40 ml of anhydrous methanol, taking the difference between the two titrations as equivalent to the water in the sample.

$$(CHOH \cdot COONa)_2{,}2H_2O \equiv 2H_2O$$

$$\therefore \quad 230.1 \text{ g } (CHOH \cdot COONa)_2{,}2H_2O \equiv 36.0304 \text{ g } H_2O$$

$$\therefore \quad 1 \text{ g } (CHOH \cdot COONa)_2{,}2H_2O \equiv 0.1566 \text{ g } H_2O$$

COGNATE DETERMINATIONS

Determination of water in the following typical materials:

Ampicillin Trihydrate
Benzylpenicillin Sodium
Betamethasone Sodium Phosphate
Calcium Acetate
Dimethyl Phthalate
Ephedrine
Ether, Anaesthetic
Fructose
Gentamicin Sulphate
Glyceryl Monostearate
Gonadorelin
Isoprenaline Sulphate
Isopropyl Alcohol
Mercaptopurine
Methyldopa
Pancuronium Bromide
Potassium Citrate
Pyrazinamide

Saccharin Sodium
Sodium Methyl Hydroxybenzoate
Sodium Nitroprusside
Sulphacetamide Sodium
Suxamethonium Chloride
Tetracosactrin
Tobramycin
Warfarin Sodium

7
Oxidation-reduction titrations

Introduction

Oxidation is (a) addition of oxygen, e.g. $SO_2 + O \rightarrow SO_3$
or (b) removal of hydrogen, e.g. $H_2S + O \rightarrow S + H_2O$
or in general
(c) increase in the ratio of electronegative to electropositive portion of the molecule:

$$Sn^{2+} + 2Cl^- + 2HgCl_2 \rightarrow SN^{4+} + 4Cl^- + Hg_2Cl_2$$
$$2Fe^{2+} + 4Cl^- + Cl_2 \rightarrow 2Fe^{3+} + 6Cl^-$$

Reduction is (a) addition of hydrogen, e.g. $C_2H_2 + 2H \rightarrow C_2H_4$
or (b) removal of oxygen, e.g. $CuO + 2H \rightarrow Cu + H_2O$
or in general
(c) increase in the ratio of electropositive to electronegative portion of the molecule.

Oxidation and reduction occur simultaneously in a reaction, one substance becoming reduced in the process of oxidising the other.

Many oxidations which are quantitive can be used as a basis for volumetric determinations, a standard solution of the oxidising agent being used for the titration. The chief oxidising agents used in volumetric work are potassium permanganate, potassium dichromate, iodine, potassium iodate and ceric sulphate.

The change in analytical practice away from expressing concentrations in terms of **normality** to the alternative method of expression in **molarity** presents an unnecessary complication in the consideration of oxidation-reduction titrations. Such methods, based on the gain or loss of electrons between oxidant and reductant, are automatically linked to the concept of equivalents and, hence, to the expression of concentrations in terms of normality. The relationship between normality and molarity is not constant, but differs for each oxidation-reduction system, as will be evident from the discussion on each reagent.

Determinations involving the use of potassium permanganate solution

The use of potassium permanganate ($KMnO_4$) as an oxidising substance in acid solution depends upon the reactions expressed by the following equations:

$$MnO_4^- + 8H^+ + 5e \rightarrow Mn^{2+} + 4H_2O$$
$$\therefore \quad KMnO_4 \equiv MnO_4^- \equiv 5e$$
$$\therefore \quad 158.0\,g\ KMnO_4 \equiv 1000\,ml\ \text{M} \equiv 5000\,ml\ \text{N}$$
$$\therefore \quad 3.160\,g\ KMnO_4 \equiv 1000\,ml\ 0.02\text{M}$$

Thus, MOLAR (M) and 5 NORMAL (5N) solutions contain the same concentrations of potassium permanganate.

The preparation of volumetric solutions of potassium permanganate

Since potassium permanganate often contains a small proportion of manganese dioxide, volumetric solutions must be made up approximately and then standardised. The intense colour of the solution makes difficult the detection of undissolved solid. The use of heat in the preparation of potassium permanganate solutions is also undesirable, since traces of grease or other contaminants on the glass vessels used can catalyse its decomposition. The following procedure is therefore recommended.

Method Weigh out approximately the appropriate amount of potassium permanganate (about 3.2 g for 1 l) on a watch-glass. Transfer to a 250 ml beaker containing cold *water* and stir thoroughly, breaking up the crystals with a glass rod, to effect solution. Decant the solution, through a small plug of glass wool supported in a funnel, into the appropriate graduated flask, leaving the undissolved residues in the beaker. Add more *water* to the beaker and repeat the process. Continue the process until all the potassium permanganate has dissolved. Make the solution up to the graduation mark and shake well to ensure thorough mixing.

Standardisation of 0.02M potassium permanganate solution by means of sodium oxalate

Potassium permanganate may be standardised with either sodium oxalate or oxalic acid. The former is preferred because it is readily available to a higher standard of purity (99.95%) and, unlike oxalic acid, it is available in the anhydrous state. The standardisation depends upon the reactions expressed by the following equations:

Oxidant $\quad MnO_4^- + 8H^+ + 5e \equiv Mn^{2+} + 4H_2O$

Reductant $\quad (COONa)_2 \overset{H^+}{\rightarrow} (COOH)_2 \rightarrow 2CO_2 + 2H^+ + 2e$

$$\therefore \quad 5(COONa)_2\ (10e) \equiv 2MnO_4^-\ (10e)$$
$$\therefore \quad 5 \times 134.0\,g\ (COONa)_2 \equiv 2000\,ml\ \text{M}\ KMnO_4$$
$$\therefore \quad 6.7\,g\ (COONa)_2 \equiv 1000\,ml\ 0.02\text{M}\ KMnO_4$$

Method Weigh out sodium oxalate (about 6.7 g) accurately into a litre graduated flask, dissolve in *water* and make up to volume. Pipette out 20 ml of this solution, add concentrated sulphuric acid (about 5 ml) and warm to about 70° (the flask can only just be held in the hand). Add the potassium permanganate solution from the burette. The first few drops result in a pink colour persisting for about 20 s. Wait until the colour disappears and then continue the titration. Formation of a brown colour during the titration is caused by insufficient acid, by using too high a temperature or by the use of a dirty flask. Clean the flask with solution of hydrogen peroxide and dilute sulphuric acid before repeating the titration. The end point is reached when a faint pink colour persists for about 30 s upon shaking the flask.

Potassium permanganate may also be standardised by means of sodium thiosulphate (p.187).

Potassium Bromide *Determination of the percentage of* KBr

This determination depends upon the reactions expressed by the following equations, in which the bromide ion is oxidised to bromine by acidified potassium permanganate:

$$MnO_4^- + 8H^+ + 5e \rightarrow Mn^{2+} + 4H_2O$$

$$2KBr \rightarrow 2Br^- \rightarrow Br_2 + 2e$$

$$\therefore \quad 10KBr \equiv 2KMnO_4$$

$$\therefore \quad 10 \times 119.0\,g\ KBr \equiv 2000\,ml\ \text{M}\ KMnO_4$$

$$\therefore \quad 0.01190\,g\ KBr \equiv 1\,ml\ 0.02\text{M}\ KMnO_4$$

Method Accurately weigh the sample (about 1.2 g) into a graduated flask (100 ml), dissolve in *water*, and dilute to volume. Pipette 10 ml into a conical flask, add dilute sulphuric acid (10 ml) and a few glass beads. Heat to boiling, and titrate (fume cupboard) with 0.02M $KMnO_4$, added dropwise, until the pink colour just persists.

Hydrogen Peroxide Solution (6%) *Determination of the percentage w/v of* H_2O_2

This determination depends upon mutual oxidation-reduction as expressed by the following equations:

$$2MnO_4^- + 6H^+ + 5H_2O_2 \rightarrow 2Mn^{2+} + 8H_2O + 5O_2$$

$$\therefore \quad 5H_2O_2 \equiv 2MnO_4^- \equiv 10e$$

$$34.02\,g\ H_2O_2 \equiv 2000\,ml\ \text{M}$$

$$\therefore \quad 0.001701\,g\ H_2O_2 \equiv 1\,ml\ 0.02\text{M}\ KMnO_4$$

Method Pipette the sample (10 ml) into a 100 ml graduated flask. Adjust to volume with *water* and titrate 20 ml portions of the solution with 0.02M $KMnO_4$ in the cold after adding sulphuric acid (50% v/v, 5 ml).

Calculation of volume strength of the solution

The 'volume strength' of the solution is the number of ml of oxygen at STP that can be obtained by complete thermal decomposition of 1 ml of solution. Under these conditions decomposition occurs according to the equation:

$$2H_2O_2 \rightarrow 2H_2O + O_2$$

$$\therefore \quad 68.04\,g\ H_2O_2 \equiv 22\,400\,ml\ O_2$$

$$\therefore \quad 1\,g\ H_2O_2 \equiv 329.2\,ml\ O_2$$

Consider a sample found to contain 5.70% w/v H_2O_2

then 100 ml sample ≡ 5.70 g H_2O_2

$$\text{and 1 ml sample} \equiv 0.0570\ \text{g}\ H_2O_2$$
$$\equiv 0.0570 \times 329.2\ \text{ml}\ O_2$$
$$\equiv 18.76\ \text{ml}\ O_2$$

$\therefore$ Volume strength of the sample is 18.76

COGNATE DETERMINATIONS

Sodium Perborate
Benzoyl Peroxide

Direct titration with iodine

Determinations involving direct titration with iodine make use of the oxidising power of iodine in aqueous solution.

$$I_2 + 2e \rightarrow 2I$$
$$\therefore \quad I_2 \equiv 2I^- \equiv 2e$$
$$\therefore \quad 2 \times 126.9\ \text{g}\ I_2 \equiv 1000\ \text{ml}\ \text{M} \equiv 2000\ \text{ml}\ \text{N}$$
$$\therefore \quad 12.69\ \text{g}\ I_2 \equiv 1000\ \text{ml}\ 0.05\text{M}$$

Thus, MOLAR (M) and 2NORMAL (2N) solutions contain the same concentration of iodine.

Because iodine is practically insoluble in water, use is made of the fact that it dissolves in solutions of potassium iodide to form KI_3, which behaves in solution as free iodine.

Preparation of approximately 0.05M iodine solution

Method Weigh the iodine (3.2 g approx.) on a rough balance. Transfer to a beaker containing potassium iodide (7.5 g) and *water* (10 ml). Dissolve (I_2 dissolves more readily in this concentrated solution of KI than in weaker solutions), transfer to a 250 ml graduated flask and adjust to volume.

Note Iodine is volatile and standard solutions should be stored in tightly stoppered (glass) bottles. Standard solutions should not be collected from bulk containers in beakers or other open vessels and, when measured from suitable containers in pipettes (or burettes), should be titrated without delay. These precautions must be observed at all times if satisfactory results are to be obtained.

Standardisation of approximately 0.05M iodine solution by means of arsenic trioxide

This standardisation depends upon the reactions expressed by the following equations:

$$As_2O_3 + 2H_2O \rightarrow As_2O_5 + 4H^+ + 4e$$
$$I_2 + 2e \equiv 2I^-$$
$$\therefore \quad As_2O_3 \equiv 2I_2$$
$$\therefore \quad 197.8\ \text{g}\ As_2O_3 \equiv 2000\ \text{ml}\ \text{M}$$

$$\therefore \quad 0.4946 \text{ g } As_2O_3 \equiv 100 \text{ ml } 0.05\text{M iodine}$$

Because of the strong reducing properties of hydriodic acid, the oxidation with iodine is a reversible reaction

$$As_2O_3 + 2O_2 + 2H_2O \rightleftharpoons As_2O_5 + 4H^+ + 4I^-$$

The reaction is made to go completely to the right (i.e. to As_2O_5) by removal of the HI with sodium bicarbonate. Sodium hydroxide or sodium carbonate cannot be used to remove HI because they react with iodine as follows:

$$6NaOH + 3I_2 \rightarrow 5NaI + NaIO_3 + 3H_2O$$
$$3Na_2CO_3 + 3I_2 \rightarrow 5NaI + NaIO_3 + 3CO_2$$

Method Accurately weigh the arsenic trioxide (0.5 g approx.) into a beaker, add sodium hydroxide solution (20%; about 2 ml) and warm to dissolve. Cool, transfer quantitatively to a 100 ml graduated flask and adjust to volume with *water*. Pipette 25 ml, acidify with dilute hydrochloric acid, testing for acidity with a little sodium bicarbonate when effervescence should occur. This acidification is necessary to remove the free NaOH which would react with the iodine. Add sodium bicarbonate to remove the excess acid, followed by a further 2 g (to remove the HI in the subsequent titration). Titrate with the approx. 0.5M iodine solution. The end point is the first permanent pale straw colour.
Note There is no need to add starch solution in this determination.

Antimony Sodium Tartrate *Determination of the percentage of* $C_4H_4O_7NaSb$

This substance behaves in solution as a mixture of sodium acid tartrate and antimony trioxide, i.e.

$$2C_4H_4O_7NaSb \equiv Sb_2O_3$$
$$Sb_2O_3 + 2H_2O \rightarrow Sb_2O_5 + 4H^+ + 4e$$
$$2[I_2 + 2e \rightarrow 2I^-]$$
$$\therefore \quad C_4H_4O_7NaSb \equiv Sb_2O_3 \equiv 2I_2$$
$$\therefore \quad 617.6 \text{ g } C_4H_4O_7NaSb \equiv 2000 \text{ ml M}$$
$$\therefore \quad 001544 \text{ g } C_4H_4O_7NaSb \equiv 1 \text{ ml } 0.05\text{M iodine}$$

Method Dissolve the sample (about 0.5 g accurately weighed) in *water* (50 ml). Add potassium sodium tartrate (5 g) and sodium tetraborate (borax) (2 g), to remove hydriodic acid as formed, and titrate with 0.05M iodine solution using starch solution as indicator. The end point is rather slow, therefore shake well, to make sure that the end persists. There is a possibility of under-titration unless this provision is observed.

Acetarsol *Determination of the percentage of* $C_8H_{10}AsNO_5$

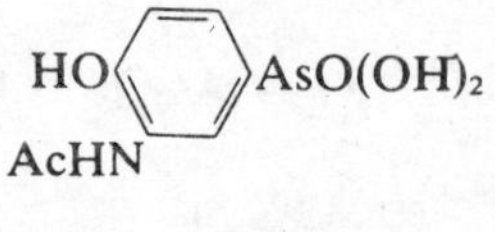

Arsenic in organic arsenicals is determined, after preliminary treatment, by oxidation of arsenic from the trivalent to the pentavalent state by means of 0.05M iodine solution.

Before the determination of As can be carried out, it must be removed from its organic combination. The following is a summary of the steps involved.

(1) The organic matter is destroyed by heating strongly with concentrated sulphuric and fuming nitric acids.

(2) The nitric acid is removed by adding ammonium sulphate and heating until all N_2O has been removed:

$$2HNO_3 + (NH_4)SO_4 \rightarrow 2NH_4NO_3 + H_2SO_4$$

$$\downarrow \text{heat}$$

$$2N_2O + 4H_2O$$

(3) Potassium iodide is added to the acid solution and the liberated HI reduces the pentavalent arsenic to the trivalent state. The solution is boiled during the reduction and most of the liberated iodine is expelled.

$$As_2O_5 + 4HI \rightarrow As_2O_3 + 2O_2 + 2H_2O$$

(4) The liquid is cooled and any remaining iodine removed by titration with 0.05M sodium sulphite solution.

(5) The solution is made just alkaline with sodium hydroxide solution and then acidified with dilute sulphuric acid to remove the free NaOH. Sodium bicarbonate is added to ensure that the reaction proceeds quantitively in the correct direction and the liquid is titrated with 0.05M iodine solution:

$$As_2O_3 + 2I_2 + 2H_2O \rightleftharpoons As_2O_5 + 4HI$$

$$\therefore \quad 2C_8H_{10}AsNO_5 \equiv As_2O_5 \equiv As_2O_3 \equiv 2I_2$$

$$\therefore \quad 2 \times 275.1\text{ g } C_8H_{10}AsNO_5 \equiv 2000\text{ ml M}$$

$$\therefore \quad 0.01375\text{ g } C_8H_{10}AsNO_5 \equiv 1\text{ ml } 0.05\text{M iodine}$$

Method Accurately weigh the sample (about 0.25 g) into a 500 ml conical flask. Add sulphuric acid (7.5 ml) and fuming nitric acid (SG 1.5; 1.5 ml) and boil gently (fume cupboard) for 45 min. Cool, add fuming nitric acid (0.5 ml) and heat until brown fumes cease to be evolved. Cool, add ammonium sulphate (5 g) in small quantities at a time and heat again (fume cupboard) until there is no further evolution of gas and the resulting liquid is colourless. Cool, dilute with water (100 ml), add KI (1 g), and boil gently (fume cupboard) until the volume is reduced to 50 ml. Cool, decolorize by the addition of a few drops of 0.05M sodium sulphite. Dilute with *water* (60 ml), add phenolphthalein solution (1 drop) and make just alkaline with sodium hydroxide solution. Acidify slightly with dilute sulphuric acid, neutralise with sodium bicarbonate, then add 4 g in excess and titrate with 0.05M iodine.

COGNATE DETERMINATIONS

Sodium Stibogluconate and Injection

Ascorbic Acid *Determination of the percentage of* $C_6H_8O_6$

The determination depends upon the quantitive oxidation of ascorbic acid to dehydroascorbic acid with iodine acid solution.

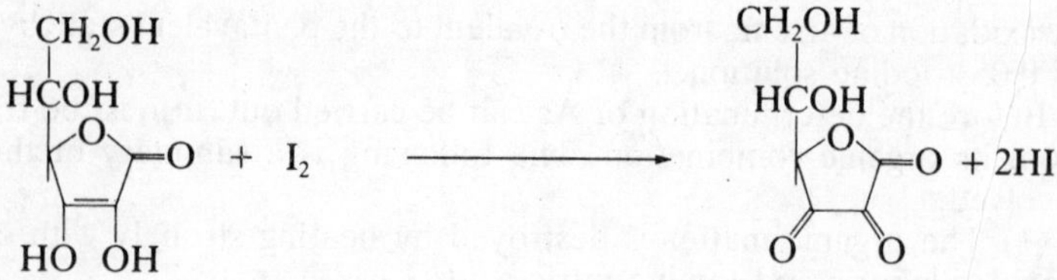

$$\therefore \quad C_6H_8O_6 \equiv I_2$$
$$\therefore \quad 176.1\ g\ C_6H_8O_6 \equiv 1000\ ml\ \text{M iodine}$$
$$\therefore \quad 0.0881\ g\ C_6H_8O_6 \equiv 1\ ml\ 0.05\text{M iodine}$$

Method Accurately weigh approximately 0.2 g into a conical flask. Dissolve in a mixture of freshly boiled and cooled *water* (80 ml) and M sulphuric acid (10 ml). Titrate with 0.05M iodine using starch solution as indicator.

Dimercaprol *Determination of the percentage w/w of* $C_3H_8OS_2$

This determination depends upon the oxidation of the thiol groups with iodine. The reaction is not reversible.

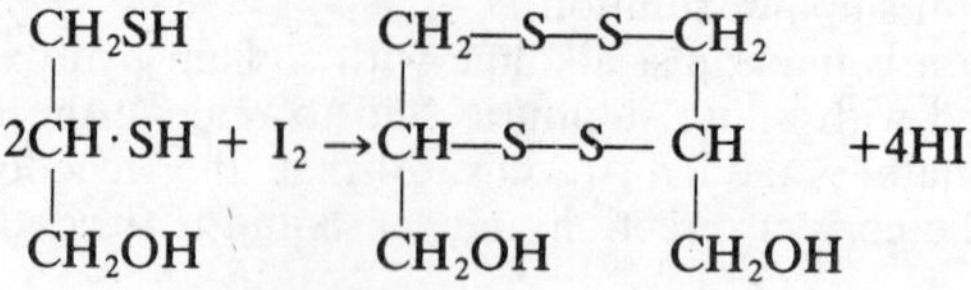

$$\therefore \quad 2 \times 124.2\ g\ C_3H_8OS_2 \equiv 2I_2 \equiv 2000\ ml\ \text{M}$$
$$\therefore \quad 0.00621\ g\ C_3H_8OS_2 \equiv 1\ ml\ 0.05\text{M iodine}$$

Method Weigh accurately 0.2 g in a small sample tube. Transfer the tube and contents to a 250 ml conical flask, add 0.1M hydrochloric acid (40 ml) and titrate with 0.05M iodine.

COGNATE DETERMINATION

Dimercaprol Injection

Iodine-sodium thiosulphate titrations

Standardisation of iodine solution by sodium thiosulphate

AnalaR sodium thiosulphate ($Na_2S_2O_3,5H_2O$) can be used to standardise iodine solution, which oxidises the former to sodium tetrathionate.

$$2S_2O_3^{2-} \rightarrow S_4O_6^{2-} + 2e$$
$$I_2 + 2e \rightarrow 2I^-$$
$$\therefore \quad 2S_2O_3^{2-} \equiv I_2$$
$$\therefore \quad 2\times\ 248.2\ g\ Na_2S_2O_3,5H_2O \equiv 2000\ ml\ \text{M}\ Na_2S_2O_3 \equiv 1000\ ml\ \text{M iodine}$$
$$\therefore \quad 24.82\ g\ Na_2S_2O_3,5H_2O \equiv 1000\ ml\ 0.1\text{M}\ Na_2S_2O_3$$
$$\equiv 1000\ ml\ 0.05\text{M iodine}$$

Method Transfer 6.025 g (accurately weighed) of AnalaR sodium thiosulphate to a 250 ml graduated flask. Dissolve in *water*, adjust to volume and shake. If large volumes of sodium thiosulphate solutions are being prepared for storage, the solution should be stabilised against decomposition by the addition of one or two drops of sodium hydroxide solution (20% w/v).

Pipette 25 ml of 0.05M iodine into a conical flask and titrate with the standard sodium thiosulphate solution until just colourless.

COGNATE DETERMINATIONS

Iodine
Sodium Thiosulphate

Back-titration of excess iodine with sodium thiosulphate

Anhydrous Sodium Sulphite *Determination of the percentage of* Na_2SO_3

This is determined by treatment with hydrochloric acid to liberate sulphurous acid:

$$Na_2SO_3 + 2HCl \rightarrow 2NaCl + H_2SO_3$$
$$SO_3^{2-} + H_2O \rightarrow SO_4^{2-} + 2H^+ + 2e$$
$$I_2 + 2e \rightarrow 2I^-$$
$$\therefore \quad Na_2SO_3 \equiv I_2$$

Cloaxacillin Sodium *Limit of iodine-absorbing substances*

Iodine-absorbing substances are present as manufacturing impurities and decomposition products in many penicillins. These are determined in phosphate buffer (to stabilize the parent substance) by addition of iodine and back-titration with sodium thiosulphate.

Dichloralphenazone

The sample in sodium acetate buffer is treated with standard iodine solution to yield a precipitate of phenazone-iodine complex. This is dissolved in chloroform, and excess iodine in the aqueous phase is titrated with 0.1M sodium thiosulphate.

A separate assay for chloral is also required to ensure that the composition of the chloral-phenazone complex is consistent from batch to batch.

Sodium Metabisulphite *Determination of the percentage of* $Na_2S_2O_5$

This substance reacts as follows:

$$Na_2S_2O_5 + 2HCl \rightarrow 2NaCl + H_2O + 2SO_2$$
$$2[SO_2 + H_2O \rightarrow H_2SO_3]$$
$$2[SO_3^{2-} + H_2O \rightarrow SO_4^{2-} + 2H^+ + 2e]$$

$$2[I_2 + 2e \rightarrow 2I^-]$$
$$\therefore \quad Na_2S_2O_5 \equiv 2I_2$$

Release of iodine and titration with sodium thiosulphate

Chloramine *Determination of the percentage of* $C_7H_7O_2NClSNa,3H_2O$

In acid solution this substance behaves as though it were a hypochlorite and yields active chlorine.

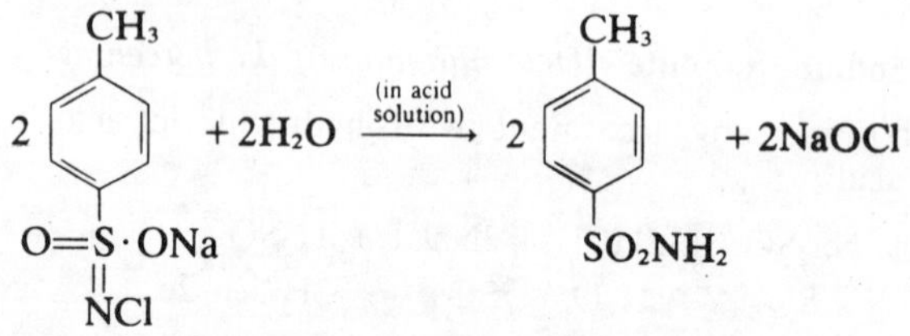

$$2NaOCl + H_2SO_4 \rightarrow Na_2SO_4 + 2HOCl$$
$$2[HO{-}Cl + I^- \rightarrow HO^- + I{-}Cl]$$
$$2[I^- + I{-}Cl \rightarrow I_2 + Cl^-]$$

$$C_7H_7O_2NClSNa \equiv I_2$$
$$\therefore \quad 281.7\text{ g } C_7H_7O_2NClSNa,3H_2O \equiv 1000\text{ ml M iodine} \equiv 2000\text{ ml M } Na_2S_2O_3$$
$$\therefore \quad 0.01409\text{ g } C_7H_7O_2NClSNa,3H_2O \equiv 1\text{ ml } 0.1\text{M } Na_2S_2O_3$$

Method Dissolve the sample (about 0.3 g), accurately weighed, in *water* (about 50 ml) in a glass stoppered bottle. Add potassium iodide (1 g) and dilute hydrochloric acid (1 ml). Shake and allow to stand for 3 min. Titrate the liberated iodine with 0.1M sodium thiosulphate solution. If a white precipitate forms during the titration, dilute with a little water so that all substances are kept in solution.

Chlorinated Lime *Determination of the percentage w/w of available* Cl_2

$$CaCl(OCl) + 2CH_3COOH \rightarrow Ca(CH_3COO)_2 + HCl + HClO$$
$$HCl + HClO \rightarrow H_2O + Cl_2$$
$$2HI + Cl_2 \rightarrow 2HCl + I_2$$
$$\therefore \quad 70.92\text{ g available } Cl_2 \equiv I_2 \equiv 1000\text{ ml M } I_2 \equiv 2000\text{ ml M } Na_2S_2O_3$$
$$\therefore \quad 0.003545\text{ g available chlorine} \equiv 1\text{ ml } 0.1\text{M } Na_2S_2O_3$$

An official sample contains about 30% Cl_2 and, therefore, about 0.4 g sample ≡ about 40 ml 0.01M thiosulphate.

Method Triturate the sample (about 4 g), accurately weighed, with *water* in a mortar to a smooth cream. This must be performed rapidly to avoid loss of chlorine. Transfer the paste to a 1 l graduated flask by means of a glass rod and funnel, washing out the mortar and funnel several times with *water*. Dilute the suspension to the 1 l mark and shake well. There must not be any large particles in the suspension. Rinse out a 100 ml graduated flask with a portion of the suspension, shaking the litre flask well before pouring out the suspension. Fill the flask to the graduation mark with the suspension.

Rinse out a conical flask containing a little dilute acid with some of the suspension from the litre graduated flask to oxidise any organic material which may be present. Then *wash out well* with *water*. Transfer the 100 ml of suspension from the graduated flask to this conical flask, washing out the suspension completely from the graduated flask by means of *water*. Add KI (3 g) and acetic acid (6M; 5 ml), and titrate the liberated iodine with 0.1M sodium thiosulphate solution.

Copper Sulphate *Determination of the percentage of* $CuSO_4,5H_2O$

The determination of copper compounds depends on the instability of cupric iodide formed by the reaction between the copper salt and potassium iodide. The cupric iodide breaks up into cuprous iodide and iodine and this iodine can be titrated with standard thiosulphate solution. The reactions are expressed by the following equations:

$$2CuSO_4 + 4KI \rightarrow 2CuI_2 + 2K_2SO_4$$

$$2CuI_2 \rightarrow Cu_2I_2 + I_2$$

$$\therefore \quad 2CuSO_4,5H_2O \equiv I_2$$

$$\therefore \quad 2 \times 249.7\text{ g } CuSO_4,5H_2O \equiv 1000\text{ ml M iodine} \equiv 2000\text{ ml } Na_2S_2O_3$$

$$\therefore \quad 0.02497\text{ g} \equiv 1\text{ ml } 0.1\text{M } Na_2S_2O_3$$

Method Accurately weigh the sample (about 1 g) into a flask. Dissolve in *water* (about 50 ml). Add a little sodium carbonate until turbid to remove any free mineral acid. Add acetic acid to clear the solution and then 5 ml in excess. Add potassium iodide (3 g), dilute to about 200 ml with *water*, and titrate the liberated iodine with 0.1M sodium thiosulphate solution. During the titration the brown colour of the iodine becomes less intense. When a pale yellowish colour has been obtained add starch mucilage (1 ml; Note 1). The solution should now be blue. Continue the titration drop by drop, and when near the end point add 2 g of potassium thiocyanate (Note 2) and shake well. Complete the titration, shaking well and allowing 10 s between each drop, because the end point is rather slow. The end point is white or flesh colour and 1 drop of 0.1M sodium thiosulphate should result in a change from the blue colour to this end point.

Note 1 The starch indicator must not be added until the end of the titration.

Note 2 Cuprous thiocyanate is less soluble than cuprous iodide. Addition of potassium thiocyanate, therefore, ensures completion of the reaction and release of iodine.

COGNATE DETERMINATION

Calcium Copperedetate

Phenol *Determination of the percentage w/w of* C_6H_5OH

The phenol is treated with excess 0.05M bromine:

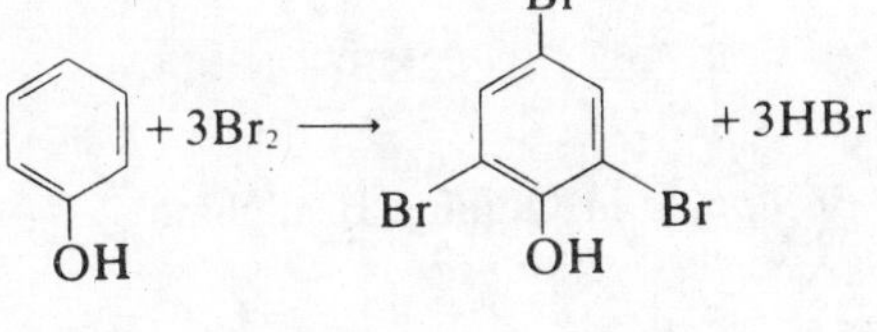

The excess of bromine not required for the formation of tribromophenol is determined by adding potassium iodide and titrating the liberated iodine with sodium thiosulphate solution.

$$Br_2 + 2I^- \rightarrow 2Br^- + I_2$$

$$\therefore \quad C_6H_5OH \equiv 3Br_2 \equiv 3I_2$$

$$\therefore \quad 94.11\text{ g } C_6H_5OH \equiv 3000\text{ ml M bromine} \equiv 6000\text{ ml M } Na_2S_2O_3$$

$$\therefore \quad 0.001569\text{ g } C_6H_5OH \equiv 1\text{ ml } 0.05\text{M bromine} \equiv 1\text{ ml } 0.1\text{M } Na_2S_2O_3$$

Solution of bromine is not stable and, therefore, a solution containing potassium bromide and potassium bromate is used. This solution on acidification liberates bromine:

$$BrO_3^- + 5Br^- + 6H^+ \rightarrow 3Br_2 + 3H_2O$$

The quantities of potassium bromide and potassium bromate used in the solution are such that, on acidification, an approximately 0.05M bromine solution is produced. The exact strength of the solution is not required because a blank determination is performed.

Method Dissolve the sample (about 2 g accurately weighed) in *water* in a litre graduated flask and adjust to volume. Pipette 25 ml (*care*) into a glass-stoppered bottle. Add bromine solution (50 ml) from a pipette and concentrated hydrochloric acid (5 ml). Shake repeatedly for half an hour and then set aside for 15 min. The shaking is to complete the bromination of the phenol to tribromophenol. Add potassium iodide (1 g), shake well, wash the stopper and titrate the liberated iodine with 0.1M sodium thiosulphate solution. When the solution is pale yellow in colour, add chloroform (10 ml), and a few drops of starch mucilage, and titrate with vigorous shaking to a colourless end point. Carry out a blank determination on the bromide-bromate solution (50 ml). The difference between the test and blank titrations is equivalent to the amount of bromine required by the sample.

COGNATE DETERMINATIONS

Liquefied Phenol
Methyl Hydrobenzoate
Propyl Hydrobenzoate
Phenindione This substance is treated with bromine in ethanol to give the dibromo addition compound. Excess bromine is removed by the addition of β-naphthol; traces of bromine vapour by means of a current of air. The bromine atom in 2-bromo-2-phenylindane-1,3-dione is labile and treatment with potassium iodide releases iodine, which can be titrated with sodium thiosulphate.

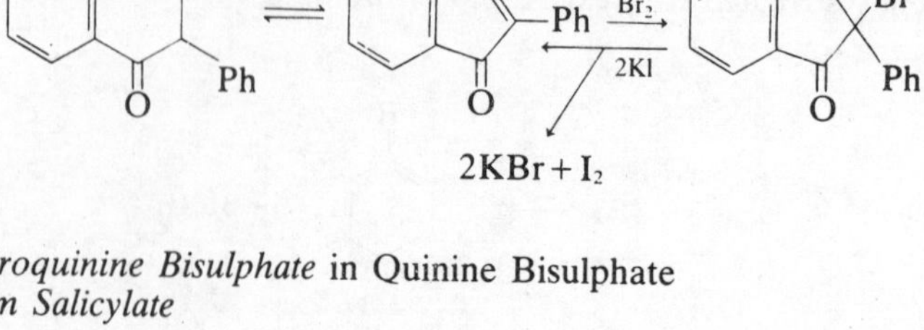

Dihydroquinine Bisulphate in Quinine Bisulphate
Sodium Salicylate

Standardisation of potassium permanganate with sodium thiosulphate

The standardisation depends upon the reactions expressed by the following equations:

$$2[MnO_4^- + 8H^+ + 5e \rightarrow Mn^{2+} + 4H_2O]$$

$$5[2I^- \rightarrow I_2 + 2e]$$

$$\therefore \quad 2KMnO_4 \equiv 5I_2$$

$$\therefore \quad 2000 \text{ ml M } KMnO_4 \equiv 5000 \text{ ml M iodine} \equiv 10\,000 \text{ ml M } Na_2S_2O_3$$

$$\therefore \quad 1 \text{ ml } 0.02\text{M } KMnO_4 \equiv 1 \text{ ml } 0.1\text{M } Na_2S_2O_3$$

Method Pipette 25 ml of the approx. 0.1M $KMnO_4$ solution into a conical flask. Add dilute sulphuric acid (10 ml) and potassium iodide (3 g). Quantitive liberation of iodine occurs immediately. Titrate the liberated iodine with 0.1M sodium thiosulphate solution, shaking continuously during the titration to prevent the acid attacking the thiosulphate. Starch solution is unnecessary because the end point is colourless.

Standardisation of 0.1M sodium thiosulphate with potassium iodate

As there is always some doubt as to the exact water content of sodium thiosulphate crystals, it is usual to standardise thiosulphate solutions by such substances as potassium iodate, potassium bromate or potassium dichromate, all of which can be obtained in a high state of purity.

The standardisation with potassium iodate depends upon the reactions expressed by the following equations:

$$IO_3^- + 5I^- + 6H^+ \rightarrow 3I_2 + 3H_2O$$

$$3[I_2 + 2e \rightarrow 2I^-]$$

$$3[2S_2O_3^{2-} \rightarrow S_4O_6^{2-} + 2e]$$

$$\therefore \quad KIO_3 \equiv 3I_2$$

$$\therefore \quad 214.0 \text{ g } KIO_3 \equiv 3000 \text{ ml M iodine} \equiv 6000 \text{ ml M } Na_2S_2O_3$$

$$\therefore \quad 3.567 \text{ g } KIO_3 \equiv 1000 \text{ ml } 0.1\text{M } Na_2S_2O_3$$

Reaction between KIO_3 and excess KI in acid solution results in liberation of iodine which can be titrated with sodium thiosulphate solution.

Method Accurately weigh about 1.3 g of AnalaR potassium iodate into a 250 ml graduated flask. Dissolve in *water* and adjust to volume. The substance is only slowly soluble. Pipette the solution (25 ml) into a conical flask, add potassium iodide (2 g) and dilute sulphuric acid (4 ml) and titrate the liberated iodine with the approx. 0.1M sodium thiosulphate solution.

COGNATE DETERMINATION

Potassium Bromate This can be determined by the addition of potassium iodide and dilute hydrochloric acid:

$$KBrO_3 + HI \rightarrow HIO_3 + KBr$$

$$IO_3^- + 5I^- + 6H^+ \rightarrow 3I_2 + 3H_2O$$

$$\therefore \quad KBrO_3 \equiv IO_3^- \equiv 3I_2$$

$$\therefore \quad 167.02 \text{ g } KBrO_3 \equiv 3000 \text{ ml M iodine} \equiv 6000 \text{ ml M } Na_2S_2O_3$$

$$\therefore \quad 0.002784 \text{ g } KBrO_3 \equiv 1 \text{ ml } 0.1\text{M } Na_2S_2O_3$$

Iodine Value of fixed oils

The Iodine Value is defined as the weight of iodine absorbed by 100 parts by weight of the substance. It is a measure of the unsaturated compounds present in the substance and is based upon the addition of halogen across a carbon-carbon double bond. Iodine itself reacts too slowly; instead, either iodine monochloride or bromine in pyridine may be employed as reagents in official determinations. A quantitive measure of the unsaturation present is obtained provided that certain experimental conditions are complied with, since there is a tendency with both reagents for substitution reactions as well as addition to take place. This is especially liable to occur where appreciable concentrations of steroid material are present and, in these instances, the so-called pyridine bromide method is obligatory.

Pure glyceryl trioleate has a theoretical iodine value of 87.62 and most edible oils have iodine values in this region, e.g. *Arachis Oil* (86–106). Such oils are often described as non-drying oils. The so-called drying oils and semi-drying oils, which have more centres of unsaturation, exhibit much higher iodine values, e.g. *Linseed Oil* (170–200); *Cod-liver Oil* (150–180). On the other hand, the solid fats, which contain a high proportion of esters of saturated fatty acids, have relatively low iodine values, e.g. *Theobroma Oil* (35–40).

Iodine monochloride method

This determination depends upon reactions expressed by the following equations:

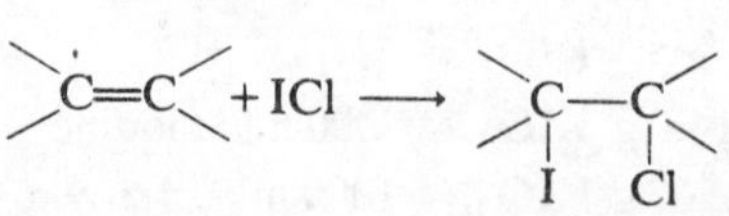

Excess iodine monochloride is converted to iodine by the addition of the potassium iodide and titrated with sodium thiosulphate in the usual way.

$$I^- + I{-}Cl \rightarrow I_2 + Cl^-$$

$$>C{=}C< \;\equiv ICl \equiv I_2$$

$$\therefore \quad 2 \times 126.9\,g \equiv 1000\,ml \text{ M iodine} \equiv 2000\,ml \text{ M } Na_2S_2O_3$$

$$\therefore \quad 0.01269\,g\ I_2 \equiv 1\,ml\ 0.1\text{M } Na_2S_2O_3$$

Method Weigh the oil (see below for calculation of the required amount) in a small specimen tube and place it in an iodine flask or a 250 ml clean, dry, glass-stoppered bottle. Add 10 ml of carbon tetrachloride to dissolve the oil. Run in 20 ml of iodine monochloride solution from an automatic pipette (see Fig 4.7, p.126), insert the stopper previously moistened with potassium iodide solution and set aside in the dark for 30 min (exactly), at a temperature of about 20°C. The reaction is performed in the dark to avoid side reactions, such as substitution, which are light-catalysed. After 30 min, add 10% aqueous potassium iodide solution (15 ml), and back-titrate the excess iodine with 0.1M $Na_2S_2O_3$ using starch mucilage as indicator. Carry out a blank determination using all the reagents, but omitting the sample. Calculate the iodine value from the formula:

$$\text{Iodine Value} = \frac{(b - a) \times 0.01269 \times 100 \times \text{molarity}}{\text{Weight of sample (in g)}}$$

where b = blank burette reading; a = test burette reading; molarity = molarity of standard sodium thiosulphate solution used.

Iodine Monochloride Solution Dissolve iodine trichloride (8 g) in about 200 ml of glacial acetic acid and dissolve iodine (9 g) in carbon tetrachloride (300 ml). Mix the two solutions and adjust to 1000 ml with glacial acetic acid. This operation should be carried out in a fume cupboard. *Iodine Bromide Solution*, 2% IBr in glacial acetic acid, may be used in place of iodine monochloride.

Calculation of the weight of oil required

Each 20 ml of iodine monochloride solution is approximately equivalent to 40 ml of 0.1M $Na_2S_2O_3$. Allowing for at least 100% excess, this leaves approximately 10 ml iodine monochloride (≡ 20 ml 0.1M $Na_2S_2O_3$) available for reaction with the oil.

$$\begin{aligned}\text{Sample of oil} &\equiv 20\text{ ml } 0.1\text{M } Na_2S_2O_3 \\ &\equiv 20 \times 0.01269\text{ g iodine} \\ &\equiv 0.2538\text{ g iodine}\end{aligned}$$

For calculation purposes the oil is regarded as equivalent to 0.2 g of iodine.

Let the expected iodine value of the sample be x.

Then x g of iodine will be absorbed by 100 g of sample

$$\therefore \quad 0.2\text{ g of iodine will be absorbed by } 100 \times \frac{0.2}{x}\text{ g of sample}$$

$$= \frac{20}{x}\text{ g of sample}$$

Hence, the weight of oil (in g) required for each determination is calculated by dividing the highest expected iodine value into 20..

If more than half the available halogen is absorbed, the test must be repeated on a smaller quantity of substance.

Pyridine bromide method

This method depends upon the reactions expressed by the following equations:

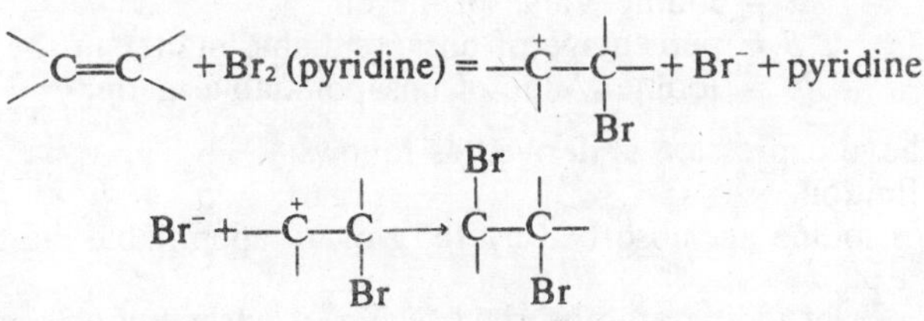

Excess bromine is destroyed by addition of potassium iodide and the iodine so released titrated with sodium thiosulphate.

$$I^- + Br_2 \rightarrow IBr + Br^-$$
$$I^- + IBr \rightarrow I_2 + Br^-$$
$$\therefore \quad >C=C< \equiv Br_2 \equiv I_2$$

Method Weigh the sample in a small specimen tube and place in a dry, glass-stoppered bottle with 10 ml of carbon tetrachloride as in the iodine monochloride method. Add 25 ml of pyridine bromide solution, stopper and allow to stand in the dark for 10 min. Add aqueous potassium iodide solution and proceed as in the iodine monochloride method.

Pyridine bromide reagent Dissolve pyridine (8 g) and sulphuric acid (5.4 ml) in glacial acetic acid (20 ml) keeping the solution cool. Add bromine (2.6 ml) dissolved in glacial acetic acid (20 ml) and dilute to 1000 ml with acetic acid.

Calculation of the weight of oil required

Each 25 ml of pyridine bromide solution contains 0.20 g of bromine. Allowing for rather more than 100% excess, this leaves 0.08 g of bromine (0.125 g iodine) for reaction with the oil.

Let the expected iodine value of the sample be x.

Then

$$x \text{ g of iodine will be absorbed by 100 g of sample}$$

$$\therefore \quad 0.125 \text{ g of iodine will be absorbed by } \frac{100 \times 0.125}{x} \text{ g of sample}$$

$$= \frac{12.5}{x} \text{ g of sample}$$

Hence, the weight of oil (in g) for each determination is calculated by dividing the highest expected iodine value into 12.5.

If more than half the available halogen is absorbed, the test must be repeated, using a smaller quantity of substance.

Halibut-liver Oil contains a high proportion of unsaponifiable matter, and two iodine values are determined, both by the pyridine bromide method. The first determination is on the oil itself and the second on the residue left in the determination of unsaponifiable matter using 1 g of oil. The iodine value of the glycerides is obtained by calculation from the two values using the formula

$$\text{Iodine Value} = \frac{100x - Sy}{100 - S}$$

where

x = Iodine Value of the oil
S = percentage of unsaponifiable matter in the oil
y = Iodine Value of unsaponifiable matter

The above expression is derived as follows:

By definition,

y g of iodine are absorbed by 100 g of unsaponifiable matter

$\therefore \frac{SY}{100}$ of iodine are absorbed by S g of unsaponifiable matter (present in 100 g of oil).

But x g of iodine are absorbed by 100 g of oil.

$$\therefore \quad \text{iodine absorbed by the glycerides in 100 g of oil} = x - \frac{Sy}{100}$$

$$= \frac{100x - Sy}{100}$$

But weight of glycerides in 100 g of oil = 100 − *S*

$$\therefore \quad \text{iodine absorbed by 100 g of glycerides} = \frac{100x - Sy}{100} \times \frac{100}{100 - S}$$

$$\therefore \quad \text{Iodine Value of glycerides} = \frac{100x - Sy}{100 - S}$$

Potassium iodate titrations

Potassium iodate is a fairly strong oxidising agent. Under suitable conditions it reacts quantitively with both iodates and iodine. Iodate titrations can be performed in the presence of alcohol, saturated organic acids, and many other kinds of organic matter.

Potassium Iodide *Determination of the percentage of* KI *by titration with potassium iodate*

If the concentration of hydrochloric acid does not exceed M, the reaction between potassium iodate and potassium iodide stops when the iodate has been reduced to free iodine:

$$IO_3^- + 5I^- + 6H^+ \rightarrow 3I_2 + 3H_2O \qquad (1)$$

In the presence of more concentrated hydrochloric acid (exceeding 4M), the iodine produced in the above reaction is oxidised by iodate to the iodine cation, I^+. The high concentration of chloride ions leads to the formation of iodine monochloride, which is stabilised against hydrolysis by the hydrochloric acid present

$$2[I_2 \rightarrow 2I^+ + 2e]$$

$$IO_3^- + 6H^+ + 4e \rightarrow 3H_2O + I^+$$

$$5I^+ + 5Cl^- = 5ICl$$

This reaction may, therefore, be summarised as:

$$KIO_3 + 2I_2 + 6HCl \rightarrow KCl + 5ICl + 3H_2O \qquad (2)$$

Thus, the complete reaction between KIO_3 and KI in the presence of 4M HCl is expressed as follows by combining equations (1) and (2)

$$KIO_3 + 2KI_2 + 6HCl \rightarrow 3KCl + 3ICl + 3H_2O \qquad (3)$$

$$2 \times 166 \text{ g KI} \equiv 1000 \text{ ml M } KIO_3$$

$$\therefore \quad 0.0166 \text{ g KI} \equiv 1 \text{ ml } 0.05\text{M } KIO_3$$

Method Dissolve the sample (about 0.5 g), accurately weighed, in *water* (about 50 ml). Add concentrated hydrochloric acid (60 ml) and titrate with 0.05M potassium iodate solution. The solution will become brown as the titration proceeds due to the liberation of iodine (eq. 1). The solution then becomes a lighter colour as titration is continued, due to the formation of iodine monochloride (eq. 2). When the solution becomes pale yellow, add 1 ml of amaranth solution and continue the titration until the red colour changes to pale yellow. Amaranth, which is an azo dye, is oxidised by excess titrant at the end point to a colourless derivative. Alternatively, chloroform can also be used as indicator. Chloroform is immiscible with aqueous solutions and gives a violet coloured solution in the presence of iodine. Vigorous shaking is necessary as the end point is approached to ensure equilibrium between the indicator and titration solutions. The end point is reached when the chloroform solution just becomes colourless.

The procedure can be modified in the presence of cyanide ions when iodine monochloride is converted to iodine cyanide, which is less susceptible than the former to hydrolysis. This procedure is described in earlier editions of this book but, although effective, is no longer acceptable because of the volatility and extreme toxicity of the hydrogen cyanide which is released.

Standardisation of potassium iodate solutions

Because potassium iodate can be purchased in a high state of purity (99.9%) a standard solution may be prepared by direct weighing. However, solutions may be standardised by one of the following methods:

(1) To a definite volume of potassium iodate solution add excess potassium iodide and dilute hydrochloric acid and titrate the liberated iodine with standard sodium thiosulphate solution. Conversely, this method can be used to standardise sodium thiosulphate solutions.

(2) Use pure potassium iodide and perform the determination as described under the determination of potassium iodide.

Weak Iodine Solution *Determination of the percentage w/v of* I *and* KI

The free iodine content is determined by titration with 0.1M sodium thiosulphate solution. The total iodine and potassium iodide is determined on a separate portion of solution by acidification and titration with potassium iodate solution.

(a) *Titration with sodium thiosulphate solution*

$$I_2 + 2Na_2S_2O_3 \rightarrow Na_2S_4O_6 + 2NaI$$

$$\therefore \quad 0.01269 \text{ g } I_2 \equiv 1 \text{ ml } 0.1\text{M } Na_2S_2O_3$$

(b) *Titration with potassium iodate solution*

$$2I_2 + HIO_3 + 5HCl \rightarrow 5ICl + 3H_2O$$

$$2HI \text{ (from KI)} + HIO_3 + 3HCl \equiv 3ICl + 3H_2O$$

$$\therefore \quad 0.02538 \text{ g I} \equiv 1 \text{ ml } 0.05\text{M } KIO_3$$

$$\therefore \quad 0.01660 \text{ g KI} \equiv 1 \text{ ml } 0.05\text{M } KIO_3$$

Calculation of I *and* KI *content in the solution.* The potassium iodate reacts with both the iodine and the potassium iodate whereas the sodium thiosulphate solution reacts only with the iodine.

From equation (a), 0.02538 g I ≡ 2 ml 0.1M $Na_2S_2O_3$
From equation (b), 0.02538 g I ≡ 1 ml 0.05M KIO_3

Therefore, using the same weight of iodine, the burette reading of 0.1M $Na_2S_2O_3$ will be twice the burette reading of 0.05M KIO_3, the solutions being calculated as exactly 0.1M and exactly 0.05M, respectively. Therefore, if equal volumes of Weak Iodine Solution are used for the sodium thiosulphate titration and for the potassium iodate titration, the volume of exactly 0.05M KIO_3 solution required by the free I_2 present is equal to half the volume of exactly 0.1M $Na_2S_2O_3$ required in the sodium thiosulphate titration.

0.05M KIO_3 required for KI in sample =
volume of 0.05M KIO_3 − $\frac{1}{2}$ volume 0.1M $Na_2S_2O_3$

Method Use 10 ml of sample for both sodium thiosulphate and the potassium iodate titrations. Drain the pipette for 30 s (ethanolic solution).

(a) *Sodium thiosulphate titration.* Dilute 10 ml with *water* (20 ml) and titrate with 0.1M sodium thiosulphate until colourless.
(b) *Potassium iodate titration.* Pipette the sample (10 ml), add *water* (30 ml) and concentrated hydrochloric acid (50 ml) and titrate with 0.05M KIO_3 as described for *Potassium Iodide* (p.191).

COGNATE DETERMINATION

Aqueous Iodine Solution. Determination of the percentage w/v of I *and* KI.

For these determinations 20 ml of the diluted sample are taken for the sodium thiosulphate titration and 10 ml of diluted sample for the potassium iodate titration. Therefore, one quarter of the 0.1M (exact) $Na_2S_2O_3$ required must be subtracted from the 0.05M (exact) KIO_3 needed, to obtain the volume of 0.05M KIO_3 required by the potassium iodide.

Sodium Diatrizoate *Determination of the percentage of* $C_{11}H_8I_3N_2NaO_4$

The iodine is in organic combination and, before it can be titrated with potassium iodate solution, it must be converted to an ionised form by boiling with sodium hydroxide and zinc powder under reflux to produce zinc iodide. Water is added and the solution filtered and cooled. Concentrated hydrochloric acid is added and the liberated HI titrated with 0.05M KIO_3, using chloroform as indicator.

COGNATE DETERMINATIONS

Iodipamide Meglumine Injection
Iodised Oil Fluid Injection
Meglumine Diatrizoate Injection
Meglumine Iothalamate Injection
Sodium Diatrizoate Injection
Sodium Iothalamate Injection

Isoniazid *Determination of the percentage of* $C_6H_7N_3O$

This is determined by the addition of potassium bromide and direct titration in acid solution with potassium bromate. Bromine is released as the titration proceeds and reacts with isoniazid as shown below. The azo dye ethoxychrysoidine used as indicator becomes decolorised by oxidation at the end point.

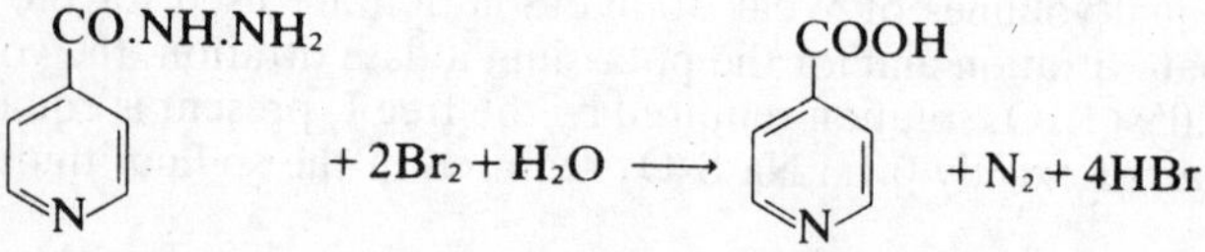

Ammonium cerium(IV) sulphate titrations

Cerium(IV) sulphate is a powerful oxidising agent in acid solution (1–8N). It is bright yellow in colour, both in the solid state and in solution, and since the corresponding cerium(IV) salt formed by reduction is colourless, strong solutions are self indicating. However, usually 0.1M solutions are used and this concentration is too dilute for observation of the end point without the addition of a suitable indicator. The oxidation reaction is represented as follows:

$$Ce^{4+} + e \rightleftharpoons Ce^{3+}$$

As a practical oxidising agent cerium(IV) sulphate possesses a number of advantages over potassium permanganate. As the solutions are not so highly coloured, observation of the meniscus in burettes, pipettes and other volumetric apparatus is less difficult. More important, the solutions are stable in air over long periods, they do not require protection from light and can even be heated for short periods without change in composition. Cerium(IV) sulphate, unlike potassium permanganate, can be used as an oxidising agent in the presence of high concentrations of hydrochloric acid.

Preparation of volumetric solutions of ammonium cerium(IV) sulphate

Ammonium cerium(IV) sulphate $Ce(SO_4)_2 2(NH_4)_2SO_4, 2H_2O$ contains not less than 95% of the pure cerium(IV) salt and volumetric solutions may be prepared by dissolving the appropriate weight (see below) in 2M sulphuric acid.

Since the oxidation reaction is given by:

$$Ce^{4+} + e \rightleftharpoons Ce^{3+}$$

$\therefore$ 632.6 g $Ce(SO_4)_2 2(NH_4)_2SO_4, 2H_2O \equiv$ 1000 ml M

$\therefore$ 63.26 g $Ce(SO_4)_2 2(NH_4)_2SO_4, 2H_2O \equiv$ 1000 ml 0.1M ammonium cerium(IV) sulphate

Standardisation of 0.1M ammonium cerium(IV) sulphate solution

The standardisation is based upon the oxidation of sodium arsenite to arsenate by the ceric component of the salt, using ferroin as indicator. The reaction is slow in the absence of a catalyst and osmic acid is used to increase the reaction rate.

$$NaAsO_2 + 2H_2O \rightarrow NaH_2AsO_4 + 2H^+ + 4e$$
$$4[Ce^{4+} + e \rightarrow Ce^{3+}]$$
$$\therefore \quad 4Ce^{4+} \equiv NaAsO_2$$
$$\therefore \quad 4 \times 632\text{ g } Ce(SO_4)_2 2(NH_4)_2SO_4,2H_2O \equiv 1000\text{ ml M } NaAsO_2$$
$$\therefore \quad 0.0632\text{ g } Ce(SO_4)_2 2(NH_4)_2SO_4,2H_2O \equiv 1\text{ ml } 0.025\text{M } NaAsO_2$$

Method Prepare a standard solution (0.025M) of sodium arsenite, by dissolving 1.2365 g of arsenic trioxide in a mixture of NaOH (5M; 5 ml) and *water* (5 ml) in a graduated flask, and dilute to 200 ml with *water*. Add hydrochloric acid (2M) until the solution is neutral. Dissolve sodium hydrogen carbonate (0.5 g) in the solution and dilute with *water* to 250 ml. Titrate the ammonium cerium(IV) solution (approx. 0.1M 25 ml) with 0.025M sodium arsenite in the presence of osmic acid solution (0.15 ml) and ferroin sulphate solution (0.1 ml) as indicator.

Ferrous Gluconate *Determination of the percentage of total iron*

The determination depends on a preliminary reduction with Zn/H_2SO_4 to convert any iron(III) which may be present to iron(II), followed by titration with ammonium cerium(IV) sulphate (or nitrate).

$$Fe^{2+} \rightarrow Fe^{3+} + e$$
$$Ce^{4+} + e \rightarrow Ce^{3+}$$
$$\therefore \quad C_{12}H_{22}O_{14}Fe \equiv Fe \equiv 1000\text{ ml M}$$
$$\therefore \quad 55.85\text{ g Fe} \equiv 1000\text{ ml M}$$
$$\therefore \quad 0.005585\text{ g Fe} \equiv 1\text{ ml ammonium cerium(IV) sulphate}$$

A separate limit test is included to control the amount of ferric iron which may be present. This method leading to the expression of the standard in terms of total iron is not satisfactory, since the effective valency state in iron therapy is iron(II) and it is generally considered desirable to express dosage in terms of this entity. This can be determined by direct titration of the sample with ammonium cerium(IV) sulphate without prior reduction of iron(III).

Method Accurately weigh the sample (1.5 g) in a conical flask and dissolve in a mixture of *water* (75 ml) and sulphuric acid (M; 15 ml). Add zinc powder. Allow to stand for about 20 min until the solution is colourless. Filter through a No. 4 sintered-glass filter covered with a thin layer of zinc powder and wash the flask and filter with CO_2-free *water* (*c.* 20 ml). Add ferroin sulphate solution (0.2 ml) to the combined filtrate and washings, then titrate with 0.1M ammonium cerium(IV) sulphate. The end point is reached when the colour changes from orange to green.

COGNATE DETERMINATIONS

Ferrous Fumarate, and *Tablets*
Ferrous Gluconate Tablets
Ferrous Succinate, and *Tablets*
Ferrous Sulphate, and *Tablets*

Titanium Dioxide *Determination of the percentage of* TiO_2

The determination depends upon:

(a) Conversion of the titanium dioxide (TiO_2) to titanic sulphate by heating with sodium hydrogen sulphate in sulphuric acid.

(b) Quantitive reduction of the titanic sulphate in sulphuric acid solution by amalgamated zinc in a Jones reductor to titanous sulphate:

$$Zn \rightarrow Zn^{2+} + 2e$$
$$Ti^{4+} + e \rightarrow Ti^{3+}$$

(c) Treatment of the resulting titanous sulphate solution with excess acidified ammonium iron(III) sulphate, which is reduced to ammonium iron(II) sulphate:

$$Ti^{3+} + Fe^{3+} \rightarrow Ti^{4+} + Fe^{2+}$$

(d) Titration of the iron(II) solution with

$$Fe^{2+} \rightarrow Fe^{3+} + e$$
$$Ce^{4+} + e \rightarrow Ce^{3+}$$
$$\therefore \quad TiO_2 \equiv Ti^{4+} \equiv Ti^{3+} \equiv Fe^{2+} \equiv Ce^{4+}$$
$$\therefore \quad 79.9\text{ g } TiO_2 \equiv 1000\text{ ml M}$$
$$\therefore \quad 0.00799\text{ g } TiO_2 \equiv 1\text{ ml } 0.1\text{M ammonium cerium(IV) sulphate}$$

The Jones reductor shown in Fig. 7. consists, essentially, of a column fitted with a sintered or porcelain disc near the base and a tap to control the outflow. The outlet tube passes through a bung into a Buchner flask which can be attached to a source of vacuum, so that the flow-rate can be increased if necessary. The column which should be almost 2 cm in diameter, is packed with amalgamated zinc to a height of about 36 cm. Amalgamated zinc is prepared from pure granulated zinc (AR) by shaking with 2% mercury(II) nitrate solution containing nitric acid and then washing with water.

Method Accurately weigh the sample (about 0.5 g) into a conical flask. Add anhydrous sodium sulphate (5 g) and *water* (10 ml). Mix, add sulphuric acid (10 ml) and boil gently until clear. Cool and slowly add 25% v/v sulphuric acid (30 ml), transfer to graduated flask and dilute with *water* to 100 ml.

Activate the Jones reductor as follows. Pass 100 ml of M sulphuric acid through the column and wash the column with *water* (100 ml). Repeat the procedure collecting both together in the receiving flask containing ammonium iron(III) sulphate solution (15% w/v in a 25% v/v solution of dilute sulphuric acid; 50 ml). Titrate the resulting solution with 0.1M ammonium cerium(IV) sulphate.

Reactivate the column with M H_2SO_4 (100 ml). Pass 20 ml of the titanium solution through the column, followed by H_2SO_4 (100 ml) and *water* (100 ml) collecting as before in 50 ml of acidified ferric alum solution. Titrate with 0.1M ammonium cerium(IV) sulphate. The difference between the two titrations gives the amount of ammonium cerium(IV) sulphate required by the titanium dioxide.

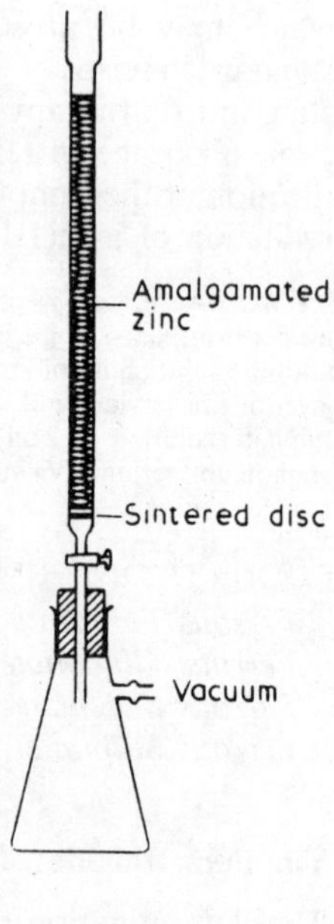

Fig. 7.1 Jones reductor

8
Precipitation and complexometric methods

Argentimetric titrations

Introduction

Quantitive precipitation can be used for volumetric determinations, provided that the point at which precipitation is complete can be determined. Thus, when silver nitrate solution is run into a solution of sodium chloride a precipitate of silver chloride is formed:

$$NaCl + AgNO_3 \rightarrow AgCl\downarrow + NaNO_3$$

The end point is the point at which all the chloride has been precipitated as silver chloride. However, detection of complete precipitation by observing the point at which addition of reagent causes no further precipitation is too tedious in practice and use is usually made of a chemical reaction to give a coloured precipitate or coloured solution at the end point. For instance, in the above example potassium chromate is added to the solution. As the silver nitrate solution is added, the chloride is precipitated as silver chloride, but when all the chloride has been precipitated, the next drop of silver nitrate causes precipitation of red chromate, indicating that the end point has been reached (p.98).

Preparation and standardisation of 0.1M silver nitrate solution

The standardisation depends upon the reactions expressed by the following equations:

$$AgNO_3 + NaCl \rightarrow AgCl\Rightarrow + NaNO_3$$
$$\therefore \quad AgNO_3 \equiv NaCl$$
$$\therefore \quad 169.9\text{ g } AgNO_3 \equiv 58.4\text{ g NaCl} \equiv 1000\text{ ml M}$$
$$\therefore \quad 0.01699\text{ g } AgNO_3 \equiv 0.00584\text{ g NaCl} \equiv 1\text{ ml } 0.1\text{M } AgNO_3$$

Method AnalaR silver nitrate contains 99.9 to 100% $AgNO_3$ and can, therefore, be used as a primary standard. A 0.1M solution can be prepared by dissolving 16.99 g of silver nitrate in *water* and adjusting to 1000 ml. A solution may be standardised against a standard solution of sodium chloride using potassium chromate as indicator (Mohr's method).

Dry some AnalaR sodium chloride at about 300° for 2 h, cool in a desiccator and use to prepare an accurate 0.1M solution of sodium chloride. Pipette 25 ml of the 0.1M sodium chloride solution into a conical flask, add potassium chromate solution (5% w/v; 1 ml) and titrate with the silver nitrate solution. *Do not dilute* the solution during the titration, but shake continuously. If these instructions are carried out, coagulation of the precipitated silver chloride usually occurs just before the end point. Continue to add the silver nitrate dropwise, shaking the flask well, until a permanent *faint reddish-brown* colour is obtained.

Mohr's method, described above, is only applicable if the solution to be titrated is neutral. Acid solutions may be determined if they are first neutralised with chloride free calcium carbonate or sodium bicarbonate.

Direct titration with silver nitrate

Sodium Chloride intravenous infusion *Determination of the percentage w/v* NaCl

The sodium chloride content is determined as in the standardisation of silver nitrate solution, using a volume of the sample equivalent to 0.2 g NaCl and titrating with 0.1M $AgNO_3$.

Chloral Hydrate *Determination of the percentage of* $CCl_3CH(OH)_2$

This determination depends upon the reactions expressed by the following equation:

$$CCl_3CH(OH)_2 + NaOH \rightarrow CHCl_3 + HCOONa + H_2O \quad (1)$$

$$\therefore \quad CCl_3CH(OH)_2 \equiv NaOH$$

$$\therefore \quad 165.4\text{ g } C_2H_3O_2Cl_3 \equiv 1000\text{ ml M}$$

$$\therefore \quad 0.1654\text{ g } C_2H_3O_2Cl_3 \equiv 1\text{ ml M NaOH}$$

Since the chloroform produced also reacts with the alkali to some extent, addition of alkali followed by back-titration does not give the correct result.

$$CHCl_3 \xrightarrow{\text{alkali}} \underbrace{HCOOH + 3HCl}_{\downarrow\ 4NaOH} \rightarrow HCOONa + 3NaCl + 4H_2O \quad (2)$$

Therefore, in reaction (2)

$$CHCl_3 \equiv 4NaOH$$

The ionised chloride produced in this side reaction can be determined by titration with 0.1M silver nitrate solution and a correction applied to the alkali titration reading to allow for this side reaction. From eq.(2)

$$CHCl_3 \equiv 4NaOH \equiv 4000\text{ ml M solution}$$

Also:

$$3NaCl \equiv 30\,000\text{ ml } 0.1M$$

Thus, $\frac{2}{15}$ of the volume of 0.1M $AgNO_3$ required ($\frac{2}{15}$ of 30 000 = 4000) will give the volume of M NaOH which reacted with chloroform according to eq.(2).

The sample is dissolved in excess M NaOH. After standing for 2 min (reaction 1), the excess alkali is back-titrated with 0.5M H_2SO_4 using phenolphthalein as indicator. Because reaction (2) has occurred to some extent, the neutral solution is titrated with 0.1M $AgNO_3$ using potassium chromate as indicator. A volume equivalent to $\frac{2}{15}$ of the amount of 0.1M (exact) $AgNO_3$ required gives the volume of M NaOH taken up by reaction (2). This value plus the amount of M H_2SO_4 required by the previous back-titration when subtracted from the volume of M NaOH originally added, gives the amount of M NaOH required by the sample in the reaction expressed by eq.(1).

COGNATE DETERMINATION

Dichloralphenazone

Potassium Iodide

Bromides and iodides cannot be titrated directly with silver nitrate by Mohr's method, since the colour of the precipitate interferes with the end point. Potassium iodide can, however, be determined by direct titration with silver nitrate using iodine and starch as indicator.

$$KI + AgNO_3 \rightarrow KNO_3 + AgI\downarrow$$

$$KI \equiv AgNO_3$$

$$\therefore\ 166\ g\ KI \equiv 1000\ ml\ M\ AgNO_3$$

$$\therefore\ 0.0166\ g\ KI \equiv 1\ ml\ 0.1M\ AgNO_3$$

The titration is simply a double decomposition. At the end point, excess silver nitrate reduces the indicator, iodine, to iodide ion, destroying the purple starch-iodide complex and leaving a pale yellow solution.

Method Dissolve the sample (about 0.3 g), accurately weighed, in *water* (10 ml). Add iodide-free starch solution (10 ml, containing 1 drop of ethanolic iodine solution in 50 ml) and titrate with 0.1M $AgNO_3$ until the blue colour is dispersed and a pale yellow solution remains.

Sodium Nitroprusside *Determination of the percentage of* $Na_2Fe(CN)_5NO$

This determination depends on the reactions expressed by the following equations:

$$Na_2Fe(CN)_5NO + 2AgNO_3 \rightarrow Ag_2Fe(CN)_5NO + 2NaNO_3$$

$$\therefore\ Na_2Fe(CN)_5NO \equiv 2AgNO_3$$

$$\therefore\ 262\ g\ Na_2Fe(CN)_5NO \equiv 2000\ ml\ M\ AgNO_3$$

$$\therefore\ 0.01301\ g\ Na_2Fe(CN)_5NO \equiv 1\ ml\ 0.1M\ AgNO_3$$

Method Accurately weigh approximately 0.35 g and dissolve in *water* (100 ml). Add M H_2SO_4 (0.1 ml) and ethanol (96%; 20 ml), and titrate potentiometrically with 0.1M $AgNO_3$ using a silver-mercury(I) sulphate electrode system.

Ammonium thiocyanate titration of silver salts

Ammonium thiocyanate reacts with silver nitrate in nitric acid solution:

$$NH_4SCN + AgNO_3 \rightarrow AgSCN\downarrow + NH_4NO_3$$

The thiocyanate solution is always used in the burette and is run into the silver nitrate solution which has been acidified with nitric acid (nitrous acid-free, because nitrous acid gives a red colour with thiocyanate ion). The end point is detected by use of ammonium iron(III) sulphate since ferric ions give a deep red colour (ferric thiocyanate) with a trace of thiocyanate ion. The temperature of the solution must be kept below 25° because at higher temperatures the colour of the ferric thiocyanate complex fades.

$$\therefore \quad NH_4SCN \equiv AgNO_3$$

Standardisation of approximately 0.1M NH_4SCN solution

Ammonium thiocyanate solution is standardised against 0.1M silver nitrate solution.

Method Pipette a standard 0.1M $AgNO_3$ solution (25 ml) into a conical flask. Add concentrated nitric acid (5 ml) and ammonium iron(III) sulphate solution (10% w/v; 3 ml). Dilute with *water* (100–150 ml) since this sharpens the end point. Titrate with ammonium thiocyanate solution. A white precipitate of silver thiocyanate is formed as the thiocyanate solution is added and then a reddish-brown colour is produced which disappears upon shaking, leaving a white precipitate of silver thiocyanate. The end point is the permanent, faint reddish brown colour which does not disappear upon shaking.

COGNATE DETERMINATIONS

Silver Nitrate
Toughened Silver Nitrate

(Volhard's method)

Ammonium thiocyanate solution is used in conjunction with 0.1M silver nitrate solutions in the assay of substances which react with nitrate but which cannot be determined by direct titration with silver nitrate solution. Excess standard silver nitrate solution is added together with concentrated nitric acid and the excess silver nitrate titrated with 0.1M ammonium thiocyanate solution. This is called Volhard's method.

Bromides and iodides can be determined by Volhard's method without filtering off the silver halide formed, but in the case of chlorides it is usual either to filter off the silver chloride or to coagulate the precipitate by means of nitrobenzene or, preferably, dibutyl phthalate, which is non-toxic, because silver chloride reacts slowly with ammonium thiocyanate:

$$AgCl + NH_4SCN \rightarrow AgSCN\downarrow + NH_4Cl$$

This makes the end point rather flat since it involves the production of red iron(III) thiocyanate complex with the thiocyanate ions.

Sodium Chloride *Determination of the percentage of* NaCl

This determination depends upon the reactions expressed by the following equations:

$$NaCl + AgNO_3 \rightarrow AgCl\downarrow + NaNO_3$$
$$AgNO_3 + NH_4SCN \rightarrow AgSCN\downarrow + NH_4NO_3$$
$$\therefore\ NaCl \equiv AgNO_3$$
$$\therefore\ 58.44\text{ g } NaCl \equiv 1000\text{ ml M } AgNO_3$$
$$\therefore\ 0.005844\text{ g } NaCl \equiv 1\text{ ml } 0.1\text{M } AgNO_3$$

Method Dissolve the sample (about 0.2 g), accurately weighed, in *water* (35 ml), add dilute nitric acid (15 ml), dibutyl phthalate (5 ml) and 0.1M silver nitrate (50 ml by pipette) and shake vigorously for 0.5–1 min. Add ammonium iron(III) sulphate solution (5 ml) and titrate with 0.1M NH_4SCN, shaking well until a reddish-brown colour, which does not fade in 5 min, is obtained.

COGNATE DETERMINATIONS

Carbromal
Potassium Chloride
Sodium Chloride Tablets
Thiamine Hydrochloride. The total Cl content is determined by Volhard's method.

Chlorbutol *Determination of the percentage of* $CCl_3\cdot C(CH_3)_2\cdot OH,\frac{1}{2}H_2O$

Organically combined chlorine is converted by hydrolysis with sodium hydroxide to ionic chloride which can be determined by Volhard's method, in the presence of nitrobenzene.

$$CCl_3\cdot C(CH_3)_2\cdot OH + 3NaOH \rightarrow C(OH_3)_2\cdot C(CH_3)_2\cdot OH$$
$$\downarrow NaOH$$
$$(CH_3)_2\cdot C(OH)\cdot COONa$$
$$3NaCl + 3AgNO_3 \rightarrow 3NaNO_3 + 3AgCl$$
$$\therefore\ C_4H_7OCl_3 \equiv 3NaCl \equiv 3AgNO_3 \equiv 3000\text{ ml M}$$
$$\therefore\ 186.5\text{ g } C_4H_7Cl_3O\tfrac{1}{2}H_2O \equiv 3000\text{ ml M}$$
$$\therefore\ 0.00622\text{ g } C_4H_{76}Cl_3O\tfrac{1}{2}H_2O \equiv 1\text{ ml } 0.1\text{M } AgNO_3$$

Lindane (Gamma Benxene Hexachloride)

A preliminary hydrolysis is carried out with ethanolic potassium hydroxide, to convert organically combined chloride to potassium chloride. The latter is determined by Volhard's method after acidification of the hydrolysate with nitric acid.

COGNATE DETERMINATIONS

Cloxacillin Sodium Initial oxygen flask combustion (p.254).
Mustine Hydrochloride
Triclofos Sodium Initial ignition with anhydrous sodium carbonate.

Ammonium thiocyanate titration of mercury compounds

The determination of mercury salts with ammonium thiocyanate depends upon the reactions expressed by the following equation:

$$Hg(NO_3)_2 + 2NH_4SCN \rightarrow Hg(SCN)_2 + 2NH_4NO_3$$

This reaction is quantitative in the presence of nitric acid and the end point can be detected by using ammonium iron(III) sulphate as indicator.

Phenylmercuric Nitrate *Determination of the percentage of* $C_{12}H_{11}Hg_2NO_4$

The mercury must be freed from organic combination before it can be titrated with ammonium thiocyanate. The sample is refluxed with formic acid and zinc dust. A zinc amalgam is obtained. The excess of zinc powder and the amalgam are filtered off, washed with water and dissolved in nitric acid. Nitrous acid and reducing substances are removed by the addition of a little urea followed by sufficient 0.02M potassium permanganate to produce a faint pink colour. The slight excess of potassium permanganate is removed with a few drops of hydrogen peroxide solution and the mercuric nitrate solution titrated with NH_4SCN using ferric ammonium sulphate as indicator.

$$C_6H_5HgOH,C_6H_5HgNO_3 \rightarrow 2Hg/Zn \rightarrow 2Hg(NO_3)_2$$

$$2[Hg(NO_3)_2 + 2NH_4SCN \rightarrow Hg(SCN)_2 + 2NH_4NO_3]$$

$$\therefore \quad 634.4\text{ g } C_{12}H_{11}Hg_2NO_4 \equiv 4000\text{ ml M } NH_4SCN$$

$$\therefore \quad 0.01586\text{ g } C_{12}H_{11}Hg_2NO_4 \equiv 1\text{ ml } 0.1\text{M } NH_4SCN$$

COGNATE DETERMINATIONS

Phenylmercuric Borate

Thiomersal Organic matter is first destroyed by heating with concentrated sulphuric acid.

Theory of complexometric analysis

Introduction

A complexing agent is any electron-donating ion or molecule, usually called a **ligand**, which, by its ability to form one or more covalent or dative bonds with the metal ion, produces a complex which has different properties from those of the free metal ion. Thus, the metal may not be precipitated from the complex by the usual metal ion precipitants. The decomposition potential of the complex at the dropping mercury cathode, also, is usually more negative than that of the free metal ion, and where the ion can exist in different valency states, the redox potential is reduced if the complexing agent combines more avidly with the cation of higher valency. Moreover, if the higher valency state of the ion is stabilised by complex formation, it will be more difficult to reduce. For example, if potassium iodide solution is acidified with acetic acid and treated with copper(II) sulphate solution, iodine is liberated and copper(I) iodide precipitated. If, however, sodium tartrate is added to the copper(II) sulphate solution first, there is no reduction to the copper(I) state and no iodine liberated. This is because the copper(II) ion is stabilised as the cupritartrate complex, which is not reduced by iodide. Similarly, an acidic solution of iron(III) chloride is not reduced by potassium iodide if excess sodium edetate (see below) is first added. Moreover, iron(II) in the presence of excess edetate is actually oxidised by iodine to the iron(III) complex.

The stability of complexes varies greatly and the greater the stability, the more marked will be the differences in properties from those of the original cations. When potassium iodide is added to a suspension of red mercuric iodide, the latter will pass into solution as the colourless complex $2K^+(HgI_4)^{2-}$, the dissociation of which is so slight that there are insufficient mercury(II) ions to precipitate as mercury(II) oxide in the presence of alkali. Potassium hexacyanoferrate(III) $K_3[Fe(CN)_6]$, in which the iron is in the iron(III) state, does not yield iron(III) hydroxide in the presence of alkali in spite of the sparing solubility of the latter (solubility product = 1.1×10^{-36}), showing that the dissociation of the complex into free iron(III) ions is negligible. Ammonium iron(III) citrate is a complex since it does not yield iron(III) hydroxide with ammonia, although it does so with sodium hydroxide. The removal by ignition of organic radicals such as citrate, tartrate and lactate in group analysis before passing to group III is essential because of their tendency to prevent precipitation of certain metal ions under the usual conditions owing to complex formation. Great difference in stability of complexes is illustrated in the group II separation of copper and cadmium, in which hydrogen sulphide is passed into a mixture of their complex cyanides, $K_3(Cu(CN)_4]$ (copper stabilised in the Cu(II) state) and $K_2[Cd(CN)_4]$. Sufficient free cadmium ions are present for the solubility product of cadmium sulphide to be reached and it is precipitated; the Cu(II) complex on the other hand is extremely stable, so that Cu(II) sulphide is not precipitated under these conditions, although its solubility product is extremely low (2×10^{-47}).

Neutral molecules with lone pairs of electrons, e.g. $\ddot{N}H_3$, will also form complexes, as illustrated by the familiar copper(II) ammonium ion, $[Cu(NH_3)_4]^{2+}$, and cobaltammine, $Co[(NH_3)_6]^{3+}$. Generally speaking, the electron-donating atoms are N, S, O and the halogens and, possibly, carbon in such groups as CO and $(CN)^-$

Bonding in complexes

The bonds are either ordinary covalent bonds in which both the metal and the ligand contribute one electron each, or co-ordinate bonds in which both electrons are contributed by the ligand. Thus, the hexacyanoferrate ion may be considered to consist of three ordinary covalent bonds and three co-ordinate bonds (Fig. 8.1a), although in the complex the bonds are identical hybrid bonds (Fig. 8.1b) which have been shown to be directed towards the apices of a regular octahedron (Fig 8.1c).

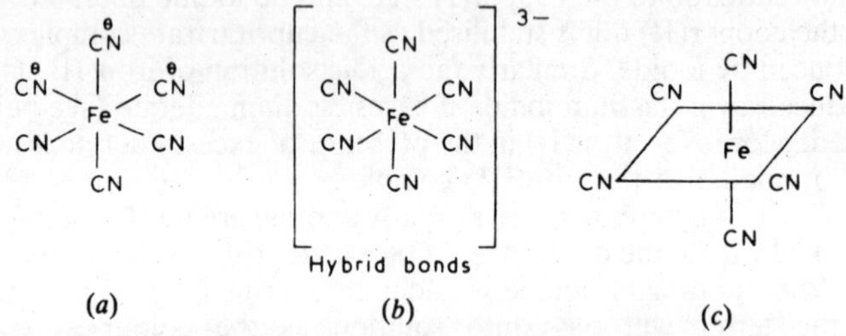

Fig. 8.1 The hexacyanoferrate iron(III) ion

The negative charge on the complex ion is equal to the total number of the negative groups *minus* the valency of the metal ion. When neutral groups only are involved, the charge on the complex is positive and is equal to the valency of the metal ion, e.g. $[Cu(NH_3)_4]^{2+}$

Werner's co-ordination number and electronic structure of complex ions

Werner (1891) first noticed that for each atom there is an observed maximum number of small groups which can be accommodated around it. This number, which is called Werner's co-ordination number, depends purely upon steric factors and is in no way related to the valency of the ion. Thus, although the valency shell of the elements of the third period is theoretically capable of expanding up to 18 electrons, and that of the fourth to 32 electrons, there is, in practice, a limit to the number of small groups which can be accommodated owing to limitations of space around the ion. In the second period, the maximum number is 4 and in the third period it is 6; with certain elements such as tungsten and molybdenum it is 8. For example, in the $[BF_4]^-$ ion, the octet is completed and the maximum co-ordination number is reached, but in the $[AlF_6]^-$ ion the outer shell contains 12 electrons and cannot expand to the maximum number of 18 electrons since the maximum co-ordination number has been reached.

Within the limits imposed by Werner's co-ordination number, there is a tendency for the metal to attain or approach inert gas structure, and this is probably the driving force for complex formation. The examples in Table 8.1 illustrate this point.

Table 8.1 Electronic configuration of metal ions and complex ions

Metal ion	Electronic configuration	Complex ion	Electronic configuration of complex ion
Co^{3+}	2, 8, 14	$[Co(NH_3)_6]^{3+}$	2, 8, 18, 8
Cu^{2+}	2, 8, 17	$[Cu(NH_3)_4]^{2+}$	2, 8, 17, 8
Cu^{+}	2, 8, 18	$[Cu(CN)_4]^{3-}$	2, 8, 18, 8
Hg^{2+}	2, 8, 18, 32, 18	$[Hg(CNS)_4]^{2-}$	2, 8, 18, 32, 18, 8
Fe^{2+}	2, 8, 14	$[Fe(CN)_6]^{4-}$	2, 8, 18, 8
Fe^{3+}	2, 8, 13	$[Fe(CN)_6]^{3-}$	2, 8, 17, 8

Chelating agents

Complexes involving simple ligands, that is, those forming only one bond are described as **co-ordination compounds**. Ligands having more than one electron-donating group are called **chelating agents** (from a Greek word meaning *claw*), since they combine with metal ions in a manner reminiscent of a crab's claw to form cyclic structures. Thus, 1,2-diaminoethane (ethylenediamine) behaves like two ammonia molecules and forms *chelates* with copper and cobalt ions:

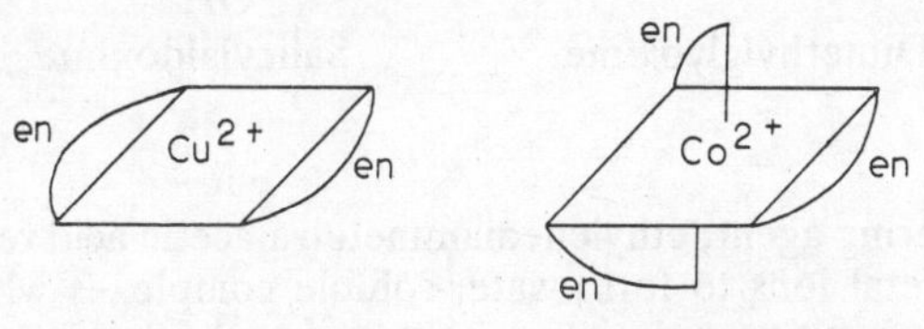

en = ethylenediamine ($H_2NCH_2CH_2NH_2$)

The copper ion is common to two five-membered rings and the cobalt ion to three five-membered rings. The latter has no plane of symmetry and, thus, exists in two optically active forms. There is no fundamental difference between co-ordination and chelate compounds except that in the latter, ring size influences stability. Five- or six-membered rings are the most stable and are readily formed, although both larger and smaller rings may also be formed.

Many organic compounds will chelate metals if they contain groups with an easily replaceable proton (−COOH, phenolic and enolic OH), or neutral groups offering a lone pair of electrons (NH_2, CO and alcoholic OH), and the structures of the molecules are such as to permit the formation of stable rings. The greater the number of rings which can be formed, the more stable the chelate is likely to be. Most rings formed in chelates involve the highest valency state of a metal since these are more stable than those involving lower valency states.

The solubility of metal chelates in water depends upon the presence of hydrophilic groups such as COOH, SO_3H, NH_2 and OH. When both acidic and basic groups are present, the complex will be soluble over a wide range of pH. When hydrophilic groups are absent, the solubilities of both the chelating agent and the metal chelate will be low, but they will be soluble in organic solvents. The term *sequestering agent* is generally applied to chelating agents which form water-soluble complexes with bi- or polyvalent metal ions. Thus, although the metals remain in solution, they fail to give normal ionic reactions. Ethylenediaminetetra-acetic acid is a typical sequestering agent, whereas dimethyglyoxime and salicylaldoxime are chelating agents, forming insoluble complexes.

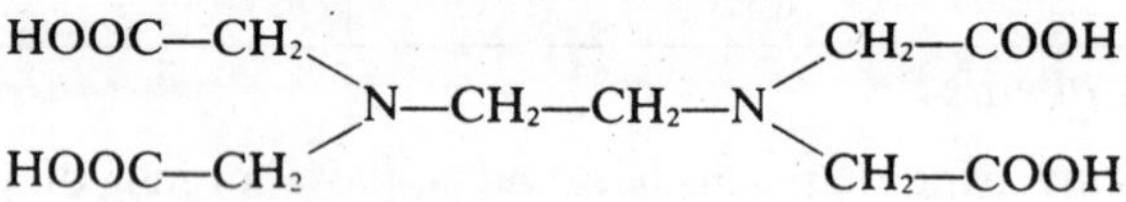

Ethylenediamine tetra-acetic acid

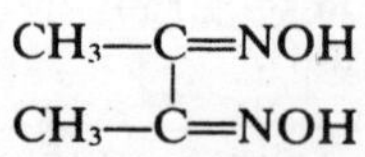

Dimethylglyoxime

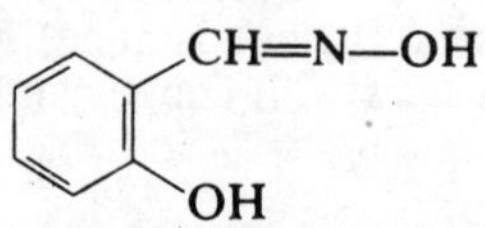

Salicylaldoxime

As a sequestering agent, ethylenediaminetetra-acetic acid reacts with most polyvalent metal ions to form water-soluble complexes which cannot be extracted from aqueous solutions with organic solvents. Dimethylglyoxime and salicylaldoxime form complexes which are insoluble in water, but soluble in organic solvents; for example, nickel dimethylglyoxime has a sufficiently low solubility in water to be used as a basis for gravimetric assay.

Nature and stability of metal complexes of ethylenediaminetetra-acetic acid

Ethylenediaminetetra-acetic acid forms complexes with most cations in a 1:1 ratio, irrespective of the valency of the ion:

$$M^{2+} + [H_2X]^{2-} \rightarrow [MX]^{2-} + 2H^+$$
$$M^{3+} + [H_2X]^{2-} \rightarrow [MX]^{-} + 2H^+$$
$$M^{4+} + [H_2X]^{2-} \rightarrow [MX] + 2H^+$$

where M is a metal and $[H_2X]^{2-}$ is the anion of the disodium salt (*disodium edetate*) which is most frequently used. The structures of these complexes with di-, tri- and tetravalent metals contain three, four and five rings, respectively:

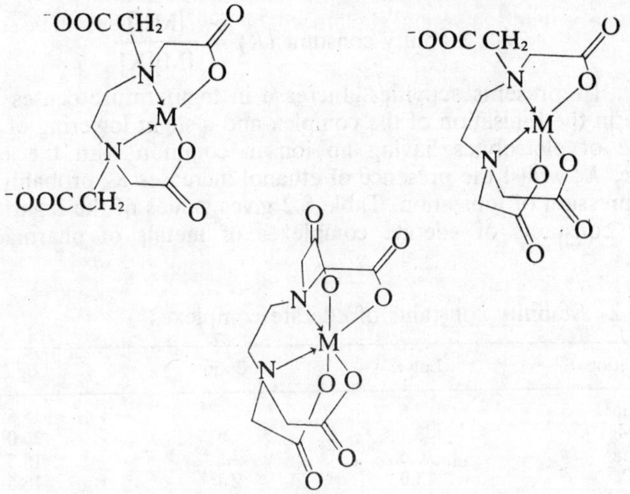

Effect of pH on complex formation

Edetic acid ionises in four stages ($pK_1 = 2.0$, $pK_2 = 2.67$, $pK_3 = 6.16$ and $pK_4 = 10.26$) and, since the actual complexing species is Y^{4-}, complexes will form more efficiently and be more stable in alkaline solution. If, however, the solubility product of the metal hydroxide is low, it may be precipitated if the hydroxyl ion concentration is increased too much. On the other hand, at lower pH values when the concentration of Y^{4-} is lower, the stability constant of the complexes will not be so high. Complexes of most divalent metals are stable in ammoniacal solution. Those of the alkaline earth metals, such as copper, lead and nickel, are stable down to pH 3 and hence can be titrated selectively in the presence of alkaline earth metals. Trivalent metal complexes are usually still more firmly bound and stable in strongly acid solutions; for example, the cobalt(III) edetate complex is stable in concentrated hydrochloric acid. Although most complexes are stable over a fair range of pH, solutions are usually buffered at a pH at which the complex is stable and at which the colour change of the indicator is most distinct.

Colour of complexes

There is always a change in the absorption spectrum when complexes are formed and this forms the basis of many colorimetric assays.

Stability of complexes

The general equation for the formation of a 1:1 chelate complex, MX, is

$$M + X \rightleftharpoons MX$$

where M is the metal ion and X the chelating ion

$$\therefore \quad \text{Stability constant } (K) = \frac{[MX]}{[M][X]}$$

where [ ] represents activities. Increase in temperature causes a slight increase in the ionisation of the complex and a slight lowering of K. The presence of eletrolytes having no ion in common with the complex decreases K, whilst the presence of ethanol increases K, probably due to the suppression of ionization. Table 8.2 gives values of the logarithms of stability constants of edetate complexes of metals of pharmaceutical interest.

Table 8.2 Stability constants of edetate complexes

Cation	Log K	Cation	Log K
Ba^{2+}	7.8	Co^{2+}	15.5
Mg^{2+}	8.7	Co^{3+}	26.0
Ca^{2+}	10.6	Al^{3+}	15.5
Cr^{2+}	13.0	Zn^{2+}	16.5
Cr^{3+}	24.0	Pb^{2+}	17.0
Mn^{2+}	13.5	Ni^{2+}	18.0
Fe^{2+}	14.3	Cu^{2+}	18.0
Fe^{3+}	25.1	Hg^{2+}	21.0

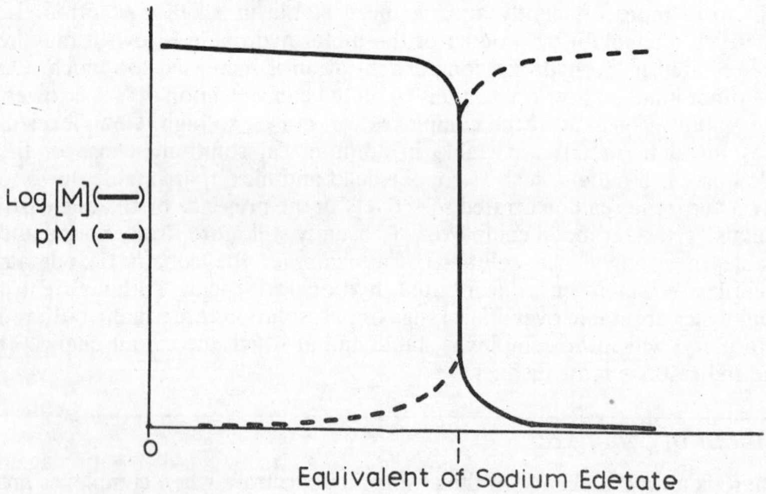

Fig. 8.2 Change of p[M] during complex formation with disodium edetate

Titration of metal ions using disodium edetate

Since edetic acid is only sparingly soluble in water (about 0.2%) the disodium salt is usually used (solubility about 10%). When disodium edetate solution is run into a solution of a metal ion buffered to promote efficient complex formation, the *rate of change* of concentration of metal ion is slow at first, but increases very rapidly as the amount of disodium edetate added approaches one equivalent, in the same way as hydrogen ion concentration changes during the titration of a strong acid with a strong base. Fig. 8.2 shows a plot of $\log[M^{n+}]$ and pM$\{\log 1/[M^{n+}]$ or, $-\log[M^{m+}]\}$ against equivalents of disodium edetate added.

End point detection in complexometric titration using pM indicators

The end point of complexometric titrations is shown by means of pM indicators. The concept of pM arises as follows:

If K is the stability constant,

$$K = \frac{[MX]}{[M][X]}$$

then

$$[M] = \frac{[MX]}{[X]K}$$

or

$$\log[M] = \log\frac{[MX]}{[X]} - \log K$$

and

$$pM = \log\frac{[X]}{[MX]} - pK$$

Therefore, if a solution is made such that $[X] = [MX]$, $pM = -pK$ (or $pM = pK'$, where K' = dissociation constant). This means that, in a solution containing equal activities of metal complex and free chelating agent, the concentration of metal ions will remain roughly constant and will be buffered in the same way as are hydrogen ions in a pH buffer. Since, however, chelating agents are also bases, equilibrium in a metal-buffer solution is often greatly affected by a change in pH. In general, for chelating agents of the amino acid type (e.g. edetic acid and ammoniatriacetic acid), it may be said that when $[X] = [MX]$, pM increases with pH until about pH 10, when it attains a constant value. This pH is, therefore, usually chosen for carrying out titrations of metals with chelating agents in buffered solution.

The pM indicator is a dye which is capable of acting as a chelating agent to give a dye-metal complex. The latter is different in colour from the dye itself and also has a lower stability constant than the chelate-metal complex. The colour of the solution, therefore, remains that of the dye complex until the end point, when an equivalent amount of sodium edetate has been added. As soon as there is the slightest excess of edatate, the metal-dye complex decomposes to produce free dye; this is accomplished by a change in colour.

The colours of dyes and of the metal complexes vary with pH. This fact together with complex stability, must be considered when deciding at which pH to carry out a titration. It is also essential to use a buffer solution to maintain the required pH during the titration.

pM indicators

Alizarin fluorine blue (alizarin complexone)

This may be used in acid solution at pH 4.3 for the titration of lead, zinc, cobalt(II), mercury(II) and copper(II) when the colour changes from red to yellow.

It is used in the determination of **fluorine** in *Triamcinolone* and *Fluocinolone Acetonides* and in *Fluocortolone Hexanoate* and *Pivalate*. The determination depends upon the fact that the red cerium(II) complex changes to blue in the presence of fluoride ions. The organically combined fluorine is converted to inorganic fluoride by burning the compound in oxygen and absorbing the gases in water (p.000). The complex is formed by the addition of alizarin fluorine blue and cerium(II) nitrate in the presence of a sodium acetate–acetic acid buffer and its intensity measured spectrophotometrically at 610 nm.

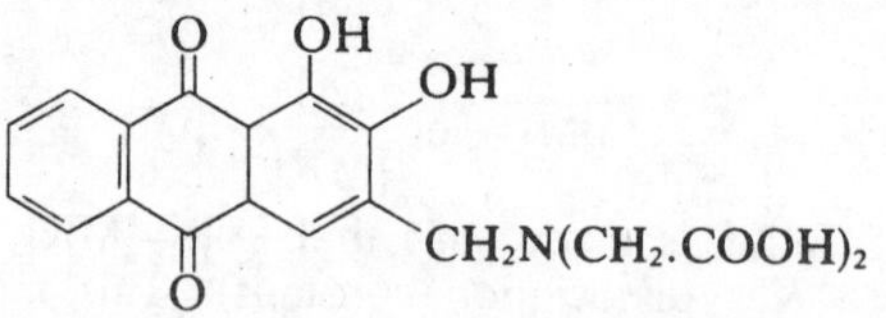

Calcon (mordant black 17)

This gives a reddish purple colour with calcium in alkaline solution and a blue colour in the absence of free (uncomplexed) calcium ions. It is used in the assay of *Calcium Carbonate*, *Calcium Chloride*, *Calcium Gluconate* and *Calcium Sodium Lactate*.

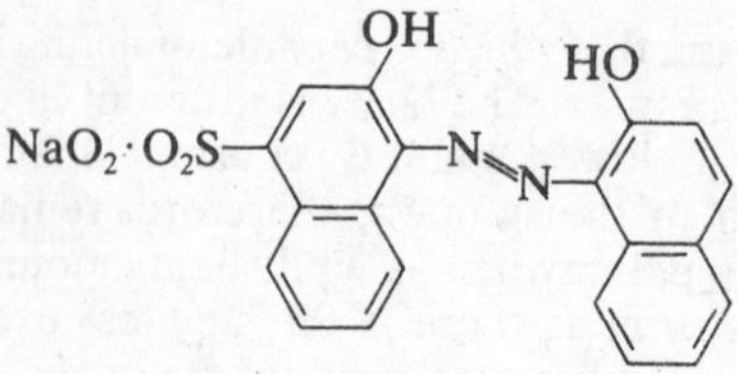

Calcon

Calcon carboxylic acid

This is also sensitive to uncomplexed calcium ions in alkaline solution, giving a reddish purple colour, changing to blue in their absence. It is used in the determination of *Calcium Lactate*.

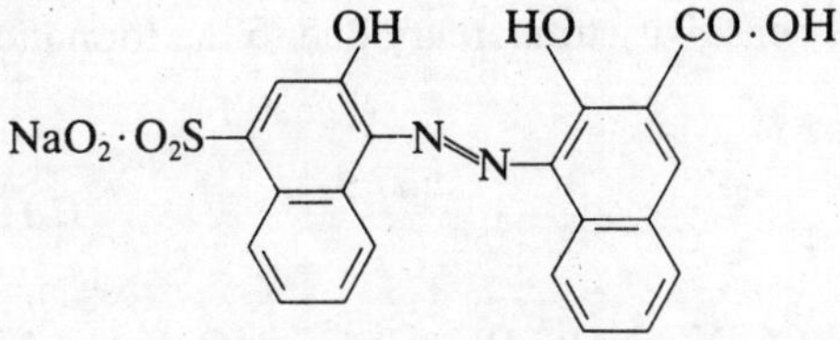

Calcon carboxylic acid

Catechol violet

The catechol groups of the molecule form weak, highly coloured complexes with a wider range of metals than mordant black II or murexide in neutral, acid and alkaline solution. These complexes may contain one or two metal atoms; all are blue in alkaline solution and the more stable ones are also blue in acid solution, except at low pH values when they change to purple or red. For example, the thorium(IV) complex is blue down to pH 3 and the bismuth(III) complex is blue down to pH 1.5. The indicator itself is red in acid solution below pH 1.5, when it changes colour to yellow owing to loss of proton from the sulphonic acid group, thence to violet at pH 6 owing to loss of a further proton and, finally, to reddish-purple at pH 9 owing to the loss of a third proton. In alkaline solution, clear end points are shown with magnesium, manganese, iron(II), cobalt, nickel, zinc, cadmium and calcium. Between pH 4 and 7, copper, lead and lanthanum give clear end points, whilst bismuth, thorium and antimony can be titrated in the pH range 1.5 to 4 and, hence, selective titrations may be carried out. Catechol violet also has the advantage that its aqueous solutions are highly stable.

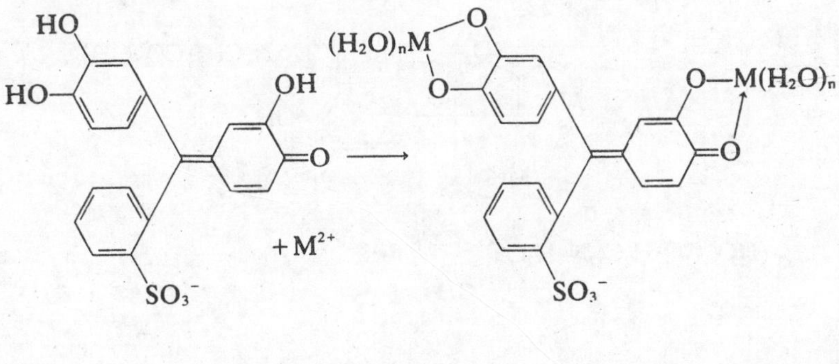

Diphenylcarbazone

This is used as an indicator for mercury in the determination of *Propylthiouracil* (p.253). It forms a violet-coloured complex with mercury-(II) ions, but the reaction is only specific for mercury in 0.2M nitric acid, provided that chromates and molybdates are absent. In neutral or slightly acid solution, copper, iron, cobalt and other ions give coloured complexes. Chlorides reduce sensitivity to mercury due to the formation of $(HgCl_4)^{2-}$ ions.

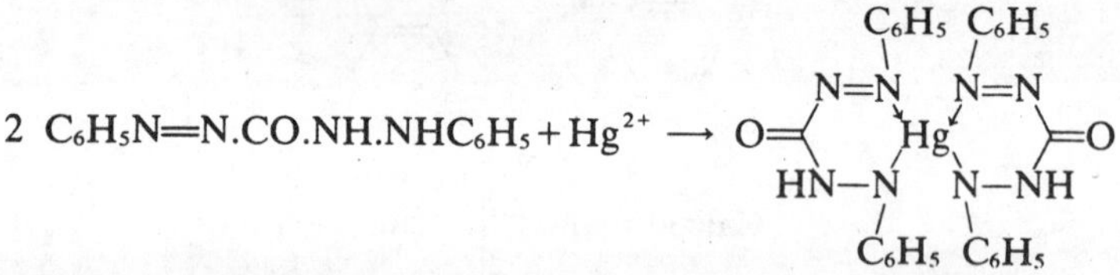

Dithizone (1,5-Diphenylthiocarbazone; Ph·N = N·CS·NH·NH·Ph)

This is a sensitive reagent for the detection and determination of lead(II), with which it forms a red, chloroform-soluble complex, and zinc, with which it forms a reddish-violet complex. It is used in the complexometric titration of *Dried Aluminium Phosphate*, in which aluminium is complexed with excess disodium edetate and the excess reagent back-titrated with zinc chloride solution.

Methyl thymol blue

This is derived from thymol blue. It can be used for bismuth in strongly acid solution (M nitric acid) and for lead, zinc, cadmium, mercury(II) and cobalt(III) in weakly acid solution (hexamine buffer). In alkaline solution, lead, zinc, cadmium, cobalt(II), magnesium and maganese(II) can be determined using an ammonium chloride–ammonia buffer at pH 10, and in still stronger alkali (0.05M sodium hydroxide) calcium, strontium and barium can be titrated. A 1% aqueous solution of the dye is stable and the colour changes from blue to yellow in acid solution and blue to grey in alkaline solution. It is used in the limit test for calcium in *Sulphobromophthalein Sodium*.

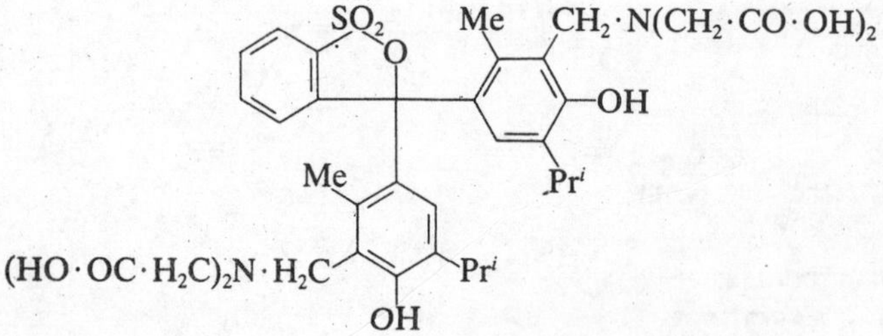

Mordant black II (Erichrome Black T, Solochrome Black T)

This indicator has wide application. It is blue at about pH 10 and most of its complexes are reddish. Below pH 6.3 and above 11.5, the dye itself is reddish like the complexes. It is, therefore, necessary to carry out the titration in the presence of a buffer at pH 10. The reaction with a divalent metal cation may be expressed as follows:

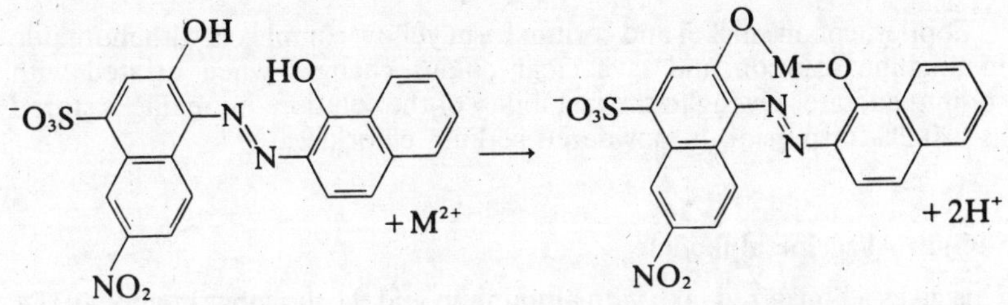

Blue (pH 10) Pink

Magnesium, calcium, cadmium, zinc, manganese, lead, lanthanum and mercury(II) may be titrated directly using this indicator. For barium and strontium, it is better to add excess sodium edetate and back-titrate with a standard magnesium chloride solution.

Cobalt, nickel, copper, aluminium, silver, titanium and platinum form more stable complexes with the dye than with the edetate and hence, the former cannot be used as an indicator. Iron(III), cerium(IV) and vanadate ions oxidise the dye, whilst Sn(II) and Ti(III)-ions reduce it. All these ions interfere and must be removed or masked (see below). For the same reason as murexide, mordant black II is used as a freshly prepared 0.2% dispersion in sodium chloride.

Mordant blue 3 (Solochrome cyanine R)

This gives a deep purple colour with aluminium ions in acidic solution and a pale pink colour in the absence of free (uncomplexed) aluminium ions.

Murexide (ammonium purpurate)

This is mainly used for the titration of calcium at pH 12. Calcium is bound in an eight-membered ring in the murexide chelate. The larger ring confers a lower stability than the usual five- and six-membered ring chelates. Since magnesium-murexide is less stable than the calcium complex, calcium can be titrated in the presence of magnesium. This is applied in water analysis (p.222).

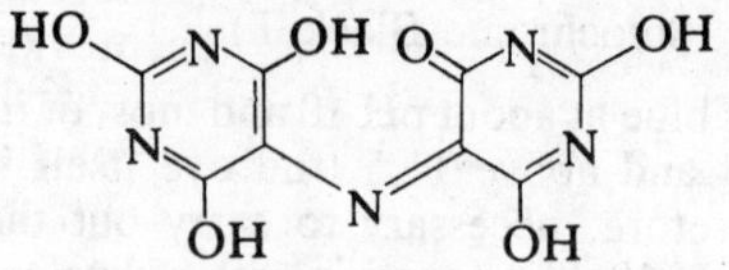

Murexide (violet at pH 12)

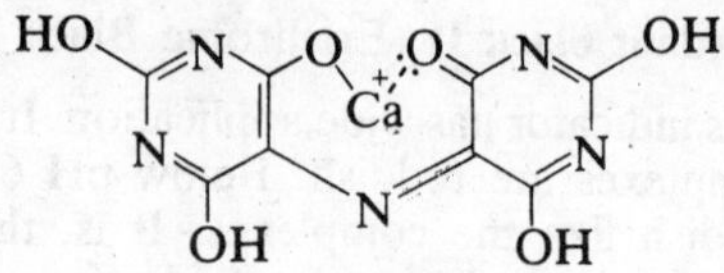

Calcium-murexide

Copper, cobalt, nickel and cerium form yellow complexes with murexide in alkaline solution and give clear colour changes when titrated with sodium edetate. Owing to the instability of the solution, murexide is stored as a 0.2% dispersion in powdered sodium chloride.

Sodium Alizarinesulphonate

This gives a bluish-red lake with aluminium and thorium ions at about pH 4 and is yellow at this pH in the absence of these ions. It has been used as an indicator for the determination of fluorine in *Triamcinolone* by titration with standard thorium nitrate solution.

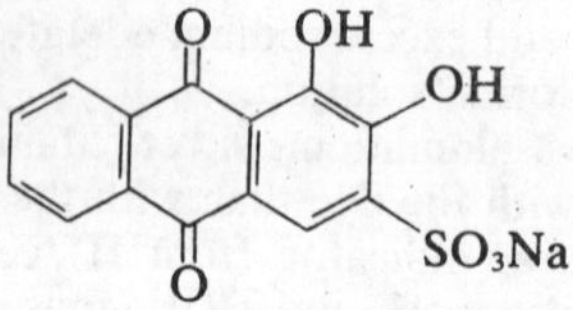

Tiron (disodium, 1,2-dihydroxyphenol-3,5-disulphonate)

This forms a blue colour with iron(III) between pH 2 and 5, a violet colour between pH 5.7 and 7 and a red complex in alkaline solution. The blue complex in acid solution is a suitable indicator for titration with disodium edetate, the colour changes being greenish-blue through green to yellow. Titrations are best carried out at 40 to 50°C. The red iron(III) complex in alkaline solution is stable and specific for ferric iron; it is sensitive to 1 in 200 million. It is not used as an indicator on edetate-iron titrations, but it can be used as the basis of a colorimetric assay for ferric iron.

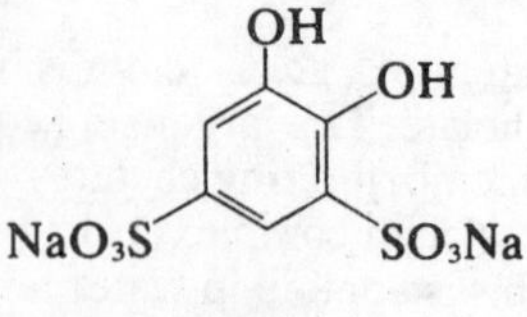

Tiron

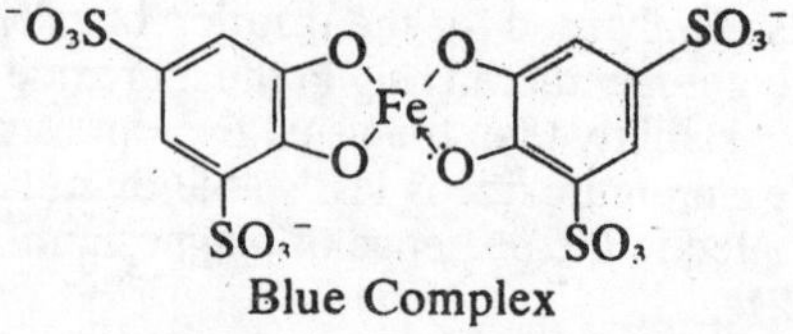

Blue Complex

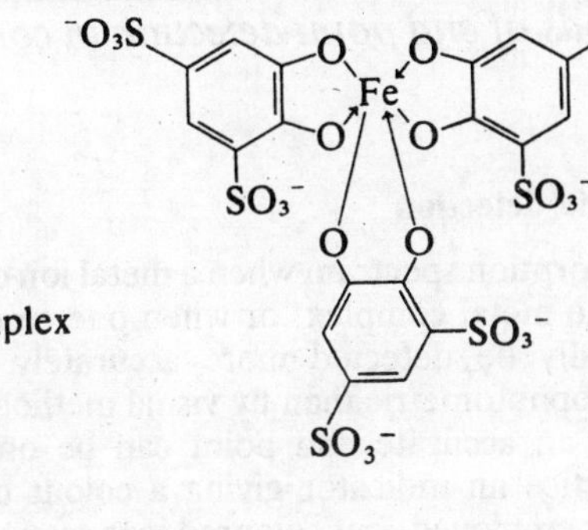

Red Complex

Xylenol Orange

This is an acid-base indicator in which two imino-acetic acid groups have been introduced into each cresol ring *ortho* to the phenolic hydroxyls to promote the formation of the complex. As an acid-base indicator, xylenol orange has the same characteristics as cresol red, i.e. lemon-yellow in acid solution and red in alkaline solution. The metal complexes are red. Hence it is restricted to titrating metals whose edetate compexes are stable in acid solution, e.g. bismuth, thorium, lead, lanthanum, cadmium and mercury. The stability of the complexes, however, varies with pH. Thus, from pH 1 to 3 bismuth and thorium can be titrated, from pH 4 to 5 lead and zinc, and from pH 5 to 6 cadmium and mercury can be titrated. Hexamine alone or with varying amounts of nitric, hydrochloric or acetic acid is a suitable buffering agent.

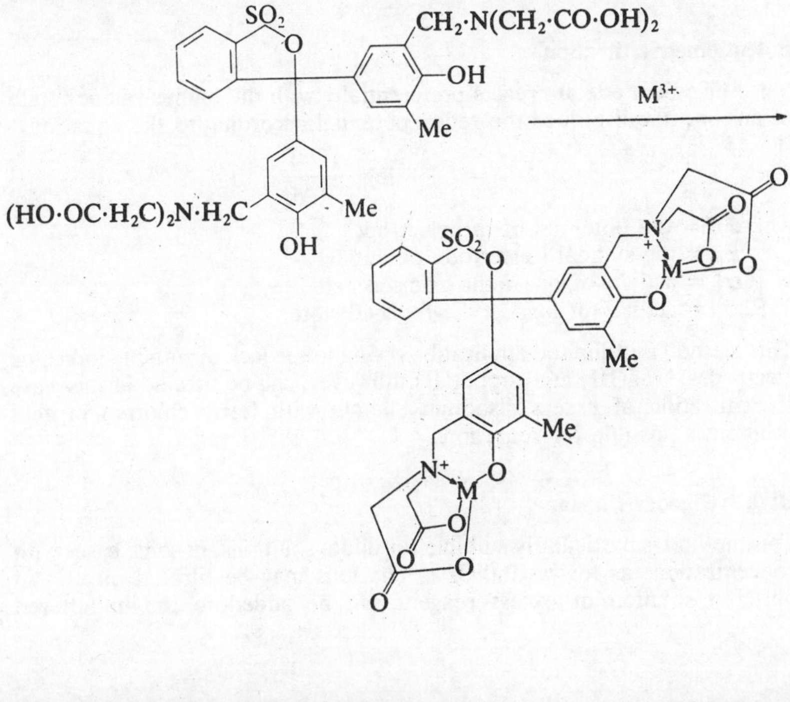

Physical methods of end point detection in complexometric titrations

Spectrophotometric detection

The change in absorption spectrum when a metal ion of a complexing agent is converted to the metal complex, or when one complex is converted to another can usually be detected more accurately and in more dilute solution by spectrophotometric than by visual methods. Thus, in disodium edetate titrations an accurate end point can be obtained using 0.001M solutions. In practice an indicator giving a colour change in the visible region is generally employed, but coloured ions may be titrated without an indicator using spectrophotometric methods. Also it is sometimes possible to use an end point in the ultraviolet region for ions and complexes which are colourless in the visible region.

Amperometric titration

The effect of complex formation on the half-wave potential of an ion is to render it more negative. If the electrode potential is adjusted to a value between that of the half-wave potential of the free cation and that of the complex, and disodium edetate solution is added slowly, the diffusion current will fall steadily until it equals the residual current, that is, until the last trace of free cation has been complexed. This is the end point and the amount of standard disodium edetate solution added is equivalent to the amount of metal present.

Potentiometric titration

Since disodium edetate reacts preferentially with the higher valency state of an ion, it will reduce the redox potential according to the equation

$$E = E_0 + \log_e \frac{[\mathrm{Ox}]}{[\mathrm{Red}]}$$

where E = the potential of the electrode
E_0 = the standard electrode potential
[Ox] = activity of ions in the oxidised state
[Red] = activity of ions in the reduced state.

This method is of limited application owing to the lack of suitable indicator electrodes. Iron(III) and copper(II), however, can be titrated in this way. Back-titration of excess disodium edetate with ferric chloride in acid solution is possible for some ions.

High frequency titrator

This method is particularly suitable for dilute solutions, in some cases with concentrations as low as 0.0002M. The ions may be titrated directly in buffered solution or excess reagent can be added to the unbuffered solution and the liberated protons titrated with standard alkali. Since buffer solution and other extraneous electrolytes reduce the sensitivity of the titration, their concentration must be kept to a minimum.

General principles involved in disodium edetate titrations

Direct titration

A suitable buffer solution and indicator are added to the metal ion solution and the solution titrated with standard disodium edetate until the indicator just changes colour. A *blank* titration may be performed, omitting the sample as a check on the presence of traces of metallic impurities in the reagents. Greater accuracy may be obtained if the amount of indicator used and volumes of solution are controlled and the colour of the solution at the end point is matched with that from a blank titration or a slightly over-titrated sample.

Back-titration

This procedure is necessary for metals which precipitate as hydroxides from solution at the pH required for titration, for insoluble substances (e.g. lead as sulphate, calcium as oxalate), for substances which do not react instantaneously with disodium edetate and for those metal ions which form more stable complexes with disodium edetate than with the desired indicator. Excess of standard sodium edetate and a suitable buffer solution is added to the metal solution or suspension. The solution is heated to effect complex formation, cooled and the disodium edetate not required by the sample back-titrated with magnesium or zinc chloride (or sulphate) using a suitable indicator.

Replacement of one complex by another

When direct titration or back-titration do not give sharp end points, the metal may be determined by the displacement of an equivalent amount of magnesium or zinc from a less stable edetate complex according to the equation:

$$M^{2+} + MgX^{2-} \rightarrow MX^{2-} + Mg^{2+}$$

Calcium, lead and mercury can be determined satisfactorily using Mordant Black II by this method.

Alkalimetric titration of metals

In this method, protons from disodium edetate are displaced by a heavy metal, and titrated with standard alkali, according to the equation:

$$M^{n+} + [H_2X]^{2-} \rightarrow [MX]^{(n-4)} + 2H^+$$

The titration is carried out in unbuffered solution. A visual pH indicator may be used, but a potentiometric method of detecting the end point is also suitable, especially when the colour of the complex would mask that of a pH indicator.

Masking and demasking agents

Owing to the range of cations complexed by disodium edetate, the selectivity of the method is poor and metal impurities may be titrated with the ion it is desired to determine. When it is required to assay selectively one or more ions in a mixture of cations and to eliminate the effects of possible impurities which would add to the titre, **masking agents** are used. These act either by precipitation or by formation of complexes more stable than the interfering ion-edetate complex. It is important that any colour due to auxiliary complexes or precipitates should not obscure the end point.

Masking by precipitation

Many heavy metals (e.g. cobalt, copper and lead) can be separated either in the form of insoluble sulphides using sodium sulphide or as insoluble complexes using thioacetamide. These are filtered, decomposed and titrated with disodium edetate. Other common precipitating agents are sulphate for lead and barium, oxalate for calcium and lead, fluoride for calcium, magnesium and lead, ferrocyanide for zinc and copper and cupferron (ammonium *N*-nitrosophenylhydroxylamine) and 8-hydroxyquinoline for many heavy metals. Thioglycerol ($CH_2SH \cdot CHOH \cdot CH_2OH$) is used to mask copper by precipitation in the assay of lotions containing copper and zinc. If it is desired to mask a *trace* of ions, which interfere, filtration is usually unnecessary since the rate of reaction of sodium edetate with the insoluble complexes is very slow.

Masking agents

pH control. A very simple method of masking is based upon the fact that the alkaline earth metals form edetate complexes below pH 7, whilst most transition elements form edetate complexes stable down to pH 3. Certain metal ions, such as tin(IV), iron(II), cobalt(III) and thorium(IV), can be selectively titrated at still lower pH values.

Ammonium fluoride will mask aluminium, iron and titanium by complex formation.

Ascorbic acid is a convenient reducing agent for iron(III) which is then masked by complexing as the very stable hexacyanoferrate(II) complex. This latter is more stable and less intensely coloured than the hexacyanoferrate(III) complex.

Dimercaprol (2,3-Dimercaptopropanol); ($CH_2SH \cdot CHSH \cdot CH_2OH$).
Cations of mercury, cadmium, zinc, arsenic, antimony, tin, lead and bismuth react with dimercaprol in weakly acid solution to form precipitates which are soluble in alkaline solution. All these complexes are stronger

than the corresponding edetate complexes and are almost colourless. Cobalt, copper and nickel form intense yellowish-green complexes with the reagent under the above conditions. Cobalt and copper, but not nickel, are displaced from their edetate complexes by dimercaprol.

Potassium cyanide reacts with silver, copper, mercury, iron, zinc, cadmium, cobalt and nickel ions to form complexes in alkaline solution which are more stable than the corresponding edetate complexes, so that other ions, such as lead, magnesium, manganese and the alkaline earth metals, can be determined in their presence. Of the metals in the first group mentioned, zinc and cadmium can be *demasked* from their cyanide complexes by aldehydes, such as formaldehyde or chloral hydrate (due to the preferential formation of a cyanohydrin), and selectively titrated.

Potassium iodide is used to mask the mercury(II) ion as $(HgI_4)^{2-}$ and is specific for mercury. It can be used in the assay of mercury(II) chloride.

Tiron (disodium catechol-3,5-disulphonate) will mask aluminium and titanium as colourless complexes. Iron forms highly coloured complexes and is best masked as its hexacyanoferrate(II) complex.

Triethanolamine $[N(CH_2 \cdot CH_2 \cdot OH)_3]$ forms a colourless complex with aluminium, a yellow complex with iron(III), the colour of which is almost discharged by adding sodium hydroxide solution, and a green manganese(III) complex which oxidises mordant black II. For these reasons, if murexide is used in the presence of iron and manganese, it is best to mask them with triethanolamine; similarly, mordant black II can be used in the presence of triethanolamine-aluminium complex.

Complexometric methods

Purification of disodium edetate

Commercial samples of disodium edetate may be purified for use as a *primary standard* by adding ethanol to a saturated aqueous solution until the first permanent precipitate appears; filter and add an equal volume of ethanol; filter the precipitated disodium edetate, wash with acetone and ether, and dry to constant weight at 80°C. Drying may require four days. The official material contains not less than 98 per cent of the dihydrate and is assayed as described under disodium edetate.

Standardisation of approximately 0.05M disodium edetate

The standardisation is based on the titration of the disodium edetate solution with a standard zinc chloride solution prepared from a known weight of granulated zinc.

$$Zn + 2HCl \rightarrow ZnCl_2 + H_2$$

$$ZnCl_2 + C_{10}H_{14}N_2Na_2O_8 \rightarrow C_{10}H_{14}N_2O_8Zn + 2NaCl$$

$$\therefore \quad 65.38 \text{ g Zn} \equiv 1000 \text{ ml M}$$

$$\therefore \quad 0.003269 \text{ g Zn} \equiv 1 \text{ ml } 0.05\text{M disodium edetate}$$

Method Accurately weigh approximately 0.125 g of granulated zinc. Disolve in the minimum volume of hydrochloric acid (7M), add bromine water (0.2 ml) to ensure oxidation of trace iron impurity to iron(III), which forms a much less stable edetate complex than iron(II). Gently boil to remove excess bromine. Cool, add sodium hydroxide (2M) until almost neutral, dilute to 250 ml and add ammonia buffer pH 10 until the precipitate is just dissolved and then 5 ml in excess. Add a mixture of mordant black II and sodium chloride (1:99; 50 mg) as indicator and titrate with 0.05M disodium edetate solution until the solution becomes green.

Ammonia buffer solution pH 10 Dissolve ammonium chloride (5.4 g) in ammonia (5M; 70 ml) and dilute with *water* to 100 ml.

Direct titration with disodium edetate

Bismuth Subcarbonate *Determination of the percentage of* Bi

The determination depends upon the reactions expressed by the following equation:

$$Bi^{3+} + [H_2X]^{2} \rightarrow [BiX]^- + 2H^+$$
$$Bi^{3+} \equiv Na_2H_2X_22H_2O$$
$$\therefore \quad 208.96 \text{ g Bi} \equiv 1000 \text{ ml M}$$
$$\therefore \quad 0.01045 \text{ g Bi} \equiv 1 \text{ ml } 0.05\text{M disodium edetate}$$

Method Accurately weigh approximately 0.2 g of sample. Dissolve in the minimum volume of hot nitric acid (2M), add *water* (50 ml) and adjust to pH 1–2 by the dropwise addition of either nitric acid (2M) or ammcnia (5M). Add a mixture of xylenol orange and potassium nitrate (1:99; 30 mg) as indicator and titrate with 0.05M disodium edetate until the colour becomes yellow.

Calcium Carbonate *Determination of the percentage of* $CaCO_3$

$$Ca^{2+} + [H_2X]^{2-} \rightarrow [CaX]^{2-} + 2H^+$$
$$\therefore \quad CaCO_3 \equiv Ca^{2+} \equiv Na_2H_2X,2H_2O$$
$$\therefore \quad 100.1 \text{ g } CaCO_3 \equiv 1000 \text{ ml M}$$
$$\therefore \quad 0.005005 \text{ g } CaCO_3 \equiv 1 \text{ ml } 0.05\text{M disodium edetate}$$

Method Accurately weigh approximately 0.1 g of sample. Dissolve in hydrochloric acid (2M; 3 ml) and *water* (10 ml). Boil for 2 min, cool and dilute to 50 ml with *water*. Titrate with 0.05M disodium edetate almost to the expected end point. Add NaOH (10M; 4 ml) and a mixture of calcon and anhydrous sodium sulphate (1:99; 0.1 g) as indicator and continue the titration until the colour becomes full blue.

COGNATE DETERMINATIONS

Calcium Acetate
Calcium Chloride
Calcium Gluconate
Calcium Lactate and *Tablets*
Sodium Calcium Lactate, Injection and *Tablets*

Cobalt Oxide *Determination of the percentage of* Co

Cobalt oxide, which is used in veterinary practice, consists of a mixture of cobalt(II,III) oxide (Co_3O_4) and cobalt(III) oxide (Co_2O_3). It is reduced to the cobalt(II) state and determined according to the relationship shown in the following equations:

$$Co^{2+} + [H_2X]^{2-} \rightarrow [CoX] + 2H^+$$
$$\therefore \quad 58.93 \text{ g Co} \equiv 1000 \text{ ml M}$$
$$0.002946 \text{ g Co} \equiv 1 \text{ ml } 0.05\text{M disodium edetate}$$

Method Accurately weigh approximately 0.1 g of sample. Dissolve in hydrochloric acid (20 ml), if necessary by heating, evaporating to dryness and reheating with a further 20 ml of acid. Add *water* (300 ml), hydroxylamine hydrochloride (4 g) and strong ammonia solution (13.5M; 25 ml). Warm to 80°, add a mixture of *methyl thymol blue* and potassium nitrate (1:100) as indicator and titrate with 0.05M disodium edetate until the solution becomes purple.

Magnesium Sulphate *Determination of the percentage of* $MgSO_4$

This determination depends upon the reactions expressed by the following equation:

$$Mg^{2+} + [H_2X]^{2-} \rightarrow [MgX]^{2-} + 2H^+$$
$$\therefore \quad MgSO_4 \equiv Mg^{2+} \equiv Na_2H_2X,2H_2O$$
$$\therefore \quad 120.38 \text{ g } MgSO_4 \equiv 1000 \text{ ml M}$$
$$\therefore \quad 0.006019 \text{ g } MgSO_4 \equiv 1 \text{ ml } 0.05\text{M disodium edetate}$$

Method Accurately weigh approximately 0.3 g sample. Dissolve in *water* (50 ml). Add ammonia buffer pH 10 (p.000) and a mixture of mordant black II and sodium chloride (1:99; 50 mg) as indicator and titrate with 0.05M disodium edetate until the solution becomes full blue.

COGNATE DETERMINATIONS

Disodium Edetate The method is essentially the same, but with magnesium sulphate as titrant.

Magnesium Acetate
Magnesium Carbonate (*Heavy* and *Light*)
Magnesium Chloride
Magnesium Hydroxide
Magnesium Oxide (*Heavy* and *Light*)
Magnesium Trisilicate and *Compound Tablets*

Zinc Sulphate *Determination of* $ZnSO_4,7H_2O$

This determination depends upon the reactions expressed by the following equation:

$$Zn^{2+} + [H_2X]^{2-} \rightarrow [ZnX]^{2-} + 2H^+$$
$$\therefore \quad ZnSO_4,7H_2O \equiv Zn^{2+} \equiv Na_2H_2X,2H_2O$$
$$\therefore \quad 287.5 \text{ g } ZnSO_4,7H_2O \equiv 1000 \text{ ml M}$$

$$\therefore \quad 0.01438 \text{ g } ZnSO_4,7H_2O \equiv 1 \text{ ml } 0.05\text{M disodium edetate}$$

Method Accurately weigh approximately 0.2 g of sample. Dissolve in acetic acid (2M; 5 ml) and dilute to 50 ml with *water*. Add a mixture of *xylenol orange* and potassium nitrate (1:99; 50 mg). Add sufficient hexamine to produce a violet-pink colour and 2 g in excess. Titrate with 0.05M disodium edetate until the solution becomes pure yellow in colour.

COGNATE DETERMINATIONS

Bacitracin Zinc for Zn
Zinc Chloride
Zinc Oxide
Zinc Undecanoate

Determination of the hardness of water

This involves the determination of the calcium and magnesium ions present, the **total hardness** being expressed as parts per million or grains per gallon of calcium and magnesium salts together calculated as calcium carbonate. This is determined by titration with 0.01M disodium edetate in buffered alkaline solution using mordant black II as indicator since it forms chelates with both Ca^{2+} and Mg^{2+}. To determine calcium hardness only, murexide is used as indicator in strongly alkaline solution since, under these conditions, it chelates with Ca^{2+} only. If the water contains the alkaline-earth metals only (as is usually the case), the buffer used in the above procedure is satisfactory for total hardness; if, however, heavy metals, such as Cu, Fe, Sn, Zn, Pb and Al, are present in traces, they will lead to errors and an alternative buffer containing sulphide is recommended.

Total hardness

Buffer solution Dissolve sodium edetate (0.93 g) and magnesium sulphate (0.616 g) in strong ammonia–ammonium chloride solution (1000 ml).
Method Add buffer solution (2 ml) to the sample (100 ml). Add 6 drops (from a dropper) of mordant black II indicator solution and titrate with 0.01M disodium edetate until there is no further colour change. The indicator changes colour from pink through mauve to blue and titrant should be added dropwise towards the end point. Add a few drops of titrant in excess and repeat the titration, taking the colour of the over-titrated sample as the end point.

$$\text{Total hardness} = \frac{\text{Volume of 0.01M disodium edetate}}{\text{Volume of Sample}} \times 1000 \text{ ppm } CaCO_3$$

Calcium hardness

Buffer solution Potassium cyanide solution (10%; 1 ml); sodium sulphide solution (10%; 1 ml); dilute with M NaOH to 100 ml.
Indicator Murexide (0.2%) in finely powdered AnalaR sodium chloride.
Method Add M NaOH (5 ml) and murexide indicator (0.2 g) to the sample (100 ml). Titrate with 0.01M disodium edetate until the pink colour changes to purple, and addition of excess titrant causes no further change. Repeat and obtain an accurate end point by matching.

Back-titrations with disodium edetate

Alum *Determination of the percentage of* $KAl(SP_4)_2,12H_2O$

The determination depends upon the reaction expressed by the following equations:

$$Al^{3+}[H_2X]^{2-} \rightarrow [AlX]^- + 2H^+$$
$$\therefore \quad KAl(SO_4)_2,12H_2O \equiv Al^{3+} \equiv Na_2H_2X,2H_2O$$
$$\therefore \quad 474.4\text{ g } KAl(SO_4)_2,12H_2O \equiv 1000\text{ ml M}$$
$$\therefore \quad 0.02372\text{ g } KAl(SO_4)_2.12H_2O \equiv 1\text{ ml } 0.05\text{M disodium edetate}$$

Method Accurately weigh about 0.6 g of sample. Dissolve in hydrochloric acid (M; 2 ml) and *water* (50 ml). Add standard disodium edetate (0.05M; 50 ml) and neutralise with M NaOH to methyl red. Heat the solution to boiling on a boiling water bath, allow to remain for 10 min and cool. Add hexamine (5 g) and a mixture of xylenol orange and potassium nitrate (1:99; 50 mg) as indicator and titrate with 0.05M lead(II) nitrate. The colour will change at the end point from yellow to reddish-purple.

Note The alums are highly hydrated and sampling errors are reduced by making a solution and taking a one-fifth aliquot part for the assay. The aluminium edetate complex is stable ($\log K = 15.5$) but it is only formed slowly. Hence, the sample is heated with excess disodium edetate to ensure complete complex formation. Hexamine acts as a buffer, stabilising the pH between 5 and 6, the optimum pH for the titration of the disodium edetate not required by the aluminium with 0.05M lead nitrate using xylenol orange as indicator.

COGNATE DETERMINATIONS

Aluminium Glycinate
Dried Aluminium Hydroxide
Aluminium Sulphate

Calcium Phosphate *Determination of the percentage of* $Ca_3(PO_4)_2$

Since calcium phosphate is insoluble, the sample is dissolved with the aid of heat in excess hydrochloric acid; an aliquot of the solution is treated with excess 0.05M disodium edetate, ammonium buffer added to bring the pH to 10 and the disodium edetate not required by the sample is back-titrated with 0.05M zinc chloride using mordant black II as indicator. This procedure illustrates an alternative way of avoiding the indistinct colour change with calcium alone and mordant black II.

COGNATE DETERMINATIONS

Aluminium Phosphate (Dried)
Calcium Hydrogen Phosphate

Sodium Fluoride *Determination of the percentage of* NaF

The determination depends upon the reaction expressed by the following equations:

$$2F^- + Pb^{2+} \rightarrow PbF_2\downarrow$$
$$Pb^{2+} + [H_2X]^{2-} \rightarrow [PbX]^{2-} + 2H^+$$

Method To the sample (80 mg), accurately weighed, in *water* (45 ml) add sodium chloride (0.2 g) and ethanol (96%; 20 ml). Heat to boiling and add 0.05M lead nitrate, initially dropwise and then more rapidly, with constant stirring. Heat to coagulate the precipitate, cool to room temperature, filter and wash the residue with small volumes of alcohol (20%). To the filtrate and washings, add hexamine (1 g) and titrate the excess lead nitrate with 0.05M disodium edetate using xylenol orange solution as indicator. The end point is yellow.

Displacement of one complex by another

Sodium Calcium Edetate *Determination of the percentage of* $C_{10}H_{12}N_2O_8CaNa_2$

The determination depends upon the reactions expressed by the following equations:

$$[CaX]^{2-} + 2H^+ \rightarrow Ca^{2+} + [H_2X]^{2-}$$
$$[H_2X]^{2-} + Pb^{2+} \rightarrow [PbX]^{2-} + 2H^+$$
$$\therefore \quad Na_2[CaX] \equiv [H_2X]^{2-} \equiv Pb^{2+}$$
$$\therefore \quad 374.3\text{ g } C_{10}H_{12}N_2O_8CaNa_2 \equiv 1000\text{ ml M}$$
$$\therefore \quad 0.01871\text{ g } C_{10}H_{12}N_2O_8CaNa_2 \equiv 1\text{ ml } 0.05\text{M lead nitrate}$$

Method Dissolve the sample (about 0.5 g), accurately weighed, in *water* (90 ml). Add hexamine (7 g) and dilute hydrochloric acid (5 ml), and titrate with 0.05M lead nitrate using xylenol orange as indicator.

Note The hexamine–hydrochloric acid buffer has a pH of about 4 and at this pH the calcium edetate complex breaks down. The liberated edetate can then be titrated with lead ions, which form a stable complex at this pH.

Miscellaneous complexometric methods

Volumetric determination of sulphate

Many methods for the determination of sulphate have been described, the most satisfactory being titration with barium perchlorate in aqueous ethanol buffered to pH 3.7, using alizarin red S as indicator.

Determination of lead using dithizone

Method Place the sample containing 1–100 µg of lead in a separating funnel, add 20 ml of dilute ammonia solution, 1 ml of 10% KCN solution and 5 ml of 0.002% dithizone in chloroform and shake vigorously. Allow to separate; shake the lower layer with a mixture of 5 ml of dilute ammonia and 1 ml of the KCN solution to remove excess of dithizone. The intensity of the red colour in the chloroform layer may be compared visually or spectrophotometrically with that produced by known amounts of lead under the same conditions. The absorption maximum of lead dithizonate is at 525 nm which is close to the position of minimum absorption of dithizone itself.

Gravimetric methods

Introduction

Gravimetric analysis is a procedure for isolating and weighing an element or compound in as pure a form as possible; the element or compound is separated from a definite portion (weight or volume) of the substance being examined, and the weight of the constituent in the sample calculated from the weight of the product.

In pharmaceutical analysis, the product to be weighed is obtained by one of the following procedures:

(i) volatilisation or ignition,
(ii) solvent extraction,
(iii) precipitation from solution.

Procedures involving volatilisation, ignition and solvent extraction are described in detail in Chapters 1, 4 and 9. Gravimetric procedures involving precipitation from solution are based on the quantitive precipitation of the anion or cation to be determined, either in the form of an insoluble compound of definite composition, or as an insoluble compound which leaves a residue of definite composition upon ignition. The techniques involved, i.e. precipitation, filtration, washing of the precipitate and drying or ignition of the residues to constant weight, are described in Chapter 4.

Gravimetric methods, based on precipitation techniques, are time-consuming. For this reason, they are now seldom used where more rapid alternative methods are available.

Sodium Aurothiomalate *Determination of the percentage of* Au

Sodium Aurothiomalate is of undefined constitution and, moreover contains a small but variable proportion of moisture. The percentage of gold is calculated with reference to the substance dried over phophorus pentoxide at a pressure not exceeding 5 mm of mercury.

The method depends upon the oxidation of organic matter to volatile products by digestion with sulphuric acid. Oxidation occurs at the expense of the sulphuric acid, which is reduced to a sulphur dioxide and hydrogen sulphide. The latter reduces the gold salts to metallic gold, which is precipitated. The subsequent addition of nitric acid and digestion ensures the complete decomposition of organic matter and oxidation of inorganic substances, such as sulphur (from thiomalic acid), which might otherwise be weighed as gold.

Method Accurately weigh about 0.2 g of sample into a 100 ml Kjeldahl flask. Add concentrated sulphuric acid (10 ml) and heat gently to boiling (fume cupboard). Boil gently until the liquid is clear and pale yellow in colour. Cool, add nitric acid (1 ml) dropwise, and then boil again for 1 h. Cool, dilute with *water* (70 ml), boil for 5 min and filter through a Whatman No. 42 filter paper. Wash with hot water, allow to drain, and dry the paper and precipitate, still in the funnel, in an oven at 120°. Transfer the paper and precipitate to an ignited and weighed silica crucible. Heat carefully to burn off the paper and then ignite to constant weight.

COGNATE DETERMINATIONS

Sodium Aurothiomalate Injection
Aurothioglucose

Thenium Closylate *Determination of the percentage of* $C_{21}H_{24}CoNO_4S_2$

This method depends upon the precipitation of insoluble thenium reineckate, which is formed on addition of ammonium reineckate to an acidified solution of the sample.

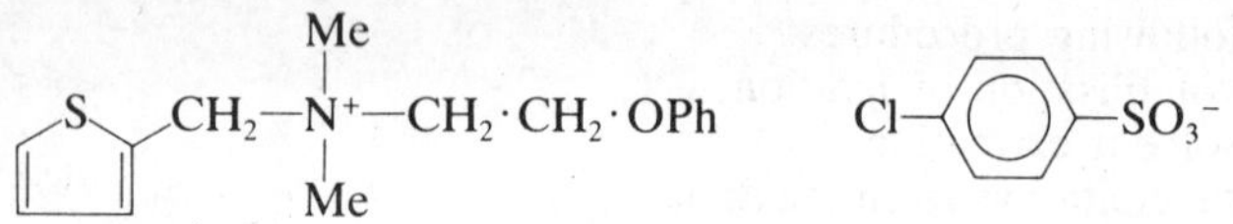

$$(C_{14}H_{20}\overset{+}{N}OS)(C_6H_5ClO_3S) + \overset{+}{N}H_4[Cr(NH_3)_2(SCN)_4]^-$$
$$= (C_{14}H_{20}\overset{+}{N}OS)[(Cr(NH_3)_2(SCN)_4]^-$$
$$\therefore \quad 458\text{ g } C_{21}H_{24}ClNO_4S_2 \equiv 585.9\text{ g of residue}$$
$$\therefore \quad 0.7817\text{ g } C_{21}H_{24}ClNO_4S_2 \equiv 1\text{ g of residue}$$

Method Accurately weigh the sample (about 0.2 g) into a 250 ml beaker. Dissolve in *water* (50 ml), add dilute sulphuric acid (8 ml) and boil. Slowly add freshly prepared ammonium reineckate solution (1%; 60 ml), stirring continuously. Set aside for 1 h, allowing the solution to cool to room temperature. Cool to 0° and allow to stand for 1 h. Filter through a No. 4 sintered glass crucible, previously dried to constant weight. Wash the precipitate successively with 20, 15, 15 and 15 ml of water at 0°, dry at 80° for 1 h and then dry to constant weight at 105°.

COGNATE DETERMINATION

Pentolinium Injection

Thiamine Hydrochloride *Determination of the percentage of* $C_{12}H_{17}ON_4SCl,HCl$

This method depends upon the precipitation of insoluble thiamine silicotungstate which is formed on the addition of a solution of silicotungstic acid to a slightly acidified solution of the sample. The precipitating reagent is a complex silicate, $SiO_2,12WO_3,xH_2O$ of somewhat variable composition in respect of the degree of hydration. The quality of the reagent, however, is controlled by the requirement that it should contain $\nless 85\%$ of WO_3 and $\nless 1.85\%$ of SiO_2. Thiamine silicotungstate, on the other hand, is of constant composition:

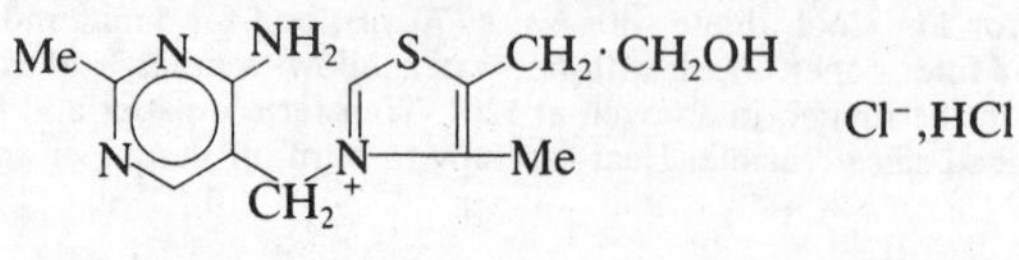

$$2C_{12}H_{17}N_4OSCl,HCl + [SiO_2,12WO_3] + 6H_2O \rightarrow (C_{12}H_{17}N_4OSCl)_2,[SiO_2(OH)_2,12WO_3]4H_2O$$

$$\therefore \quad 674.6\ g\ C_{12}H_{17}N_4OSCl,HCl \equiv 3858\ g\ \text{of residue}$$

$$\therefore \quad 0.1929\ g\ C_{12}H_{17}N_4OSCl,HCl \equiv 1\ g\ \text{of residue}$$

Method Accurately weigh the sample (about 0.05 g) and dissolve it in *water* (50 ml) in a 250 ml beaker provided with stirring rod and clock-glass cover. Add hydrochloric acid (2 ml) (Note 1), heat to boiling and add silicotungstic acid solution (10%; 4 ml) in a rapid stream of drops (Note 2). Boil the solution for 4 min (Note 3) and filter through a No. 4 sintered glass crucible, previously dried to constant weight at 105°. Wash with a boiling mixture of hydrochloric acid (1 volume) and *water* (19 volumes) (Note 4), then with *water* (10 ml) and finally with 2 portions of 5 ml each of acetone. Dry the residue to constant weight at 105°.

Notes on the determination of Thiamine Hydrochloride as thiamine silicotungstate

Note 1 Excess of hydrochloric acid is necessary to produce a readily filterable precipitate.

Note 2 The rate of addition does not influence the result if the thiamine hydrochloride is reasonably pure, but slow addition in the presence of appreciable impurity may give poor results.

Note 3 Boiling for less than 4 min gives low results. Longer periods of boiling do not affect the result.

Note 4 The volume of wash liquid should be restricted to about 50 ml. Prolonged washing gives low results.

Wool Alcohols *Determination of Cholesterol*

This method depends upon the fact that all 3β-hydroxyls, of which cholesterol is an example, form an insoluble molecular addition complex with the steroidal saponin, digitonin. The complexes have very low solubilities, and are stable and strongly crystalline.

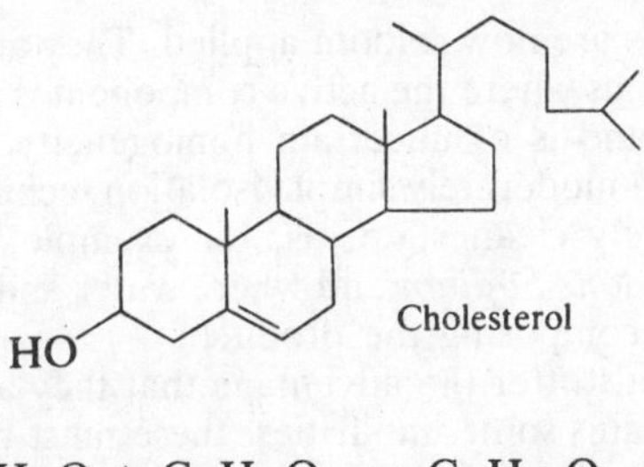

Cholesterol

$$\underset{\text{Cholesterol}}{C_{27}H_{46}O} + \underset{\text{Digitonin}}{C_{56}H_{92}O_{29}} \rightarrow \underset{\text{Complex}}{C_{83}H_{138}O_{30}} \downarrow$$

$$\therefore \quad 386.3\ g\ \text{Cholesterol} \equiv 1616\ g\ \text{of complex}$$

$$\therefore \quad 0.239\ g\ \text{Cholesterol} \equiv 1\ g\ \text{of complex}$$

Method Melt about 20 g of sample on a water-bath, mix thoroughly and cool. Accurately weigh about 1 g, dissolve in warm ethanol (90%; 25 ml) and filter immediately through a warmed No. 2 sintered glass crucible. Wash the residue on the filter with warm ethanol (90%; 50 ml). Cool the combined filtrate and washings, transfer to a 100 ml graduated flask and make up to 100 ml. To 10 ml of the solution, add a solution of digitonin (0.5% w/v; 40 ml) and warm to 60° to ensure that complex formation is complete. Filter through a No. 2 sintered glass crucible, previously dried to constant weight at 105°. Wash the precipitate successively with ethanol (90%; 15 ml), acetone (15 ml) and hot carbon tetrachloride (15 ml), allow to drain and dry to constant weight at 105°.

9
Solvent extraction methods

Theory

Introduction

Solvent extraction methods were originally evolved for the determination of alkaloids in crude drugs and galenicals. Such methods, however, are equally applicable to synthetic organic bases, their salts and pharmaceutical preparations (e.g. injections and tablets). They are also applicable to organic acids and their salts, and to non-basic substances which, by virtue of their solubilities, lend themselves to extraction with an organic solvent.

The extraction procedure is primarily a method for separating the active ingredient from extraneous matter and effecting a degree of purification sufficient to permit quantitive determination. The analytical method applied to the isolate may be either gravimetric, volumetric or spectrophotometric.

End determination methods

Gravimetric methods are now seldom applied. They are usually limited in application to materials where the active component is present in relatively high concentration and is of uncertain homogeneity. Such methods are suitable only where a moderately simple isolation technique yields the acid or base in a relatively clean state, as, for example, in *Phenobarbitone Sodium* and *Thiopentone Sodium* and where water and solvent are readily removed without decomposing the product.

Volumetric methods offer the advantage that they are applicable when the residue still contains some impurities; these must be neutral and must not be contaminated with large quantities of coloured impurities which would mask a visual end point. A back-titration method is always used because of the insolubility of organic acids and bases in water. Care is necessary if good results are to be obtained. The forward titration is always small and the use of a dilute volumetric solution to give a larger forward titration defeats its own object since the end point is less clearly defined. For example, 0.01M and 0.02M solutions are used for solanaceous alkaloids because of the small percentage of alkaloid present but the end points are noticeably flat and difficult to judge accurately. Alkaloids are, for the most part, weak bases, hence, methyl red is the indicator of choice. Phenolic alkaloids, such as cephaeline in ipecacuanha and morphine in opium, which are amphoteric, tend to exert a buffering action and give a flat end point. This buffer-like action is caused by competition between the phenolic hydroxyl group of the alkaloid and the indicator for sodium hydroxide. Non-aqueous titrations are now widely used whenever conditions allow, as, for example, in *Quinine Bisulphate*, *Carbenoxolone Sodium* and *Levallorphan Tartrate*.

Spectrophotometric methods are frequently quicker and simpler than volumetric and gravimetric methods and can usually be applied with greater accuracy to smaller quantities. Not all substances amenable to solvent extraction methods exhibit the absorption characteristics suitable for spectrophotometric determination. *Chlorpheniramine Tablets*, *Bisocodyl Tablets*, *Levorphanol Tablets* and many similar substances are assayed by solvent extraction and measurement of the ultraviolet absorption at the wavelength of maximum absorption of the active principle. A number of steroids, exemplified by cortisone acetate in *Cortisone Injection*, are isolated by extraction, and converted to a coloured complex, the absorption of which can be measured at specified wavelengths of maximum absorption in the visual spectrum. Colorimetric assays, employing comparison methods, are also specified in a few special cases where other methods are inapplicable. For example, in the determination of morphine in *Camphorated Opium Tincture*, the quantity of alkaloid is too small to be determined by volumetric or gravimetric methods. A specific colour reaction between morphine and sodium nitrite is used and the colour produced compared with the standard colour. Since it is often difficult to exclude traces of colouring matter from the alkaloid isolated in determinations of this type, it may be necessary to compare colours which are not exactly the same shade. The eye is not very sensitive to the fine distinctions of colour, shade and depth of colour and accurate comparisons are difficult. Much more satisfactory results can be obtained with the use of a photoelectric colorimeter or absorptiometer, when measurements may be made at selected wavelengths. The principles underlying these determinations are discussed fully in the appropriate chapter in Part 2.

Determination of the salts of organic acids and bases

Since these substances contain little extraneous matter, the extraction process is less complicated than that used for crude drugs and galenicals. The following general principles are involved.

(1) Most organic acids and bases are sparingly soluble in water, but readily soluble in organic solvents such as ether and chloroform. Phenolic alkaloids, such as morphine, are exceptions in that they are relatively insoluble in both water and organic solvents; they can, however, be extracted from water by mixed organic solvents such as chloroform-ethanol.

(2) Salts of organic bases, such as hydrochlorides and sulphates, are water-soluble but insoluble in organic solvents.

(3) Organic bases and their salts are readily interconvertible by reaction with either acid or alkali. Ammonia is preferable to sodium hydroxide as

alkali as it is volatile and any excess can be readily removed during the final drying of the residue.

(4) Most organic bases and all salts are non-volatile.

The following scheme of extraction for the isolation of an organic base from its salt can be evolved by taking advantage of these properties.

Solution containing salt of organic base **(aqueous or aqueous-ethanolic)**

(1) Make alkaline with ammonia.
(2) Extract by shaking with an immiscible solvent (chloroform or ether) and allow to separate into two phases.

Organic solvent **containing dissolved organic base** | ***Aqueous ammoniacal solution*** **Reject**

(3) Wash by shaking with a small quantity of *water* and allow to separate into two phases.

Organic solvent | ***Aqueous washings*** **Reject**

(4) Remove the solvent by distillation.
(5) Dry the residue by heating in an oven at 100° for 30 min. This also volatilises the last traces of ammonia.

Residue
Base

A similar scheme can be devised for the isolation of an organic acid from its salt in which the salt is decomposed with dilute mineral acid at Stage (1).

Distribution law and partition coefficients

When a substance is shaken with two immiscible solvents, it distributes itself between them in a constant ratio which is characteristic of the substance and the solvents employed at a given temperature. That is,

$$\frac{\text{Concentration in solvent A}}{\text{Concentration in solvent B}} = \frac{C_A}{C_B} = K$$

The constant K is known as the partition coefficient. It varies with temperature.

The law applies strictly only to dilute solutions, for completely immiscible solvents and for solutes which have the same molecular weight in both solvents and which do not affect the miscibility of the solvents. For concentrated solutions, the limiting case occurs when the two solvents are shaken with excess solute. Two saturated solutions are obtained and the partition coefficient is then equal to the ratio of the solubilities of the solute in the two solvents. This provides a rough approximation from which approximate distribution coefficients can be calculated. This is not strictly valid at lesser concentrations, but a useful guide where accurate distribution coefficients are not available; moreover, it provides for conditions frequently met in practice insofar as the solubility in one or other of the solvents is usually low.

In practice, the amount of material extracted depends on the amount of material present, the partition coefficient and the volumes of the two solvents. The efficiency of the process can be increased by increasing the volume of the extracting solvent and by increasing the number of extractions. This is illustrated by the following examples of the distribution between chloroform and water of 1 g of a solid of partition coefficient, $K = 9$.

Example 1 *Extraction using equal volumes of solvents*
Consider the partition of 1 g of solid between 20 ml volumes of chloroform and water.

$$\frac{\text{Concentration in chloroform}}{\text{Concentration in water}} = K = \frac{9}{1}$$

then, if x g dissolves in the chloroform

$$\frac{x/20}{(1-x)/20} = \frac{9}{1}$$

$$\therefore \quad \frac{x}{(1-x)} = 9$$

$$\therefore \quad x = 0.9$$

Hence, after the first extraction, one-tenth of the substance (0.1 g) remains in the aqueous phase. A second extraction with the same volume of solvent will reduce the weight of substance remaining in the aqueous phase to one-hundredth (0.01 g) of that originally present.

Example 2 *Extraction with unequal volumes of solvents*
Consider the partition of 1 g of solid between 40 ml of chloroform and 20 ml of water. Again,

$$\frac{\text{Concentration in chloroform}}{\text{Concentration in water}} = \frac{9}{1}$$

then if x g dissolves in the chloroform

$$\frac{x/40}{(1-x)/20} = \frac{9}{1}$$

$$\therefore \quad x/2 = 9(1-x)$$

$$\therefore \quad x = 18/19$$

Hence, after the first extraction, one-nineteenth of the substance (0.0526 g) remains in the aqueous phase. A second extraction with a further 40 ml of chloroform will reduce the weight of substance remaining in the aqueous phase to 1/361 of that originally present.

Choice of solvent

The choice of solvent depends on the solubility characteristics of the substance to be extracted, characteristics on which the partition coefficient also depends.

Chloroform has the advantages that it is non-inflammable and heavier than water. The latter property facilitates separation of the organic phase. Ether, on the other hand, is both inflammable and lighter than water, so that not only are the separation techniques more complicated, but care must also be exercised in the subsequent removal of the solvent. Mixed solvents are sometimes necessary. Morphine is insoluble in both ether and chloroform and is best extracted with a mixture of choroform and ethanol (or isopropanol).

Extraction with chloroform

Quinine Bisulphate *Determination of the percentage of*
$C_{20}H_{24}N_2O_2,H_2SO_4$

Chloroform is used as the extracting solvent and since it is heavier than water, only two separators are required. For convenience of description these two separators are designated A and B, respectively. Before commencing the extraction check that the taps are properly greased and that there is no possibility of leakage from either taps or stoppers.

Method Accurately weigh the sample (0.4 g) into separator A. Add *water* (15 ml) and dissolve. Make alkaline by the addition of exactly 25 ml of 0.1M sodium hydroxide solution (Note 1). Add chloroform (30 ml), stopper the separator and shake gently. Invert the separator and cautiously open the tap to release the pressure. Close the tap and shake vigorously for at least 2 min. Return the separator to an upright position in the stand, when the contents will separate rapidly into two layers, of which the chloroform is the lower (Note 2). Remove the stopper, washing its surface with a few drops of chloroform (teat pipette). Carefully open the tap and run off the whole of the chloroform solution into separator B without allowing any aqueous layer to follow. It is most important to make a clean separation between the two phases and any residual emulsion should be allowed to break.

Whilst the stem of separator A is still inside the neck of separator B, pour about 1 ml of chloroform into A, allow the chloroform to sink straight to the bottom and then run into B as before (Note 3). Finally, wash the outside to the stem of A with a few drops of chloroform (teat pipette) allowing the washings to run directly into B (Note 4). Wash the chloroform solution in separator B by shaking vigorously for 2 min with *water* (20 ml). Allow to separate and run off the lower chloroform layer into a clean dry alkaloid flask (150 ml). *Wash through* with chloroform, running the washings into the flask. Retain the aqueous liquor in separator B as this is required for washing successive chloroform extracts.

Repeat the extraction of the aqueous layer in A by shaking with a further 20 ml of chloroform. Allow to separate into two phases and run off the lower layer into B. *Wash through* with chloroform. Shake the chloroform solution now in B with the aqueous wash liquor, already present, as described above. Allow to separate and run off into the same alkaloidal flask.

Complete the extraction of the base from the alkaline solution in A, by shaking with two further 10 ml portions of chloroform. Transfer each extract in turn to B, wash and run into the alkaloid flask. Test the remaining alkaline liquor in A as follows to ensure that the base has been completely extracted. Shake with 2–3 ml of chloroform and run off the latter into a test tube. Add an equal volume of dilute sulphuric acid, heat in a boiling water-bath until all the chloroform has volatilised. Add a few drops of Mayer's reagent (K_2HgI_4) when there should be no opalescence or precipitate, indicating the absence of base. If base is still present, continue the extraction with further 10 ml portions of chloroform as before until the extraction is complete. Combine the aqueous solutions for use in the test for titratable cation.

Fit the alkaloid flask containing the chloroform solution for distillation (Fig. 9.1), and heat in a water-bath until most of the solvent has been recovered (Note 5). Do not on any account allow solid matter (base) to separate from solution. Add ethanol (96%; 5 ml) and complete the evaporation to dryness (Note 6) on a water-bath, gently rotating the flask to give a thin layer of base.

Dry the base in an oven at 100° for half an hour (Note 7). Care should be taken to place the flask on the shelf of the oven in such a position as to avoid any possibility of charring which may occur in close proximity to the heating elements. Cover the top of the flask to prevent specks of dirt falling into the flask.

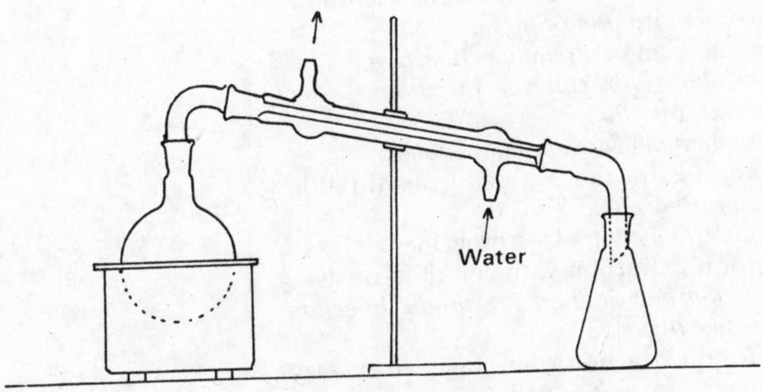

Fig. 9.1. Chloroform recovery apparatus

Dissolve the residue in glacial acetic acid (50 ml) and titrate with 0.1M perchloric acid using crystal violet as indicator (Chapter 7).

$$C_{20}H_{24}N_2O_2,H_2SO_4 \equiv 2000 \text{ ml M } HClO_4$$
$$\therefore \quad 422.6 \text{ g } C_{20}H_{24}N_2O_2,H_2SO_4 \equiv 20\,000 \text{ ml } 0.1\text{M}$$
$$\therefore \quad 0.02113 \text{ g } C_{20}H_{24}N_2O_2,H_2SO_4 \equiv 1 \text{ ml } 0.1\text{M}$$

Notes on the determination of quinine in Quinine Bisulphate

Note 1 This decomposes the quinine bisulphate, liberating the free base. A volumetric solution of sodium hydroxide is used to neutralise the liberated acid, thereby providing the basis for the test for titratable cation. This is necessary to control possible contamination with quinine sulphate. For other salts of organic bases which are free from such complications, precipitation of the organic base should preferably be carried out with ammonia rather than sodium hydroxide, since the former can be volatilised prior to titration.

Note 2 Quinine base, which is very much more soluble in chloroform than in water becomes distributed between the two phases in proportion to its solubility in each of the two solvents. A larger volume of solvent is used at this stage than is stipulated in the official process, so that any unavoidable loss of solvent results in the smallest possible loss of base.

Note 3 This small volume of chloroform displaces any of the solution from the tap and stem of A.

Note 4 Chloroform does not run cleanly away from the end of the separator stem, so that small quantities of the base are deposited on the outside of the stem, owing to creeping and subsequent evaporation of the solvent.

The operations covered by Notes 3 and 4 comprise a *wash through* technique which should become standard procedure in all manipulations involving transfer from one separating funnel to another.

Note 5 Chloroform is expensive and, therefore, should always be recovered.

Note 6 The addition and evaporation of ethanol helps to remove the last traces of chloroform and water.

Note 7 This removes traces of volatile bases, which are liable to be present.

COGNATE DETERMINATIONS

Chloroform extraction with non-aqueous titration

Benztropine Mesylate
Carbenoxolone Sodium
Chloroquine Phosphate Injection
Chloroquine Sulphate and *Injection*
Dextropropoxyphene Napsylate and *Capsules*
Hydroxychloroquine Sulphate
Lignocaine and *Adrenaline Injection*
Lignocaine Hydrochloride Injection
Pethidine Injection
Phenindamine Tartrate and *Tablets*

Chloroform extraction with aqueous titration

Codeine Phosphate Tablets
Procaine Penicillin – for procaine

Chloroform extraction with colorimetric assay

Betamethasone Sodium Phosphate Injection
Cortisone Injection

Chloroform extraction with gravimetric assay

Thiopentone Sodium and *Injection*
Methohexitone Injection

Extraction with chloroform-alcohol mixtures

Mixed solvents are not often used in the determination of salts and simple preparations. They are useful, however, for the extraction of amphoteric bases, such as morphine and nalorphine, from their salts and preparations.

The use of a mixed solvent is also often very valuable in the extraction of biological materials and of ion-pair complexes. One per cent of amyl alcohol or octanol in chloroform materially assists recovery of drugs and metabolites from plasma and counteracts adsorption of coloured complexes on glass surfaces.

Morphine Sulphate Tablets *Determination of the weight of* $(C_{17}H_{19}NO_3)_2,H_2SO_4,5H_2O$ *per tablet of average weight*

Method Weigh and powder 20 tablets. To an accurately weighed quantity of powder, equivalent to about 0.1 g of morphine sulphate in a separator, add *water* (25 ml), and M sodium hydroxide (5 ml). Mix well, add ammonium sulphate (1 g), shake to dissolve and add ethanol (95%; 20 ml). Extract with a mixture of chloroform/ethanol 95% (3/1; 40, 20, 20 and 20 ml), wash each extract with water (5 ml) and filter through a small plug of cotton wool. Evaporate the solvent and dissolve the residue in 0.05M hydrochloric acid (10 ml). Boil and cool the solution, add *water* (15 ml) and titrate the excess of acid with 0.05M sodium hydroxide, using methyl red as indicator.

$$0.01897 \text{ g } (C_{17}H_{19}NO_3)_2,H_2SO_4,5H_2O \equiv 1 \text{ ml } 0.05\text{M HCl}$$

COGNATE DETERMINATION

Nalorphine Hydrobromide and *Injection* Chloroform-isopropanol extraction.

Extraction with ether

Phenobarbitone Sodium *Determination of the percentage of* $C_{12}H_{11}N_2NaO_3$

Phenobarbitone is only sparingly soluble in chloroform, so ether is used as the extraction solvent. Since ether is less dense than water, the actual separation technique is slightly more complex than that using chloroform. Three separators are required and these are designated A, B and C in the following description.

Method Accurately weigh the sample (0.5 g approx.), introducing it via a funnel directly into separator A containing *water* (15 ml) and mix well to dissolve. Add 2M HCl (5 ml) (Note 1) and ether (50 ml), stopper and shake to dissolve the precipitated phenobarbitone; continue shaking for 2 min (NB release pressure). Allow the solutions to separate. Remove the stopper carefully and wash it and the neck of the separator with about 1 ml of ether from a teat pipette, allowing the washings to run into the separator (Note 2). Run off the lower aqueous solution into separator B. Wash through the tap and stem of A with 2–3 ml of *water*, running this off into B.

Repeat the extraction of the aqueous liquor now in separator B, shaking with a further 25 ml of ether. Allow to separate as before, remove the stopper and wash the latter and the neck of the separator with 1 ml of ether. Run off the aqueous liquor into separator C, washing through the tap and stem with 2–3 ml of *water*. Combine the ether extracts by running the second extract into A, washing the tap and stem with 2–3 ml of ether.

Continue the extraction of the aqueous solution now in separator C, with a further 25 ml of ether. Allow to separate and run off the aqueous layer into B. Transfer the ether layer to separator A. Complete the extraction of the aqueous solution now in separator B with a final

25 ml of ether. Allow to separate and run the aqueous layer into C. Transfer the ether layer to A. Four extractions should be sufficient in this and most other cases, but if necessary a fifth or even more extractions should be made to ensure complete extraction of the base. Wash the combined ether extracts with successive quantities of *water* (5 ml) until the washings are no longer acidic (Note 3). Shake the combined washings with ether (2 × 10 ml) and add the ether to the combined ether extracts. Evaporate the ether to low volume (2–3 ml), add absolute ethanol (2 ml), evaporate to dryness and dry the residue to constant weight at 105° (Note 4).

$$1 \text{ g residue} \equiv 1.095 \text{ g of } C_{12}H_{11}N_2NaO_3$$

Notes on the determination of Phenobarbitone Sodium

Note 1 Acidification of the solution liberates the free acid, which is insoluble in water.

Note 2 Ether readily evaporates about the stopper and the neck of the separator with the possibility of losses.

Note 3 The washings remove excess acid, which would interfere with the titration later.

Note 4 Ether is inflammable and distillation must be carried out well away from any naked flames.

COGNATE DETERMINATIONS

Ether extraction with aqueous titration

Dimenhydrinate and *Injection* Determination of diphenhydramine.

Emetine Hydrochloride and *Injection* Emetine is precipitated from the sample by the addition of sodium hydroxide, which retains the phenolic alkaloid cephaëline in the aqueous phase. The emetine is extracted by shaking out with successive volumes of solvent ether. The combined ether solutions are washed repeatedly with water until, after being re-extracted with ether, the washings are neutral to litmus. The emetine is re-extracted with a known volume of standard acid and then with water. The excess acid is back-titrated with standard sodium hydroxide solution using methyl red as indicator.

Methadone Injection and *Tablets*

Ether extraction with non-aqueous titration

Levallorphan Tartrate

Ether extraction with absorptiometric assay

Levallorphan Injection

Levorphanol Injection

Determination of alkaloids in crude drugs and galenicals

Separation of non-alkaloidal matter

In addition to their active principles, crude drugs and galenical preparations obtained from them contain large quantities of extraneous vegetable matter, which must first be removed if the alkaloid is to be extracted in a pure state. The non-alkaloidal matter present includes substances of many different chemical types, which can be classified as follows.

Water-soluble matter, such as sugars, glycosides, starches, proteins, gums, mucilages, tannins and saponins. These substances are present in both crude drugs and in galenicals. Although soluble in ethanol and water (the solvents most frequently used in galenical preparations), these substances are completely insoluble in chloroform and ether. They are easily removed during the course of the assay process since they are retained in the aqueous phase during extraction of the alkaloid from alkaline solution by shaking with ether or chloroform.

Resins, fats, oils and colouring matters are all readily soluble in organic solvents, but insoluble in acid, neutral and, in most instances, in alkaline aqueous solutions. They may be removed by a preliminary extraction of an acidified aqueous solution of the galenical with ether or chloroform. The alkaloids readily form water-soluble salts in acid solution and are retained in the aqueous phase whilst the unwanted material is removed in the organic solvent.

Organic acids, although often insoluble in water, all form water-soluble ammonium salts, so that in the final extraction of the alkaloid base from ammoniacal solution the acid is retained in the latter solvent, whilst the alkaloid passes into the organic phase.

Organic bases other than the required alkaloid

Since the solubility properties of such bases are similar to those of the alkaloids, their separation from the alkaloids often presents considerable difficulty and special methods are usually adopted to fit a particular case as in the following examples.

(a) **The solanaceous drugs** In addition to the active alkaloids these contain a number of volatile bases which can be removed from the alkaloidal residue by heating in an oven under specified conditions.

(b) **Nux Vomica** contains prinicipally two alkaloids, brucine and strychnine, though only the latter is determined in the official assay process. Both alkaloids are isolated together by solvent extraction. Brucine can be converted to an alkali-soluble product by oxidation with nitric and nitrous acids. The resulting solution is made alkaline with sodium hydroxide and strychnine extracted with chloroform (see also p.240). Strychnine can, however, be determined more readily in the presence of brucine by a differential light absorption method.

(c) **Opium** contains a mixture of alkaloids of which the most important is the phenolic base morphine. The bulk of the remaining alkaloids are non-phenolic. Opium is assayed on its content of phenolic alkaloids (calculated as morphine) which are separated from the non-phenolic group by conversion to water-soluble calcium salts in a preliminary extraction with calcium hydroxide.

(d) **Ipecacuanha** also contains phenolic and non-phenolic alkaloids. These are determined together to give a total alkaloid content by a direct extraction process.

General method for liquid galenicals

(A) Dilute the liquid extract with *water* and acidify to convert the alkaloids to their water-soluble salts. Extract the solution by shaking the chloroform. Allow to separate into two phases. Repeat the chloroform extraction procedure several times and combine the extracts.

Aqueous acid solution

(B) *Chloroform solution*
containing colouring matter, fat, resin and possibly traces of alkaloid.
Extract by shaking with dilute acid.

Chloroform solution
Reject

(C) *Combined aqueous-acid solution*
Make alkaline with ammonia to liberate the alkaloids from their salts. Extract the alkaloids with successive portions of chloroform.

Aqueous ammoniacal solution
Reject

Chloroform solution of alkaloid
Wash with *water*, evaporate the solution to dryness and dry the residue. Determine the alkaloid content as described previously.

Note Vigorous and continuous shaking is necessary at each extraction to ensure attainment of equilibrium in the distribution of alkaloid between two immiscible solvents. Unfortunately the presence of vegetable extractives renders the formation of emulsions likely and in these cases gentle shaking for longer periods reduces the likelihood of their formation. Compatible with the time available, it is essential to allow the two phases to separate as completely as possible. Ethanol may on occasion be used to break up or prevent troublesome emulsions (Hyoscyamus), though it should be used judiciously to avoid upsetting the partition ratio of alkaloid and other substances between the two phases. In some cases the application of gentle heat is effective (i.e. under running hot water) though great care should be exercised if the solvent is readily volatile and inflammable (ether). Coarse emulsions can be broken by scratching with a piece of wire at the emulsified interface.

General method for crude drugs and dry extracts

Preliminary treatment with a suitable organic solvent is used to extract the alkaloid, the solvent being so chosen that only the minimum of other unwanted vegetable material is also extracted. This part of the process resembles the preparation of a liquid galenical, the main steps in the process being as follows.

(a) Moisten with the solvent and allow to stand so that penetration of the tissues by the solvent can occur.

(b) Make alkaline by moistening with a small quantity of ammonia and macerate for half to one hour.

(c) Extract the alkaloid either by percolation or, where suitable, by a process of continuous extraction with an organic solvent. There are various means of carrying out a continous extraction and a Soxhlet apparatus (Fig. 9.2*a*) or the pharmacopoeial system (Fig. 9.2*b*) are two common methods for powders. Continuous liquid-liquid extraction is also used and extractors for use with solvents heavier and lighter than water are illustrated in Figs, 9.2(*c*) and 9.2(*d*), respectively. The advantage of continuous extraction is that only a small volume of solvent is necessary and tedious evaporation of large volumes of solvent can be avoided. It is also a much more rapid extraction process than percolation. The principal objection to the continuous process is that certain alkaloids cannot withstand the combined effect of heat and alkali. This applies especially to the solanaceous alkaloids.

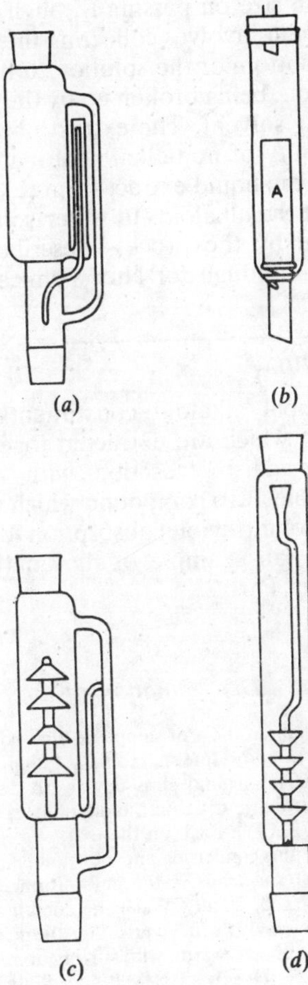

Fig. 9.2
Continuous extraction apparatus

For liquid/solid extraction the sample is held in a paper thimble (Soxhlet apparatus) or in the tube A (pharmacopoeial system) fitted with a base of calico or other suitable fabric to retain the powder.

For liquid/liquid extraction two modes are possible, one using a solvent heavier than water (downward displacement, Fig. 9.2*c*) and one using a solvent lighter than water (upward displacement, Fig. 9.2*d*). Extraction by downward displacement involves condensation of the refluxing solvent which falls through the liquid solution or suspension to the bottom of the extractor and back to the flask holding the boiling solvent. To assist rapid extraction the condensed solvent passes over baffles where the droplets are reduced in size on passing through the solution to be extracted. Upward displacement involves collecting the condensate in such a way as to convey it to the bottom of the solution to be extracted. The solvent rises through the solution, being broken up by the baffles and returns to the flask holding the boiling solvent. The extract obtained in this way contains a considerable quantity of non-alkaloidal matter and requires further purification.
(d) Treat the liquid extract from (c) with successive portions of dilute acid to convert the alkaloids to water-soluble salts. The acid solution may then be treated by the process described in the general method for liquid galenicals although for Nux Vomica the solution is examined directly.

Nux Vomica

The principal alkaloid constituents of Nux Vomica are brucine and strychnine, which are extracted together in the assay. Formerly, brucine was converted by reaction with nitrous and nitric acids to a red, alkali-soluble nitro compound which was separated from strychnine. Use is now made of ultraviolet absorption to determine strychnine in the presence of brucine, an example of the analysis of a two-component mixture as described in Part 2.

Nux Vomica *Determination of the percentage of strychnine*

Method Mix about 0.4 g of sample in fine powder accurately weighed with ethanol (70%; 2 ml) (Note 1). Add dilute ammonia solution (5M; 5 ml) and mix thoroughly. Transfer the powder to a 60 ml ground glass downward displacement liquid-liquid extractor (Fig. 9.2*c*) fitted with four baffle discs on the distributor, with the aid of water (25 ml) and chloroform (20 ml) (Note 2). The flask on the extractor apparatus should be about 100 ml volume.

Attach a reflux condenser and boil under reflux for 4 h with occasional swirling of the contents. Cool and transfer the chloroform extract to a separator and extract with 0.5M H_2SO_4 (20, 20, 20, 20 ml). Wash the combined acid extracts with chloroform (10 ml) and reject the lower chloroform layer. Filter the acid solution through a small plug of cotton wool into a shallow dish and warm, with stirring, to remove traces of chloroform. Cool and dilute to 100 ml with 0.5M H_2SO_4. Mix well and dilute 20 ml to 100 ml with 0.5M H_2SO_4.

Measure the extinction in a 1 cm cell at 262 nm and 300 nm. Calculate the percentage of strychnine from the formula

$$\text{per cent strychnine} = 5(0.321a - 0.467b)/w$$

where a is the absorbance at 262 nm, b is the absorbance at 300 nm and w is the weight of powder taken.

Note 1 The addition of ethanol (70%) is to assist thorough wetting by aqueous solvent.
Note 2 In order to avoid carry-over of suspended powder into the reflux flask the extractor must contain chloroform (60 ml) before adding the suspension. Care should be taken to avoid leaving powder on the distributor.

COGNATE DETERMINATION

Nux Vomica Liquid Extract and *Tincture*

Solanaceous Alkaloids

Belladonna Tincture *Determination of the percentage w/v of alkaloids calculated as hyoscyamine*

Method The tincture is concentrated tenfold by evaporation on a water-bath and then submitted to a preliminary process.

Make alkaline with ammonia and extract the alkaloids with chloroform

- *Chloroform solution of alkaloids*, together with chloroform-soluble colouring matter, fat and resin. Shake with dilute acid.
  - *Acid solution* Use the general process in liquid galenicals (p.238)
  - *Chloroform solution* Reject.
- *Aqueous phase* containing colouring matter. Reject.

COGNATE DETERMINATIONS

Stramonium Tincture

Hyoscyamus Tincture. The tincture is concentrated to one twenty-fifth of its volume by careful evaporation. A low temperature of evaporation is necessary to prevent decomposition of the hyoscyamine. The subsequent procedure is the same as that described for Belladonna Tincture.

Belladonna Dry Extract

In addition to desiccated extracted substance, the preparation contains a proportion of the corresponding powdered leaf which has been added to adjust the alkaloid content of the preparation to the required standard figure. This added leaf may or may not contain alkaloids. On the assumption that it does, the early stages of the extractions are suitably modified to ensure extraction of the alkaloid from the leaf tissue.

Belladonna Dry Extract is macerated with ethanol (96%) for 30 min. After making alkaline with ammonia to liberate the alkaloids from their salts, the solution is extracted with successive portions of chloroform, the latter being filtered after separation to remove pieces of leaf tissue. This preliminary process is completed by extraction of the alkaloids from the chloroform into dilute acid in the presence of ethanol to reduce emulsification.

Belladonna Herb; Prepared Belladonna Herb; Hyoscyamus Leaf; Prepared Hyoscyamus; Stramonium; Prepared Stramonium

These are drugs which consist essentially of vegetable tissue and the first step in the determination of their alkaloid content is the isolation of the alkaloids. Following the general method for crude drugs (p.229), the sample (10 g) in fine powder is moistened with ammonium solution to liberate the alkaloids from their salts and macerated with a mixture of ether and ethanol (96%) for 4 h. The mixture is transferred to a percolator and the extraction continued by percolation first with ether-chloroform (3:1) and afterwards with ether. Concentrate that percolate to 50 ml and dilute with ether to reduce the density below that of water. The alkaloids are extracted from the percolate by shaking with dilute acid and this solution is submitted to the general method for liquid galenicals (p.238).

Ipecacuanha

This drug contains a number of alkaloids of which the most important are emetine and cephaëline. Earlier Pharmacopoeias required both phenolic and non-phenolic alkaloids to be determined, but now only total alkaloids are determined, these being calculated as emetine.

Ipecacuanha *Determination of total alkaloids, calculated as emetine*

The general method for crude drugs is used, but with modifications in the later stages of the process. The alkaloids are extracted by maceration with solvent ether after the addition of ammonia to liberate the alkaloids from their salts. Maceration is effected by shaking the mixture frequently for 1 h. Solid matter is removed by the addition of water, shaking and filtration through a plug of cotton wool. The volume of the filtrate is measured by the addition of ether (5 ml) and evaporation of the solvent, followed by repetition of the procedure. The residue is dissolved in ethanol (90%), the solution diluted with freshly boiled and cooled water, and titrated with 0.1M hydrochloric acid using methyl red–methylene blue as indicator (end point purple). The alkaloid content is calculated using the ratio of the filtrate volume to that of the original total extraction mixture.

COGNATE DETERMINATION

Prepared Ipecacuanha

Ipecacuanha Liquid Extract *Determination of the percentage w/v of total alkaloids, calculated as emetine*

Total alkaloids are determined by the general method for liquid galenicals (p.238).

(a) Particular care should be exercised to remove as much colour as possible in this initial chloroform extraction.

(b) A mixture of acid and ethanol is used to wash the chloroform.

Ethanol reduces emulsification, which can also be modified if only moderate shaking is used.

(c) At this stage 5 ml of dilute ammonia solution is sufficient to make the solution alkaline. Avoid the use of any large excess of ammonia and use only moderate shaking to avoid emulsions.

The alkaloids are extracted with chloroform and determined volumetrically.

COGNATE DETERMINATION

Ipecacuanha Tincture is determined in the same way as the liquid extract using a larger volume, but without any preliminary concentration, since the latter would cause considerable darkening of colour with consequent difficulties at the end point.

Determination of non-phenolic alkaloids

These are no longer specifically determined in the official assay processes, but may be determined in the following way. After the extraction and volumetric determination of total alkaloids, transfer the solution to a separator with dilute acid and make alkaline with sodium hydroxide solution. Phenolic alkaloids form water-soluble sodium salts, whilst the non-phenolic alkaloids are precipitated and separated by extraction with ether. Titrate the alkaloidal residue remaining after evaporation of the ether.

Opium

Although several alkaloids are present in opium and its preparations only the principal one, morphine, is determined. Morphine is insoluble in chloroform and most other organic solvents, but its isolation is facilitated by the fact that the molecule is phenolic and hence soluble in solutions made alkaline with sodium or calcium hydroxide.

Opium *Determination of the percentage w/v of morphine, calculated as anhydrous morphine.*

Method Triturate the sample (8 g) with *water* (30 ml) and calcium hydroxide (2 g) in a mortar. This liberates the alkaloids from their salts, and converts morphine and narcotine into their water-soluble calcium salts. Transfer the mixture (grease the lip of the mortar) to a tared flask and make up to exactly 90 g with *water*. Shake occasionally during half an hour and filter through a dry Buchner funnel into a dry Buchner flask. Transfer exactly 52 ml of the filtrate to a clean dry conical flask and add ammonium chloride (2 g), ether (25 ml) and 90% ethanol (5 ml). The function of the ammonium chloride is to decompose the calcium salts of the alkaloids.

$$(RO)_2Ca + 2NH_4Cl \rightarrow 2ROH + 2NH_3 + CaCl_2$$

Ether acts as a solvent for narcotine and the ethanol facilitates the crystallisation of the morphine which separates from solution. Stopper the flask, shake for 5 min and occasionally during half an hour, so that the total shaking time is about 15 min. Allow to stand overnight. Decant the ethereal solution as completely as possible through a funnel with a tight plug of cotton wool retaining the crystals in the flask as far as possible. Wash the contents of the flask with a further quantity of ether (10 ml) and decant through the filter. Wash the filter with solvent ether (5 ml) added slowly and in small quantities. Pour the aqueous liquor from the flask onto the filter and wash the flask and filter with morphinated water (a saturated solution of morphine in chloroform water), until the filtrate is free from chloride. Displace the air in the flask gently with air from a bellows to ensure removal of ammonia fumes. Transfer the contents of the funnel back to the flask with *water* from a wash bottle. Add 0.05M H_2SO_4 (30 ml), boil gently to remove CO_2 and back-titrate the excess acid with 0.1M NaOH using methyl red as indicator. The end point is flat, due to the buffering effect of the phenolic hydroxyl group of morphine, and is indicated by the last shade of orange.

A correction of +0.052 g must be added to the calculated amount of anhydrous morphine, due to the solubility of morphine in the ether-water mixture. The correction as well as the general process is empirical and both depend upon 52 ml of the filtrate corresponding to exactly 5 g of the sample taken; these calculations are based on the average sample of morphine containing some definite percentage of water. It is assumed therefore that all are average samples.

COGNATE DETERMINATIONS

Powdered Opium

Opium Tincture only contains about 1% of morphine and is therefore concentrated by evaporation to dryness before proceeding as for Opium.

Camphorated Opium Tincture *Determination of the percentage w/v of morphine, calculated as anhydrous morphine*

The percentage of morphine is very low (0.5% w/v) and a colorimetric method is adopted, depending upon the fact that nitrous acid reacts with morphine to convert it into an intensely yellow nitroso derivative. The colour is intensified and becomes darker on the addition of ammonia.

Method Evaporate the sample (5 ml) to dryness on a water-bath and extract the residue with calcium hydroxide solution (10 ml). Filter into a separator and wash the evaporating dish and the filter with a further quantity of calcium chloride solution (10 ml). Add ammonium sulphate (0.1 g) to liberate the morphine and extract twice with alcohol-free chloroform (10 ml) to remove camphor and anise oil. Wash the chloroform extracts with the same 10 ml of *water*, and reject the chloroform. Add ethanol (30 ml) and chloroform (30 ml) to the combined aqueous layer and washings and shake. Separate the lower chloroform-ethanol solution of morphine and wash by shaking with 96% ethanol (5 ml) and *water* (10 ml). Repeat the extraction with two further quantities (45 ml) of chloroform(2)-ethanol(1), washing with the same liquid as before. Evaporate the combined chloroform-ethanol solution of morphine to dryness, dissolve the residue in M HCl (10 ml) and dilute with water to 50 ml.

To 10 ml of the solution (representing 1 ml of sample), add freshly prepared 1% w/v sodium nitrite (8 ml) allow to stand for 15 min and add dilute ammonia solution (12 ml). Compare the yellowish-brown colour which is produced with that obtained under the same conditions from standard amounts of morphine solution (see Part 2).

Determination of acids in crude drugs and galenicals

Extraction with ether

Balsams

Balsams are exudates from the trunks of trees and are composed of resins (both acidic and neutral), alcohols, esters of balsamic acids and the free balsamic acids themselves (benzoic and cinnamic acids).

Balsamic esters are hydrolysed by refluxing with ethanolic potassium hydroxide. Under these conditions, the potassium salts of benzoic, cinnamic and resin acids dissolve and the suspended insoluble matter consists of neutral resins and high molecular alcohols. Addition of magnesium sulphate precipitates the resin acids as insoluble magnesium salts and assists filtration. The clear aqueous filtrate is acidified and the balsamic acids extracted into ether. Purification is effected by extraction of the acids into aqueous sodium hydrogen carbonate solution, followed by acidification of the aqueous solution and re-extraction into chloroform.

Sumatra Benzoin *Determination of the percentage of total balsamic acids*

Method Reflux the sample, about 1.5 g, accurately weighed, with ethanolic KOH (0.5M; 25 ml) for 1 h. Remove the ethanol by distillation and warm with *water* (50 ml) until the residue is evenly diffused. Cool, add *water* (80 ml) and ammonium sulphate (1.5 g) dissolved in *water* (50 ml). Mix, allow to stand (10 min), filter and wash the residue with hot *water* (20 ml). Acidify the mixed filtrate and washings to liberate the free acids and extract with four successive volumes each of 40 ml of ether. Reject the aqueous solution. Shake the mixed ethereal solutions with successive volumes (20, 20, 10, 10 and 10 ml) of 5% aqueous sodium bicarbonate, separating and washing each aqueous solution with the same 20 ml of solvent ether. Acidify the mixed alkaline solutions with hydrochloric acid to liberate the acids and extract with successive portions of chloroform (30, 20, 20 and 10 ml). Dry the chloroform extract by filtering through a plug of cotton wool supporting a layer of anhydrous sodium sulphate. Remove the chloroform using a current of air, ceasing the process as soon as the last trace of chloroform has been removed to avoid volatilisation of the aromatic acids. Dissolve the residue in neutral 96% ethanol (10 ml) and titrate with 0.1M NaOH, using phenol red as indicator.

1 ml 0.1M NaOH ≡ 0.01482 of total balsamic acids, calculated as cinnamic acid

COGNATE DETERMINATIONS

Compound Benzoin Tincture
Siam Benzoin
Tolu Balsam

Soft Soap *Determination of the percentage of fatty acids*

Method Dissolve a sample of the soap (30 g) in water (100 ml), transfer the solution to a separator and acidify with dilute sulphuric acid to precipitate the fatty acids. Extract the latter by shaking with three successive portions of *water* until the latter are free from mineral acid. Evaporate the ether solution in a tared flask and dry the residue to constant weight at 80°.

The fatty acids obtained in the assay process are further examined to ensure that they comply with certain characteristics.

Acid value ≯205.

Iodine value (iodine monochloride method) ≮83.
Solidifying point ≯31°.

Limit test for Resin Shake 0.5 ml with 2 ml of acetic anhydride until a clear solution is obtained. Cool to 15.5° and transfer one drop to a white porcelain tile. Mix with one drop of 50% H_2SO_4, when there should be no transient violet colour.

Extraction with buffer solution

Cochineal

Cochineal is a colouring agent and to maintain a reasonable standard in this respect, a value is obtained by extraction of the colour, carminic acid, with a buffer solution of pH 8.0, which gives a stable colour for measurement of the extinction (Part 2).

Method Weigh exactly 0.5 g of cochineal in moderately fine powder and add solution to standard pH 8.0 (60 ml). Heat on a water-bath for 30 min, cool and dilute to 100 ml with more buffer solution. Filter and dilute 5 ml of filtrate to 100 ml with the pH 8.0 buffer. Measure the extinction of the solution at the maximum at about 530 nm. A limit of ≮0.25 in a 1 cm cell is imposed.

Determination of unsaponifiable matter

Unsaponifiable matter is determined in the Vitamin D-containing oils *Cod-liver Oil* and *Halibut-liver Oil*, and also in *Emulsifying Wax* where it forms not less than 88.0%, calculated with reference to the anhydrous substance, and consists almost entirely of the major component, cetostearyl alcohol. By contrast, the unsaponifiable matter in fixed oils is the small non-glyceride fraction, consisting chiefly of steroids. Unsaponifiable matter is calculated as the percentage w/w of ether-soluble material remaining after complete saponification. Glycerides are saponified to water-soluble soaps and glycerol and the unsaponified matter is the small fraction of the original material which is then soluble in ether. There is a tendency for the soaps to hydrolyse in solution, and traces of fatty acid so formed may dissolve out in the ether. Except in the examination of Emulsifying Wax the residue is always titrated with alkali to ensure that large quantities of such acids are not present.

Method Accurately weigh between 2 and 2.5 g of oil and reflux for 1 h with 0.5M ethanolic potassium hydroxide. Transfer the contents of the flask to a separator with *water* and, whilst still warm, extract vigorously with three successive portions of 50 ml of ether. Take great care to release the pressure in the separator. Separate the ethereal solution and mix with 20 ml of *water*. Very gently rotate the separator for a few minutes. Avoid violent shaking. Allow to separate and run off the aqueous layer. Continue washing, shaking vigorously with two successive portions of 20 ml of *water*. Shake vigorously with 0.5M aqueous KOH (20 ml) and wash again with 20 ml of *water*. Repeat the dual alkali-water washing procedure twice more. Wash with successive portions of *water* until the latter are no longer alkaline to phenolphthalein. Evaporate the ether solution in a tared flask. Evaporate the residue with a small volume of acetone (3 ml). Dry to constant weight at a temperature not exceeding 96° and titrate with 0.1M alcoholic NaOH. If a titre of more than 0.1 ml is obtained reject the results and repeat the determination.

Indicator extraction titrations

The formation of highly coloured complexes between cations and inorganic or organic compounds is well known. It forms the basis of sensitive tests for many metallic ions as the complexes may often be extracted by organic solvents, for example zinc or lead with dithizone, copper with diethyldithiocarbamate and iron(III) with thiocyanate (Chapter 8). Similarly, the formation of a blue, chloroform-soluble base cobaltithiocyanate forms the basis of a test for the identification of *Poldine Methylsulphate*. Amine salts and quaternary ammonium compounds may also be identified and determined by formation of a salt or ionic complex with complex metal ions or indicators and extraction of the base-complex into an appropriate solvent. Thus, Prudhomme (1938) found that quinine formed a chloroform-soluble complex with acid dyes, such as eosin, and Auerbach (1943) described the use of bromophenol blue for the determination of some quaternary ammonium compounds. Under carefully controlled conditions the absorbance (p.18; see also Part 2) of the organic extract may be used as a measure of the amine salt or quaternary ammonium compound. Similar principles apply to the determination of anionic compounds such as alkylaryl sulphonates, but methylene blue must be used instead of the acid indicators.

Although absorptiometry is widely used and is particularly advantageous for small concentrations of active material, titrimetric procedures can also be applied as, for example, in the titration of surface-active materials. Anionic and cationic surface-active agents are mutually incompatible and it is possible to titrate one against the other in the presence of a suitable indicator. The indicator forms a chloroform-soluble complex with one of the substances, and the end point is indicated either by a transfer of colour from chloroform to the aqueous phase or by a change of colour in the chloroform phase.

In these titrations, many amine hydrochlorides act in the same way as cationic surface-active agents and this is evident in the following technique, which is closely akin to indicator extraction methods. When an excess of potassium iodide solution is shaken with a quaternary ammonium compound and chloroform, the excess iodide may be titrated with potassium iodate. If the extraction is made from slightly alkaline solution, only quaternary ammonium compounds are measured, whereas if the aqueous layer is slightly acid, non-quaternary cationic amine impurities are also included.

It is not possible to apply one particular method to all compounds and several factors contribute to this failure. These include:

(a) *The stability of the complex*, which may be so great that the titrant fails to liberate the indicator at the end point, e.g. the hexamethonium-dimethyl yellow complex is very stable.
(b) *The relative solubilities of the compound in water and chloroform* It is important for a sharp end point that the difference in solubility in the two phases is significant.
(c) *Molecular weight* Compounds of low molecular weight often differ

from those of high molecular weight in the readiness with which complexes are formed.

(d) *The pH of the aqueous phase* Different complexes may require different pH values for extraction into organic phases.

The official methods described below are based on those of Carkhuff and Boyd (1954, *Dicyclomine Hydrochloride*) and Brown (1954, *Benzalkonium Chloride*). Indicator extraction methods based on titration with sodium tetraphenylborate are described in Chapter 11 (p.281).

Dicyclomine Tablets *Determination of the weight of* $C_{19}H_{35}NO_2$ *per tablet of average weight*

The determination depends upon the reaction of dyclomine hydrochloride with a standard sodium dodecyl sulphate solution, in the presence of chloroform, using dimethyl yellow as the extraction indicator.

Method Weigh and powder 20 tablets. To an accurately weighed quantity of the powder equivalent to about 30 mg of dicyclomine hydrochloride in a 100 ml glass-stoppered cylinder, add *water* (20 ml) and shake the mixture. Add dilute sulphuric acid (10 ml), dimethyl yellow solution (1 ml) and chloroform (40 ml) and shake well to obtain a bright yellow chloroform layer on allowing to stand. Titrate the mixture with 0.004M sodium dodecyl sulphate, shaking vigorously and allowing the layers to separate after each addition, until a permanent orange-pink colour appears in the chloroform layer.

$$0.001384 \text{ g } C_{19}H_{35}NO_2,HCl \equiv 1 \text{ ml } 0.004\text{M sodium dodecyl sulphate}$$

COGNATE DETERMINATIONS

Dicyclomine Elixir
Procyclidine Tablets

Sodium Lauryl Sulphate *Determination of the percentage of sodium alkyl sulphates calculated as* $C_{12}H_{25}OSO_3Na$

Method Dissolve the sample (about 0.25 g accurately weighed) in *water* and dilute to 100 ml in a volumetric flask. Transfer the solution (10 ml) to a 100 ml glass-stoppered cylinder, add chloroform (40 ml), dilute sulphuric acid (10 ml) and dimidium bromide–sulphan blue solution (1 ml) and titrate with 0.004M benzethonium chloride, shaking vigorously and allowing the layers to separate after each addition, until the pink colour in the chloroform layer changes to greyish-blue (Note).

$$0.001154 \text{ g } C_{12}H_{25}OSO_3Na \equiv 1 \text{ ml } 0.004\text{M benzethonium chloride}$$

Note The determination of the end point is often assisted by observing the appearance of the drops that form on the surface of the aqueous layer during separation.

COGNATE DETERMINATIONS

Sodium Dodecyl Sulphate
Cetrimide Emulsifying Ointment
Emulsifying Wax

Benzhexol Tablets *Determination of the percentage of* $C_{19}H_{29}NO,HCl$

Benzhexol Hydrochloride is extracted from the powdered tablets by heating with methanol and *water*. The base is extracted from an aliquot of a cooled solution (appropriately buffered) as the chloroform-soluble bromocresol purple complex, and determined by absorptiometry at 408 nm.

COGNATE DETERMINATION

Neostigmine Injection The quaternary ammonium base is extracted as the hexanitrodiphenylamine complex in dichloromethane and the absorbance measured at 420 nm.

Benzalkonium Chloride Solution *Determination of the percentage w/v benzalkonium chloride calculated as* $C_{22}H_{40}ClN$

In this determination, the quaternary salt is extracted as the quaternary ammonium iodide into chloroform. Excess potassium iodide in the aqueous phase is then determined by titration with potassium iodate.

Method Accurately weigh a sample (Note 1) containing about 0.5 g of anhydrous benzalkonium chloride and transfer with the aid of *water* (35 ml) to a 500 ml glass-stoppered conical separator containing chloroform (25 ml). Add M sodium hydroxide (0.5 ml) (Note 2) followed by solution of potassium iodide (5% w/v, exactly 10 ml) (Note 3). Stopper the separator and shake well. Allow the two layers to separate and run off the lower chloroform layer through a loosely packed plug of absorbent cotton wool (about 0.5 g) or through phase separating filter paper placed in a small glass funnel to absorb traces of entrained aqueous liquid. Discard the chloroform layer and repeat the extraction with chloroform (3 × 10 ml) running the chloroform layer each time through the same filter. Finally, wash the filter with a further 5 ml of chloroform and allow to drain.

Add to the separator, hydrochloric acid (40 ml), previously chilled (Note 4) and titrate the mixture in the separator with 0.05M potassium iodate until the solution becomes light brown in colour; add chloroform (5 ml), stopper and shake. Continue the titration, with shaking, until the chloroform becomes colourless and the supernatant liquid is clear yellow. Remove the cotton wool or paper from the funnel and add directly to the contents of the separator. Wash the glass funnel with *water* (2 or 3 ml), receiving the washings in the separator. Stopper, shake and complete the titration if necessary.

Perform a blank by titrating a mixture of *water* (20 ml), solution of potassium iodide (exactly 10 ml) and hydrochloric acid (40 ml), adding chloroform (5 ml) when the mixture becomes light brown in colour. The differences between the two titrations represents the 0.05M potassium iodate equivalent to the anhydrous benzalkonium chloride in the amount of sample taken for assay.

$$0.0354 \text{ g } C_{22}H_{40}ClN \equiv 1 \text{ ml } 0.05\text{M } KIO_3$$

Note 1 The quantity stated is for solutions containing 30–100% of benzalkonium chloride. For solutions containing less than 30% the volume of liquid used should not exceed 35 ml. If the volume is less, add *water* to 35 ml.

Note 2 The amount of alkali used is normally sufficient to liberate all the amines that may be present as impurities, but more alkali should be added if it is suspected that contamination is excessive. If the procedure is carried out in the presence of 0.1M hydrochloric acid (0.5 ml, or sufficient to ensure amine impurities are present as hydrochlorides) the method will give a result which includes amine impurities.

Note 3 The solution of potassium iodide should be measured by pipette.

Note 4 The addition of chilled hydrochloric acid prevents undue warming-up of the solution on mixing.

COGNATE DETERMINATIONS

Cetrimide
Cetylpyridinium Bromide
Domiphen Bromide

10
Miscellaneous methods

Determination of cineole

Cineole, like most ethers, is relatively unreactive and its determination in volatile oils depends upon the formation of an addition compound with an equimolecular proportion of *o*-cresol. When pure, this addition compound melts at 55.2° and its melting point is depressed by the presence of impurities. Table 10.1 (reproduced here from the British Pharmacopoeia) correlates the freezing point of the cineole *o*-cresol mixture with various percentages of cineole in *Eucalyptus Oil*.

Table 10.1 Freezing points of cineole — *o*-cresol mixtures

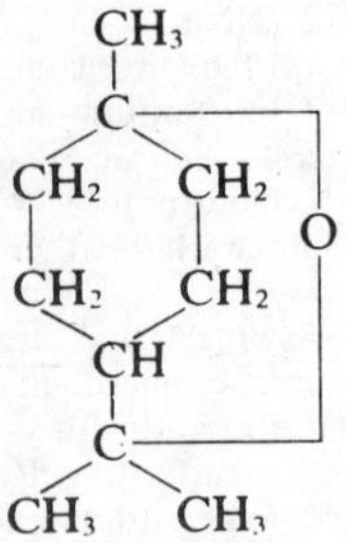

Freezing point(°)	% w/v of cineole	Freezing point(°)	% w/v of cineole
41	68.5	49	84.0
42	70.0	50	86.0
43	72.5	51	88.5
44	74.0	52	91.0
45	76.0	53	93.5
46	78.0	54	96.0
47	80.0	55	99.0
48	82.0	55.2	100.0

Intermediate values can be obtained by interpolation.

Method Dry the oil by shaking with anhydrous sodium sulphate. Weigh precisely 3 g of the oil into a hard glass test tube (80 mm × 15 mm) supported vertically on the balance pan by inserting it in a cork with a hole as in the diagram (Fig. 10.1). It is convenient to add the oil

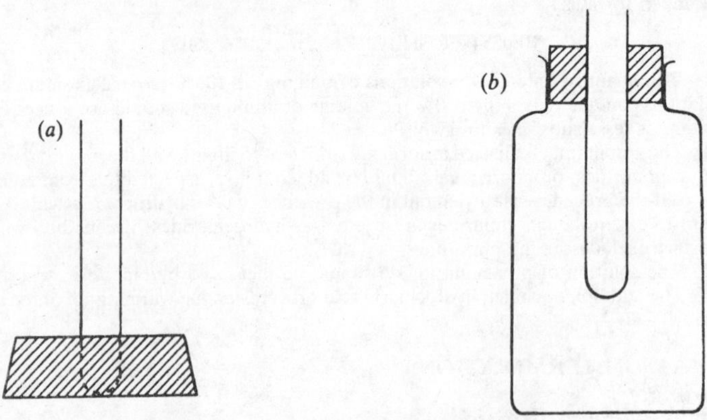

Fig. 10.1 Cineole determination apparatus

from a dry teat pipette. Add precisely 2.1 g of melted *o*-cresol. Warm the tube gently until the contents are completely molten and insert the tube in a bored cork in a wide-mouthed bottle (Fig. 10.1b). Allow to cool slowly, stirring continuously by rubbing the inside of the tube with the bulb of a thermometer graduated in fifths of a degree. Note the highest temperature recorded on the thermometer whilst crystallisation is occurring. Remelt and repeat the process until two consecutive identical results are obtained.

Determination of ethanol in liquid galenicals

Distillation procedure

The traditional process for the determination of ethanol in liquid galenicals by distillation and physical measurement is no longer official in the British Pharmacopoeia, but is still used in the examination of some galenical preparations. This process consists of the following steps.

(1) Removal of volatile impurities if any.
(2) Distillation of all the ethanol and some of the water.
(3) Dilution of the distillate to four times the original volume and determination of the specific gravity of the resulting solution.
(4) A test on the distillate to determine whether the refractive index is in agreement with the recorded specific gravity. The distillate must comply with the limit test for absence of methanol.

The alcohol content is calculated from the specific gravity using the table (quadruple bulk) in the British Pharmacopoeia, 1968.

Three general methods are available for the determination of ethanol galenicals by distillation, the choice of method depending upon the nature of the other substances also present.

Method 1 *No other volatile substances present*
Measure 25 ml of the galenical in a graduated flask at a temperature of 20°. Transfer to a distillation flask (500 ml) with about 100 to 150 ml of *water*. Add pumice powder to prevent bumping and distil the liquid through a short fractionating column, collecting about 90 ml of distillate in a 100 ml graduated flask. Adjust the temperature of the distillate to 20° and dilute to 100 ml with *water* at the same temperature. Determine the specific gravity at 20° and the refractive index of the solution at 20°. Refer to the alcohol (quadruple bulk) table. The refractive index calculated on the basis of the observed specific gravity should not differ from the observed refractive index by more than 0.00007. If the distillate complies with this requirement, read off the ethanol content from the table. If it does not comply, some impurity is present and 75 ml of the distillate should be treated with sodium chloride and light petroleum as in Method 2. Distil about 70 ml and dilute to 75 ml, then re-check the specific gravity and refractive index for mutual agreement.

Acidic solutions should be neutralised with sodium hydroxide before distillation.

Method 2 *Simple solutions containing other volatile substances*
Measure 25 ml of the galenical at 20° in a graduated flask. Transfer to a separator with about 100 ml of *water*. Saturate the solution with sodium chloride and shake vigorously with an equal volume of light petroleum (b.p. 40 to 60°) for 2–3 min. The sodium chloride helps to salt out unwanted substances which dissolve in the light petroleum. This solvent is used because it is completely insoluble in water and does not extract ethanol from aqueous solutions. Allow

to separate for 15–30 min and run the lower aqueous layer into a distillation flask. Wash the organic solution in a separator by shaking vigorously with saturated sodium chloride solution (25 ml) and run the aqueous phase into the distillation flask. In some cases a double separation is used, in which the aqueous liquors are extracted with a second volume of light petroleum. The same volume of wash liquor is used for both petroleum extracts. Make the mixed aqueous solutions just alkaline with 0.1M NaOH (use solid phenolphthalein as indicator). Distil as in Method 1.

This method may be applied to such preparations as Tinctures, Chloroform Spirit, and Aromatic Waters.

Method 3 *Complex solutions containing other volatile substances*

Measure 25 ml of the galenical as before, and transfer to a 500 ml distillation flask with water (100 to 150 ml). Add pumice powder and distil about 100 ml. Treat the distillate by Method 2.

This method may be used for galenicals such as Coal Tar solution, which contain other volatile substances or material which precipitates on dilution with water; these combinations would complicate the extractions by light petroleum.

Gas-liquid chromatographic procedure

The gas chromatographic method overcomes all the problems of the now outdated distillation method, and additionally is simple and easy to operate (Part 2).

The GLC method is carried out on a 1.5 m column, 4 mm internal diameter, of porous polymer beads (Porapak Q or Chromosorb 101) at 150° with nitrogen as carrier gas and flame ionization detector. *n*-Propanol is used as an internal standard and the column calibrated with a solution containing *n*-propanol (5% v/v) and dehydrated ethanol (5% v/v) in *water*. This is compared with a second chromatogram prepared from the sample suitably diluted with *water* to contain between 4% and 6% of ethanol, to ensure that no impurity is present with the same retention time as the internal standard. Ethanol content is then determined from a third chromatogram prepared from the sample, diluted as before, containing the internal standard (5% v/v).

Determination of Industrial Methylated Spirit

The GLC method is applied to the determination of both ethanol and methanol in Industrial Methylated Spirit and preparations prepared with it. The method is essentially that used for determination of ethanol except that a fourth solution containing the same internal standard, *n*-propanol (5% v/v), together with methanol (0.25% v/v) is used, and instrument amplifier gain is increased when the methanol peak is recorded. The total alcohol content must be within the range specified in the monograph for the preparation under examination and the methanol content must reasonably conform with that calculated on the basis that Industrial Methylated Spirit has been used.

Isopropanol in denatured ethanol can be determined together with methanol by a similar GLC method using an appropriate reference standard solution.

Mercurimetric titrations

Sulphydryl compounds react quantitatively with mercury(II) salts in either a 2:1 or a 1:1 molar ratio and, hence, are capable of being titrated mercurimetrically in either a one- or a two-step reaction, according to the following general equations.

$$2RSH + HgX \rightarrow (RS)Hg + 2HX$$
$$(RS)Hg + HgX \rightarrow 2RSHgX$$

Propylthiouracil *Determination of the percentage of* $C_7H_{10}ONS$

This determination depends upon the conversion of methylthiouracil to a water-soluble disodium derivative with sodium hydroxide and titration of the latter in an acetate buffer system with mercury(II) acetate:

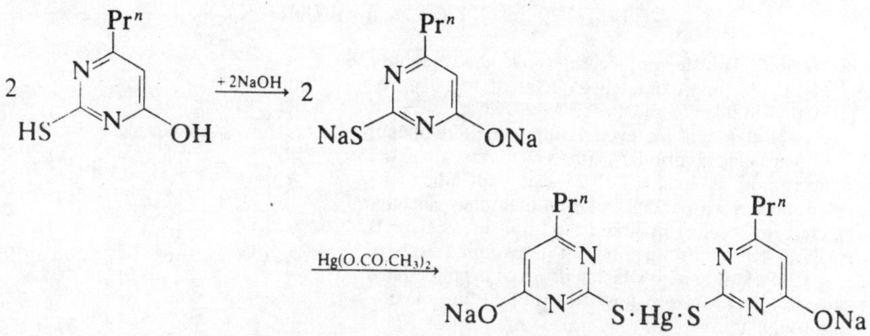

$$\therefore \quad 2 \times 170.2\,g\ C_7H_{10}ON_2S \equiv Hg(O \cdot COCH_3)_2 \equiv 1000\,ml\ \text{M}$$
$$\therefore \quad 0.01702\,g\ C_7H_{10}ON_2S \equiv 1\,ml\ 0.05\text{M mercury(II) acetate}$$

Method Accurately weigh the sample (about 0.35 g) and warm with 0.1M NaOH (50 ml) and *water* (200 ml) to dissolve. Cool and add sodium acetate (10 g) and acetic acid until the solution is just acid to litmus. Titrate with 0.05M mercury(II) acetate solution using a freshly prepared solution (1 ml) of 1,5-diphenylcarbazone (0.5% w/v) in ethanol (96%) as indicator. The end point is indicated by the formation of the mercury-diphenylcarbazone complex which is rose-violet in colour. This colour should persist for 2 to 3 min.

COGNATE DETERMINATIONS

Penicillamine, Capsules and *Tablets*
Propylthiouracil Tablets

β-Lactam antibiotics

Mercurimetric titrations are now increasingly widely used to control β-lactam antibiotics. The antibiotics themselves do not react directly with mercury(II) nitrate, but their decomposition products, the related penicilloic and penicillenic acids, can be titrated with this reagent. The use of mercurimetric preparations in the assay and control of degradation products in penicillin preparations is discussed in Chapter 11.

Determination of methoxyl

Methylcellulose *Determination of the percentage of* (CH_3O)

The proportion of methoxyl radical in the sample is determined by a modification of the Zeisel method. The reaction is expressed by the following equations:

$$R \cdot OCH_3 + HI \rightarrow ROH + CH_3I$$
$$CH_3I + CH_3COOH \rightarrow CH_3COO \cdot CH_3 + HI$$
$$HI + 3Br_2 + 3H_2O \rightarrow HIO_3 + 6HBr$$
$$(Br_2 + H \cdot COOH \rightarrow 2HBr + CO_2) \text{ Removal of excess bromine}$$
$$\text{Add KI,} \quad HIO_3 + 5HI \rightarrow 3I_2 + 3H_2O$$
$$\therefore \quad (CH_3O) \equiv CH_3I \equiv HIO_3 \equiv 6I$$
$$\therefore \quad 31.034\ g\ (CH_3O) \equiv 6000\ ml\ M$$
$$\therefore \quad 0.0005172\ g\ (CH_3O) \equiv 1\ ml\ 0.1M\ Na_2S_2O_3$$

Method Transfer an accurately weighed quantity of sample, approximately equivalent to 50 mg of methyl iodide, to the flask. Add a short piece of glass rod to ensure even boiling, melted phenol (2.5 ml) and hydriodic acid (5 ml). Assemble the apparatus as in Fig. 10.2 after half-filling the scrubber S with a 25% solution of sodium acetate. Place in receiver A about 6 ml and in receiver B about 4 ml of 10% solution of potassium acetate in glacial acetic acid to which 4 drops of bromine have been added. Pass a slow stream of carbon dioxide through the capillary side arm of the flask. This sweeps out methyl iodide from the flask as it is formed. Carefully heat the flask using a mantled microbunsen or suitable electric heating mantle in such a way that the vapours of the boiling liquid rise half way up the condenser. Continue heating thus for 30 min. Wash the contents of the two receivers A and B into a conical flask containing 25% w/v aqueous sodium acetate and make up the total volume to 125 ml with *water*. Add 6 drops of formic acid and rotate the flask until the bromine colour is discharged. Add a further 12 drops of formic acid and allow to stand 1 to 2 min. Add potassium iodide (1 g), acidify with dilute sulphuric acid and titrate the liberated iodine with 0.1M $Na_2S_2O_3$. Carry out a blank determination.

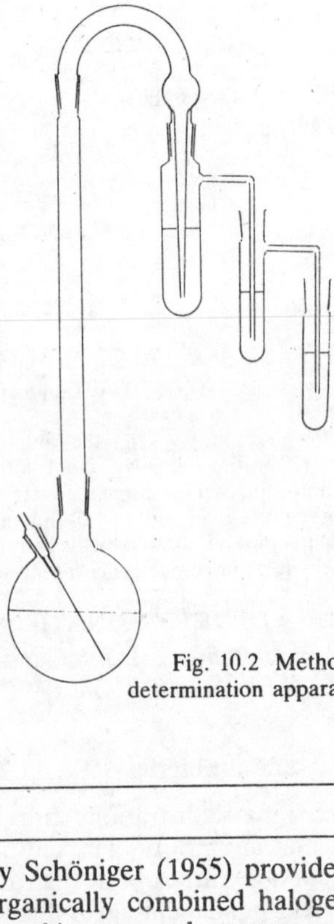

Fig. 10.2 Methoxyl determination apparatus

Oxygen flask combustion

Flask combustion analysis as developed by Schöniger (1955) provides a rapid method for the determination of organically combined halogens. Complete combustion of the organic compound in oxygen destroys organic matter, releasing the halogen, which is absorbed in the combustion flask into sodium hydroxide solution to yield, in the case of iodo-compounds, a mixture of iodide and iodate. Subsequent oxidation (with bromine) and then acidification in the presence of iodide gives a sixfold yield of free iodine which is determined by titration with sodium thiosulphate. The method is, therefore, particularly useful for expensive materials and products in which the halogen-containing material is present at low dosage.

The method is used widely in the assay of organo-iodine compounds such as *Nitroxynil, Propyliodone* and *Liothyroinine Sodium*. It is also used in the determination of chlorine in *Cloxacillin Sodium* to supplement the assay, which is only semispecific, and fluorine in *Fluocinolone Acetonide*. It is also used as an identity test to confirm the presence of chlorine in *Diloxamide Furoate* and *Chlorthalidone*.

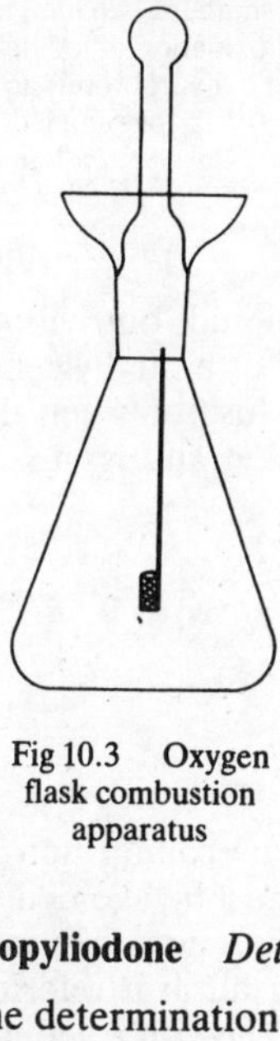

Fig 10.3 Oxygen flask combustion apparatus

Flask combustion analysis has also been adapted to the determination of elemental sulphur, which is oxidised to sulphur trioxide absorbed in water (and peroxide to ensure complete oxidation) and titrated. Other applications of flask combustion include the determination of organo-mercury compounds, and the combustion of water in animal tissue and blood samples prior to measurement of their carbon-14 and tritium content.

Apparatus

The apparatus (Fig. 10.3) consists of an iodine-flask (500 ml) conforming to British Standard Specification (BS 2735:1956), specially modified by fusion into the stopper of a platinum wire, to which is attached a platinum gauze basket of specified dimensions. The gauze complies with the dimensions of a No. 36 sieve.

Propyliodone *Determination of the percentage of* $C_{10}H_{11}I_2NO_3$

The determination depends upon the reactions expressed by the following equations:

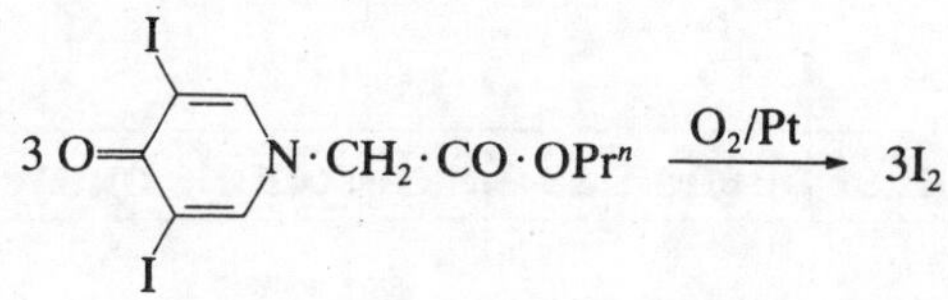

$$3I_2 + 6NaOH \rightarrow 5NasIO_3 + 3H_2O$$
$$5[NaI + 3Br_2 + 3H_2O \rightarrow NaIO_3 + 6HBr]$$
$$6[NaIO_3 + 5KI + 6HBr \rightarrow 3I_2 + NaBr + 5KBr]$$

$$I_2 + 2Na_2S_2O_3 \rightarrow 2NaI + Na_2S_4O_6$$

$$\therefore \quad 3 \times 447\text{ g } C_{10}H_{11}I_2NO_3 \equiv 6I \equiv 6NaIO_3 \equiv 36I \equiv 36\,000\text{ ml M}$$

$$\therefore \quad 0.7450\text{ mg } C_{10}H_{11}I_2NO_3 \equiv 1\text{ ml } 0.02\text{M } Na_2S_2O_3$$

Method Accurately weigh the sample (20 mg) on to a strip of filter paper (Whatman No. 1) about 5 cm by 3 cm which has been folded into three along its length. Wrap the sample in the filter paper by folding in the outer thirds and rolling up the folded strip. Place the roll in the platinum gauze cup, and insert a narrow strip of filter paper into the roll to act as a fuse. Displace air from the flask with oxygen, moisten the neck of the flask with *water*, place *water* (10 ml) and M sodium hydroxide (2 ml) in the flask and fill it with oxygen. Light the end of the fuse, insert the stopper immediately and *hold it firmly in place*. When the sample is burning vigorously, tilt the flask to prevent incompletely burned material from falling into the liquid. As soon as combustion is complete, shake the flask vigorously for 5 min. Place a few ml of *water* in the cap, carefully withdraw the stopper and rinse the stopper and wire assembly with *water*. Add an excess of acetic bromine solution (5–10 ml) and allow to stand for two min; the bromine in potassium acetate–acetic acid buffer solution completes the oxidation of iodide to iodate. Add formic acid (0.5–1 ml) to destroy excess bromine (reduction to hydrobromic acid) and sweep out any bromine vapour from the flask with a current of air. Add potassium iodide (1 g) and titrate with 0.02M $Na_2S_2O_3$ using starch mucilage as indicator. Carry out a loss on drying on a separate sample at 105°.

COGNATE DETERMINATIONS

Iophendylate Injection Weigh the sample, which is liquid, onto about 15 mg of ashless filter paper flock contained in one half of a methylcellulose capsule; seal the capsule inserting the filter paper fuse between the two parts. Place the capsule in the platinum gauze basket and ignite.

Liothyronine Sodium
Propyliodone Suspension
Propyliodine Oily Suspension
Thyroxine Sodium

Fluocinolone Acetonide *Determination of fluorine*

Organically combined fluorine is converted to fluorine by combustion in oxygen in a silica or soda glass flask. Borosilicate glass must be avoided as fluorine reacts with boron present in hard glass. Fluorine is absorbed into water and determined colorimetrically at 610 nm using alizarin fluorine blue and cerous nitrate in an acetate buffer, in comparison with a sodium fluoride standard.

Determination of phenols as alkali-soluble matter

Clove oil

Eugenol, which is present in the oil together with acetyl eugenol and other phenolic substances to the extent of 85–90% w/v, is determined by treating the oil with a solution of potassium hydroxide and measuring the material not dissolved. A Cassia flask is used. It has a capacity of about 150 ml (Fig. 10.4). The neck of the flask is long, having a capacity of 10 ml and being of such diameter that it is not less than 15 cm in length. It is graduated in tenths of a ml. Before use, the flask should be cleaned with sulphuric acid and well rinsed with *water* to free it from grease.

Method Place 80 ml of 5% aqueous potassium hydroxide in the flask and pipette in 10 ml of the sample. The phenolic eugenol dissolves in the potassium hydroxide and non-phenolic material (terpenes) forms a separate layer on the surface of the liquid. Shake thoroughly at 5 min intervals during half an hour. Gradually add more aqueous potassium hydroxide, so that the undissolved oil is slowly raised into the graduated neck of the flask. Allow to stand for not less than 24 h and record the volume of undissolved oil. The volume of undissolved oil should be between 1 and 1.5 ml.

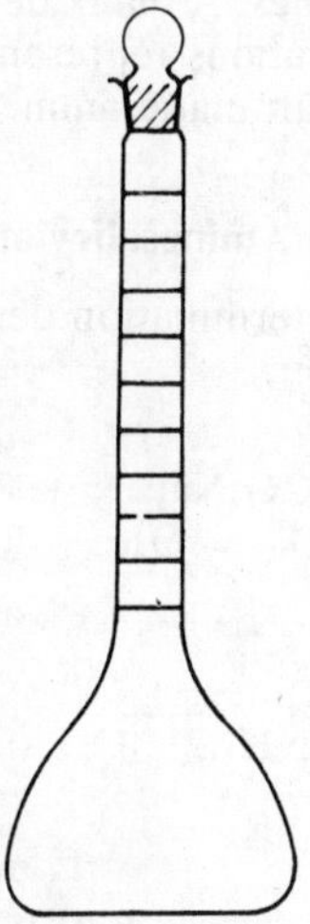

Fig.10.4 Cassia flask

Sodium nitrite titrations

Aromatic primary amines react with sodium nitrite in acid solutions (i.e. nitrous acid) to form diazonium salts:

$$C_6H_5NH_2 + NaNO_2 + HCl \rightarrow C_6H_5N_2Cl + NaCl + 2H_2O$$

Under controlled conditions the reaction is quantitative, and can be used for the determination of most substances containing a free primary amino group, as in sulphanilamide and other sulpha drugs.

Observation of the end point depends upon the detection of the small excess of nitrous acid which is then present. This can be demonstrated visually using starch-iodide paper or paste as external indicator.

$$KI + HCl \rightarrow HI + KCl$$

$$2HI + 2HNO_2 \rightarrow I_2 + 2NO + 2H_2O$$

The iodine liberated reacts with starch to form a blue colour.

Alternatively, the end point may be detected amperometrically using a pair of bright platinum electrodes immersed in the titration liquid. Electrode polarisation occurs when a small voltage (30–50 mV) is applied across the electrodes, and no current flows through the sensitive galvanometer included in the circuit, during the course of the titration. Liberation of excess nitrous acid at the end point depolarises the electrodes, current flows in the galvanometer and a permanent deflection of the galvanometer needle is observed. This is known as the dead-stop end point. The electrodes must be clean, otherwise the end point is sluggish. Cleaning can be accomplished by immersing the electrodes in boiling nitric acid containing a little iron(III) chloride for about 30 s and then washing with *water*. A blank determination is necessary; the difference between the two titrations represents the volume of sodium nitrite solution equivalent to the aromatic amine.

Sodium Aminosalicylate *Determination of the percentage of* $C_7H_6NNaO_3$

This determination depends upon the reaction expressed by the following equation:

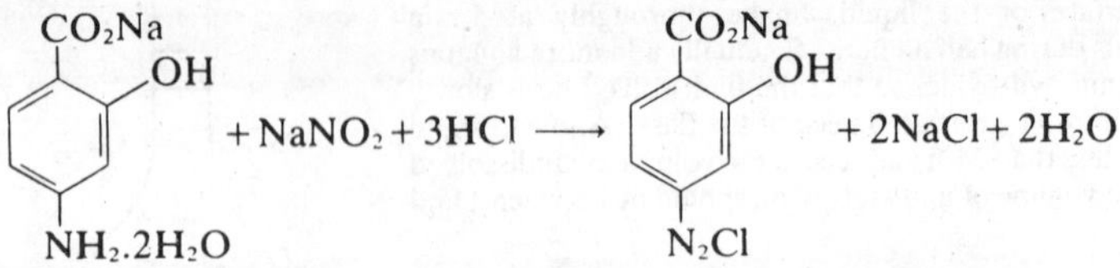

$$\therefore \quad 175.2\ g\ C_7H_6NNaO_3 \equiv NaNO_2 \equiv 1000\ ml\ \text{M}$$
$$\therefore \quad 0.01752\ g\ C_7H_6NNaO_3 \equiv 1\ ml\ 0.1\text{M}\ NaNO_2$$

Method 1 *Visual end point*
Weigh the sample (about 2.5 g), accurately, into a funnel in the mouth of a 250 ml graduated flask. Wash through with concentrated hydrochloric acid (50 ml) and *water*. Dissolve, add potassium bromide (5 g) and adjust to volume. Pipette 50 ml into a conical flask, cool to below 15° (cold tap) and titrate slowly with 0.1M $NaNO_2$ solution, shaking continuously until a distinct blue colour is produced, when a drop of titrated solution is placed on a thin film of starch-iodide paste spread on a porcelain tile 5 min after the last addition of the 0.1M $NaNO_2$ solution. The latter should be added in quantities of 0.1 ml towards the end of the titration. Note that sometimes a blue colour results even after only a few ml of sodium nitrite have been added. The reaction is slow in starting but the solution behaves normally as the titration proceeds. A blue colour is produced on the starch iodide paste before the end point is reached, but at this stage it takes time to develop. Only at the end point does the blue colour appear on the paste instantaneously.
Method 2 *The electrometric end point* (see Part 2)
Accurately weigh the sample (about 0.5 g) into a 250 ml beaker and dissolve in hydrochloric acid (10 ml) and *water* (75 ml). Insert a pair of bright platinum electrodes into the solution, connected through a sensitive galvanometer and a suitable potentiometer to a 2 V battery in such a way as to produce a potential drop of between 30 and 50 mV across the electrodes. Titrate slowly with 0.1M $NaNO_2$, keeping the lip of the bottle below the surface of the liquid, stirring continuously (mechanical stirrer), until a permanent deflection of the galvanometer is observed at the end point.

COGNATE DETERMINATIONS

Benzocaine
Dapsone
Primaquine Phosphate and *Tablets*
Procainamide Hydrochloride and *Injection*
Procaine Hydrochloride
Sulphacetamide Sodium and *Eye Ointment*
Sulphadimidine and *Tablets*
Sulphadimidine Sodium and *Injection*
Sulphadoxine
Sulphamethizole and *Tablets*

Sulphamethoxazole
Sulphamethooxypyridazine and *Tablets*
Sulphapyridine and *Tablets*
Sulphathiazole and *Tablets*

Succinylsulphathiazole *Determination of the percentage of* $C_{13}H_{13}O_5N_3S_2$ *calculated with reference to the substance dried to constant weight at* 105°

This substance is determined similarly to sulphaguanidine after a preliminary hydrolysis of the protecting acyl group. This determination depends upon the reactions expressed by the following equations:

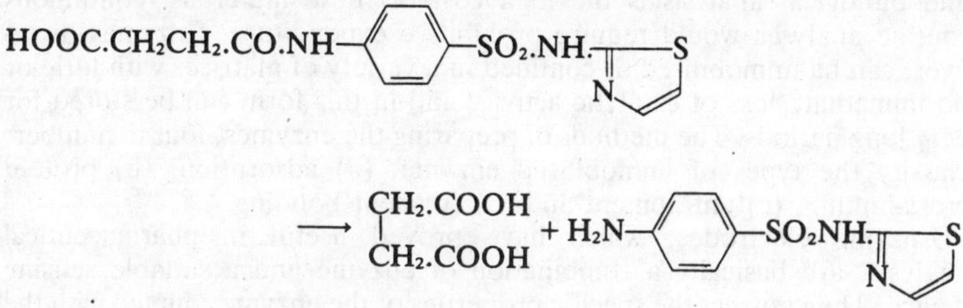

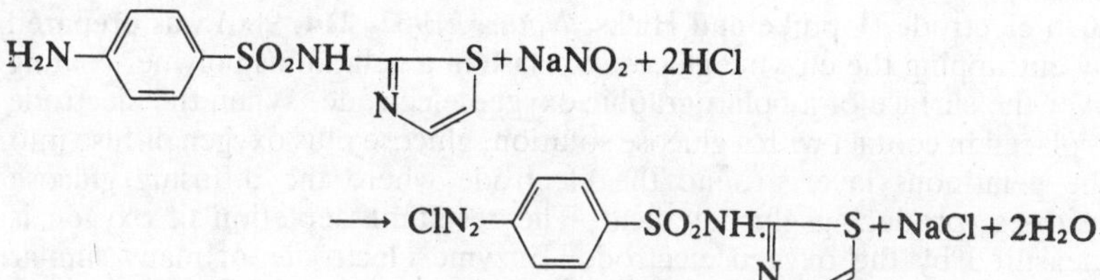

Method Dissolve the sample (about 0.5 g), accurately weighed, in a mixture of concentrated hydrochloric acid (33 ml) and *water* (66 ml). Reflux for 1 h to hydrolyse the substance as shown by the equation above, and then complete the assay as described under Sodium Aminosalicylate.

COGNATE DETERMINATION

Succinylsulphathiazole Tablets

Enzymes in pharmaceutical analysis

Enzymes are proteins which act as biological catalysts in complex chemical reactions essential to the continued existence of living systems. They have the advantage of specificity, each acting only in the presence of compounds (substrates) with a particular functional group. Additionally, it frequently happens that one member of a homologous series of compounds is active in the enzyme-catalysed reaction, while others are totally inactive or react at much slower rates. Many enzymes are also specific for one particular optical isomer of the substrate. Another advantage in the use of enzymes for analysis is the extreme sensitivity achieved, often as low as 10^{-7}M or lower of substrate concentration. Carbohydrates, amines, amino acids, organic acids, alcohols and esters plus miscellaneous drugs not in these categories are regularly determined enzymatically, e.g. penicillins by β-lactamase.

As well as the determination of substrates, the enzymes themselves, coenzymes, activators and inhibitors are routinely determined in medical situations. Such measurements, usually in plasma, serve to indicate, reliably, sensitively and specifically, deviations from the normal state with minimal risk and inconvenience to the patient.

One of the principal objections to the use of enzymes in pharmaceutical and biological analysis is the high cost of these materials; continuous routine analyses would require prohibitive expenditure. Enzymes, however, can be immobilised or confined in a variety of matrices with little or no immediate loss of catalytic activity and in this form can be stored for very long periods. The methods of preparing the enzymes, four in number, classify the types of immobilised enzyme: (a) adsorption, (b) protein cross-linking, (c) entrapment and (d) covalent bonding.

Enzyme electrodes, which have proved useful in pharmaceutical analysis, are basically a combination of enzyme and a suitable sensing device. They possess the specific properties of the enzyme coupled with the normal electronic read-out associated with electrode systems. The first such electrode (Updike and Hicks, *Nature*, 1967, **214,** 986) was prepared by entrapping the enzyme glucose oxidase in a gelatinous polymer coating over the surface of a polarographic oxygen electrode. When the electrode is placed in contact with a glucose solution, glucose plus oxygen diffuse into the gelatinous layer around the electrode where the diffusing glucose oxidises, producing gluconic acid. The resultant depletion of oxygen is measured by the oxygen electrode. Enzyme electrodes of many similar types and varieties are possible and some have been developed.

Another type of enzyme electrode system uses enzymes to quantify either a substrate or a product. Such a substrate-selective electrode has been used, for example, in the studies of cholinesterase activity in blood. Many of the experimental difficulties of using enzymes in analysis by reaction rate methods are lessened or eliminated by automating all steps in the procedure.

As with other analytical procedures, progress in the use of enzymic analytical procedures is reviewed regularly in *Analytical Chemistry*.

Microbiological assays

Many biological products are substances and preparations used in medicine that cannot, as yet, be defined completely by chemical and physical methods and include antibiotics, immunological products, natural and synthetic polypeptide hormones, and blood and related products.

Most biological assay procedures yield highly sensitive and selective measurement of the characteristic activity of the biological product. However, they are usually complex, expensive, lengthy and of poor reproducibility, requiring a sufficient number of replicates to permit statistical evaluation. There is, thus, a strong incentive to replace them with chemical or physical methods.

By comparison with chemical and physical assays, microbiological methods are limited in scope to those substances or preparations which will influence the growth of micro-organisms in such a way as to give a measurable effect. These are, therefore, antibacterial and antifungal agents such as antibiotics and preservatives which inhibit the growth of bacteria and fungi and those substances without which the micro-organisms cannot grow, i.e. essential growth factors such as amino acids and vitamins. The vitamins amenable to microbiological assay are those of the B group, such as *Thiamine Hydrochloride, Riboflavine, Pyridoxine Hydrochloride, Nicotinic Acid* and its derivatives and *Cyanocobalamin* (Vitamin B_{12}).

In assessing the relative merits of chemical, physical and microbiological methods, it is well worth bearing in mind what is expected of an assay. Thus, it should be:

(i) specific
(ii) easy to carry out and require simple materials and apparatus
(iii) sensitive
(iv) inexpensive in time
(v) reproducible and precise.

Specificity

It is extremely rare for chemical or physical methods to be specific as they are normally based on a group reaction rather than on the molecule as a whole. Microbiological assay, on the other hand, measures directly the property which the compound is expected to exert, i.e. the potency of the compound; it involves either of two procedures.

Plate diffusion technique Measured amounts of the test substance are introduced into cups cut into agar plates which have previously been inoculated with a test organism. It diffuses into the agar and produces zones of inhibition (or stimulation of growth, depending on the test substance). The diameters of the zones may be plotted against log potency for a standard and, by interpolation on this curve, the potency of the test material may be read.

Tube-dilution method Suitable dilutions of the sample are introduced into an appropriate nutrient medium and suppression or stimulation of the growth of a test organism occurs. The method is suitable for the determination of the vitamin B group as the measurement may be made by nephelometry, titration or gravimetrically, depending on the vitamin.

A distinctive feature of results obtained from biological assays generally is that they are subject to random error arising from the inherent variability of biological responses. It is, therefore, essential to assay sample material against a standard preparation so that the relative responses are measured at the same time and under the same circumstances.

Complexity and sensitivity

Microbiological assays are not simple to carry out and experience must be obtained before reliability can be achieved. This is in contrast to physical methods such as spectrophotometry which, for simple formulations, merely involves dilution and measurement of absorbance; all the B vitamins, aromatic preservatives and many antibiotics are amenable to such treatment, though careful control of pH may be important, as, for example, in the assay of *Pyridoxine* (p. 286). Chemical methods are less easy, e.g. assay of *Thiamine Hydrochloride* by oxidation to thiochrome and fluorimetric assay, or by a gravimetric method based on silicotungstic acid. When complex formulations are involved, particularly in the presence of highly coloured organic material, microbiological methods may offer some advantage by avoiding complex separation procedures.

The apparatus required for physical methods is somewhat complex and very costly, while chemical and microbiological assays require relatively simple and cheap glassware. Reagents for biological methods, however, require care in preparation and sterilisation and the tests must be carried out in a specially equipped laboratory reserved for this purpose. In addition, the time involved in microbiological methods is considerable, as it must allow for bacterial growth, and should repeat assays be necessary, the delay is longer.

The usual level of compound necessary for spectrophotometry is 5–10 μg/ml; chemical determinations require much larger quantities, but microbiological methods have the distinct advantage of being very much more sensitive and so require no more than 0.1 μg of active ingredient. Thus, a microbiological method based on the antibiotic effects of penicillin proves to be the most sensitive way of limiting undesirable levels of penicillin impurity in Penicillamine.

Reproducibility is interrelated with accuracy and precision and one particular advantage of microbiological assays in this respect is that the analyst recognises the limitations of the method and expresses his results accordingly, in statistical terms. The final result is obtained from several assays and is always a comparative one in that the figures obtained are compared with those for the standard reference substance. Nevertheless, difficulties will always arise in agreeing the actual potency of product when the results obtained in two or more laboratories are compared. This problem is accentuated with antibiotics and other poor-stability products in which potency may fall during storage. A clear margin between the manufacturing specification (release specification) and the check pharmacopoeial specification of a product is essential in setting standards for such products. This is achieved for antibiotics in the United Kingdom by setting release standards in the *Compendium of Licensing Requirements for the Manufacture of Biological Medicinal Products* (HMSO) according to the requirement that 'when a minimum potency is specified (and is less than that corresponding to 95% equivalent purity) the calculated *lower* fiducial limit of the estimated potency shall be *greater than* the specified value.'

Biological tests on animals

A distinction can be drawn between microbiological assays using micro-organisms and biological tests using animals. The latter tests very often cannot be replaced by chemical and physical tests because of lack of correlation between results for potency obtained biologically and by chemical/physical means.

Success in replacing biological by chemical tests and assays, however, has been achieved in many instances, for example, with steroids, vitamins and antibiotics, and the list of standard biological substances provided by the World Health Organization is diminishing. The structural synthesis of a biological product, although not sufficient in itself to guarantee that complete characterisation is possible by non-biological means when supported by sophisticated chromatographic data (including HPLC), improves the prospects that biological assay methods will become progressively less important.

11
Medicaments in formulations

David Watt

Introduction

The determination of single compounds in simple formulated products (dosage forms) has been discussed in earlier chapters, but the methods described therein are restricted to those which apply to both the parent drug and, with minimal modification, to its formulation. Frequently, however, the vehicle or basis of the formulation, or additional active ingredients, may interfere with the normal assay process. Separation methods or analytical methods which differentiate one substance from another are, therefore, required in order to determine the actual composition of the formulation.

In this chapter, methods are described for the analyses of a number of more complex formulations requiring special treatment. Some of the formulations considered are drawn from official compendia. Others are proprietary preparations. Since the proprietary preparations considered may not always be readily available overseas, an indication is given of likely vehicles or bases so that suitable mixtures can be prepared for instructional purposes. Many of the methods are based on basic procedures and instrumental techniques described elsewhere in either this volume or Part 2 of the book, to which readers should refer. Methods given for one type of formulation are often applicable, with minor modification, to other formulations.

Quality control of formulated products

Specifications and standards

The specifications and standards for all dosage forms depend upon a complex and variable mixture of factors already discussed in Chapter 2. These include:

(a) the nature and properties of the active ingredient(s)
(b) dosage and the route of administration
(c) the excipients used and the manufacturing process
(d) stability under normal conditions of storage.

Specifications are designed, essentially, to ensure that the correct materials and manufacturing process have been used. Elementary quality control procedures to check compliance consist of physical and chemical identity tests for the active ingredients (Chapter 1) and, where appropriate, simple descriptions of physical form (e.g. solution; suspension; emulsion). **Standards** are set with due regard to the known variability of the particular manufacturing process, the stability of the product and the limitations of the test methods. They set limits to the amount of the active ingredients by means of an assay and specify appropriate tests to limit related substances, irrespective of whether these were originally present in the drug substance or formed by degradation. Additionally, certain standards, such as, for example, those to control uniformity of content, bio-availability, suspendability of reconstituted suspensions or sterility, provide means of ensuring the efficacy and safety of the product as a whole.

Assays and tests

Ideally, the assay for drug content should be specific, though it seldom is so. More often, it consists of a simple procedure based on a particular functional group or combination of such groups in the molecule; in all such cases it is essential actually to confirm the identity of the active ingredient and not to place undue reliance on the statement on the label naming the active ingredient. It follows, too, that where two or more active ingredients are present and only non-specific assay methods are available, the components must be separated prior to assay. Similarly, tests for related substances must be able to separate, identify and estimate degradation products and, thus, be stability-indicating. Chromatographic procedures are generally applicable and widely used for this purpose.

Assay tolerances

Variation in the weight of individual tablets and in-fill volume of injections within a given batch is such that much wider limits of tolerance are prescribed than is usual for the active ingredient from which the tablets or injections are prepared. This is particularly evident in injection solutions, which may well be relatively dilute, and in tablets containing high dilutions of medicaments in inactive excipients. Even with *Sodium Chloride* a relatively innocuous substance administered in doses up to 16 g daily, the assay tolerances permitted for 300 mg tablets are 95–105% compared with 99.5–100.5% for *Sodium Chloride* itself by the same volumetric method of assay. The somewhat wider tolerances necessary for a formulated product assayed by aqueous extraction, derivatisation and weighing, is seen in the 260 mg tablets of *Piperazin Phosphate*, which have limits of only 92.5–107.5% compared with 98.5–100.5% for the parent substance. Likewise, because of their low active ingredient content, wide tolerances (92.5–107.5% of labelled strength) are adopted for *Morphine Sulphate Tablets* (20 mg). Similarly, injections in which potent active ingredients are in suspension (*Cortisone Injection*) or in oily solution (*Testosterone Propionate Injection*) and assayed spectrophotometrically are permitted wide tolerances (90–110% of the declared content).

Availability of active ingredients

Tests to control the efficacy of formulations are applied to solid dosage forms of certain compounds. Thus, although tablets and capsules of drugs which are soluble with difficultly and present at low dosage may have adequate disintegration characteristics, the drugs may not necessarily be available for absorption. Additional tests are, therefore, necessary to ensure that the formulations are clinically satisfactory. Biological tests actually demonstrating bio-availability in intact animals and clinical assessments in human patients, which are essential to establish the efficacy of the product in the first instance, are not suitable for routine quality control. In practice, therefore, physical dissolution tests are used to control the rate of solution of such drugs from their solid dosage forms, thereby ensuring a degree of batch-to-batch consistency in the degree of availability.

Policy in the application of such tests is much more conservative in Europe compared to North America, where the USP has declared its intention to apply dissolution tests to all solid dosage forms of all drugs as suitable test methods are established, irrespective of the solubility of the drug or its dose. On the other hand, the British Pharmacopoeia limits the use of dissolution tests to two types of product only. The first group consists solely of solid oral dosage forms containing relatively insoluble medicaments and administered in low dosage, such as *Digoxin Tablets*, where failure to achieve consistent dosage could have serious consequences for the patient. The second group of products for which dissolution tests are essential to ensure satisfactory manufacturing standards are preparations intended for slow or sustained release of the active ingredient.

Control of particle size of the active ingredient also provides a useful means of ensuring satisfactory and reproducible availability, as, for example, in *Griseofulvin Tablets*. Again, to ensure adequate penetration of drugs such as *Isoprenaline Sulphate* into the respiratory tract when formulated as a suspension in aerosol inhalations, the particle size of the compound should be controlled so that it is not greater than about 8 μm. These and similar controls are highlighted in the sections which follow.

Sterility testing of pharmaceutical products

Introduction

The sterilization of injections, ophthalmic preparations and products for use in contact with exposed body tissues, together with the maintenance of clean areas and disinfection of permanent and expendable equipment associated with their manufacture, are of paramount importance to the pharmaceutical industry. Adequate quality assurance procedures are essential, but no system of quality assurance can ensure 100% effective safeguards against microbiological contamination simply because sterility tests are directed against living organisms. Thus, it is not possible to claim that a batch of a parenteral preparation, for example, is completely sterile unless the entire batch is tested using a procedure providing optimum conditions for the growth and multiplication of every conceivable contaminating organism, in whatever state or condition they may be present. This is clearly impracticable and, hence, sterility testing cannot be used as the sole means of controlling sterile processing.

To overcome this problem, manufacturing processes for sterile products are required, in practice, to be conducted in approved designated areas under microbiologically clean conditions. Active ingredients, additives and solvents should have a low bioburden and the sterilisation of the final product, and, likewise, the separate sterilisation of containers where necessary, must be conducted under carefully controlled and monitored conditions.

Although bacteria, bacterial spores, yeasts and mould spores may be contaminants of medicaments, most sterilisation problems are caused by bacteria and, particularly, their spores. Microbiological sterility may be regarded as freedom from living micro-organisms. It is achieved by killing or inactivating the contaminants using moist or dry heat, a high-energy source such as ionising radiation, by chemical means, or by physically removing them during filtration. Heat sterilisation is generally regarded as the method of choice wherever the product is sufficiently stable to withstand the necessary heat treatment.

Experience has established that sterilisation of aqueous preparations can always be achieved by maintaining them at a temperature of $\nless$ 121° for 15 min. In the absence of moisture, however, much higher temperatures are required (150°C for 1 h). Other combinations of temperature and holding time which are considered to provide an equivalent amount of thermal energy are used. Such alternative procedures for the thermal sterilisation of liquids can be rationalised and expressed quantitatively in terms of D values or F_0 values. Thus, a D_{10} value or decimal reduction time (DRT) is defined as the time required to inactivate 90% of the organisms present at any moment, while the F_0 value is defined as the time in minutes required at a temperature of 250°F (121°C) to kill all viable organisms. By constructing a plot relating the thermal death time of organisms in a standard culture against temperature, the F_0 value of any particular sterilisation temperature will be the value on the time axis which corresponds with a temperature of 250°F.

Sterility testing

Because heat sterilisation methods and autoclaves can be checked instrumentally and by bacteriological challenge tests, sterility testing of every batch is unnecessary; nevertheless, sufficient testing to provide regular confirmation that the chance of any container failing the test is negligibly small is required. However, in the case of preparations containing substances of biological origin or other heat-labile compounds, where sterilisation is achieved by filtration, a sufficient number of final containers from every batch must be tested to give a similar degree of assurance.

Meaningful sterility tests can only show that organisms capable of growing in the test media used and under the conditions selected are absent from the portion of the batch that has been tested. It is obviously essential, therefore, to consider the composition of media, incubation conditions and the number of samples to be tested.

Factors affecting the media and conditions used in sterility tests

The more important factors affecting the growth of bacteria and fungi which influence the choice of media and the techniques employed include nutrition, moisture, aeration, incubation, temperature, pH and ultraviolet light.

Nutrition and moisture. Adequate sources of carbon, nitrogen, mineral salts and growth factors are essential to ensure rapid growth of contaminating organisms. Media must also contain at least 20% water to enable organisms to utilise the food sources available.

Aeration. Aerobic bacteria grow in air, anaerobes can multiply only in the absence of oxygen, while facultative anaerobes can grow with or without oxygen. Separate tests must be used to check on each class of organism. Saprophytic moulds are very aerobic and for this reason it is customary to increase the surface area exposed by incubating tubes and bottles on their sides. Yeasts can grow under aerobic and anaerobic conditions.

pH and temperature. The optimum pH for bacterial growth is around 7.4, while for fungi it is between 5 and 6. Although most bacteria grow best at 37°, some grow optimally at 30° and may fail to grow at 37°; the optimum for fungi lies between 20 and 25°.

Light. Light in the ultraviolet region in the presence of air has a damaging effect on bacteria and culture media. Hence, the use of ultraviolet light in areas where aseptic techniques are practised, and the absence of windows in incubators.

Growth inhibitors. Substances which prevent the growth of bacteria and fungi without killing them are termed bacteriostats and fungistats, respectively, whereas bactericides and fungicides have lethal properties. Such substances, when present in multidose containers, for example, must be neutralised by dilution in sterility testing.

Sampling. The principle guiding the number of samples taken to represent a batch must be such as to reduce to a minimum the risk of releasing contaminated containers. Most national authorities use one or other of two approaches in deciding how many containers must be tested:

either (a) a fixed percentage of the final group of containers
or (b) a fixed number independent of the batch size.

For heat-sterilisation products, final containers are selected at random. For aseptically processed preparations, however, it is essential that samples be taken throughout the filling operation, so that they are representative of the batch as a whole.

To obtain reliable results from sterility tests, it is obviously essential to take a sufficient number of samples, and to use sensitive culture media under suitable incubation conditions. It is equally important to avoid accidental contamination during testing to ensure that there is little risk of rejecting a sterile batch.

The general method of test

One of two techniques is usually employed, either membrane filtration or direct inoculation. Membrane filtration is the method of choice for both aqueous and oily solutions and for soluble solids. Briefly, the sample solution, diluted if necessary to about 100 ml with a sterile diluent, is filtered through a sterile filter disc of nominal pore size not greater than 0.45 μm. If substances with growth-inhibitory properties are present in the product, the filter should then be washed not less than three times with about 100 ml of the solvent or diluent. Discs so prepared with their microbial contaminant from the solution are used as inocula in the appropriate test media. The latter are then incubated for $\nless 7$ days under the optimum conditions for growth and, finally, examined for microbial contamination.

Soluble solids are treated similarly after solution in an appropriate solvent. Oils and oily solutions may be filtered directly through a dry filter membrane, but may require dilution with an appropriate sterile solvent at a concentration of 1%; if heat is required to dissolve the product, this must be minimal ($\ngtr$ 40°C).

The quantity taken for test from any one container is also specified. If the contents are less than 1 ml or 50 mg, the entire contents must be used; half the contents must be used where they are between 1 and 4 ml or 50 and 200 mg; where the contents are larger, smaller proportions of the total are used.

The alternative method, direct inoculation, is only used for surgical dressings and other products not amenable to treatment by filtration.

Microbial contamination of formulated products

A number of liquid and semisolid preparations are liable to microbial contamination. This can arise from the use of contaminated ingredients (p.9) as in *Aluminium Hydroxide Mixture*. It can also occur in creams and ointments because of the difficulty of handling and packaging such products. Contamination may occur either during manufacture, or in use where the container is such that the product may be used on more than one occasion. Most such products are formulated with appropriate antimicrobial preservatives. It is important, nonetheless, that the effectiveness of the preservative should be capable of being assessed.

Challenge tests

The efficacy of an antimicrobial preservative is usually assessed by a series of challenge tests on the product. These are conducted by the manufacturer as part of the product development. The tests should be repeated over a period of at least 28 days and, preferably, longer periods. Such tests, therefore, are not suitable for quality control. The tests require the product to be exposed to contamination by organisms likely to be found in areas where the product is manufactured, stored and used and, particularly, to such species or strains capable of growing well on the product. Typical organisms include *Aspergillus niger, Candida albicans, Escherichia coli, Pseudomonas aeruginosa, Saccharomyces rouxii* and *Staphylococcus aureus*. Typical requirements for topical preparations under appropriate test conditions are a reduction by a factor of $\nless 10^3$ in the number of bacteria within 48 h of challenge, with no organisms recovered from 1 ml (or 1 g) after 7 days, and a reduction of $\nless 10^2$ in the number of moulds or yeasts within 7 days, with no organisms recovered from 1 ml (or 1 g) after 28 days. Likewise, oral liquid preparations should show reductions of $\nless 10^3$ bacteria within 48 h and $\nless 10^2$ in the number of yeasts and moulds per ml within 14 days, with no increase thereafter.

Aerosol inhalations

Dose

Aerosols supply unit doses and contain the medicament in suspension, or solution, under pressure in a mixture of insert propellants which, for antiasthmatic sprays, is usually either dichlorodifluoromethane or dichlorotetrafluoroethane. They may contain a surface-active agent, stabilising agents and other adjuvants. Correct unit dose from the pressurised container is dependent upon satisfactory functioning of the special metering valve as well as on the correct composition of the mixture or solution. The requirements for manufacturing specifications given in Chapter 2 (p.79) provide a reasonable control of the variables. Nevertheless, the limits for the amount of active ingredient stated on the label to be available to the patient are wide, viz. 75–125%, because of the nature of the preparation and its method of use. Thus, some of the dose delivered by the metering valve is retained in the oral adapter. The amount available to the patient is, therefore, the difference between the amount delivered by the metering valve and the amount retained in the oral adapter.

Isoprenaline Aerosol Inhalation *Determination of the content of* $C_{11}H_{17}NO_3,\frac{1}{2}H_2SO_4$ *per spray*

The determination is based upon the formulation of a colour when phenols are treated with iron salts under standard conditions (Part 2, Chapter 7) and the result obtained is an average value for 10 sprays.

Determination of the amount of isoprenaline sulphate delivered by metering valve

Method Remove the oral adapter from the container and gently shake the container. Actuate the valve several times, wash the valve stem with methanol and discard the washings (Note 1). Replace the oral adapter, invert the assembled unit and direct it into a small beaker containing carbon tetrachloride (25 ml) and 0.005M H_2SO_4 (25 ml), ensuring that the mouth of the adapter is completely immersed below the carbon tetrachloride layer (Note 2). Fire 10 sprays by pressing on the base of the container, gently shaking the container and contents of the beaker between each spray. Remove the aerosol unit from the beaker, wash it carefully with 0.005M H_2SO_4 (10 ml) and transfer the contents of beaker and washings to a separator. Shake thoroughly (5 min), allow to separate and transfer the aqueous liquid to a 50 ml volumetric flask. Extract the carbon tetrachloride layer with 0.005M H_2SO_4 (5 ml portions) and add the extracts to the 50 ml flask. Adjust to volume with more acid and mix well.

To 20 ml of extract in a 25 ml volumetric flask, add freshly prepared ferrous sulphate–citrate solution (0.3 ml) and aminoacetate buffer solution (3 ml), mix, allow to stand for 15 min and dilute to volume with water. Measure the absorbance of the solution in 2 cm cells at about 530 nm using a blank of reagents. By reference to a previously prepared calibration curve, calculate the average amount of isoprenaline sulphate delivered per spray.

Note 1 This is a clean-up procedure and check to ensure smooth working of metering valve.

Note 2 Complete immersion of the adapter in an absorbing liquid avoids loss of medicament as the spray emerges. Depending on the formulation, some aerosol inhalations give turbid or opalescent solutions on spraying directly into aqueous acid; the carbon tetrachloride dissolves this fatty material.

Ferrous sulphate–citrate solution Dissolve sodium metabisulphite (0.1 g) in *water* (20 ml). Add hydrochloric acid (M; 0.1 ml), ferrous sulphate (0.15 g) and sodium citrate (1 g) and shake to dissolve.

Aminoacetate buffer solution Mix sodium bicarbonate (4.2 g) and potassium bicarbonate (5 g) with *water* (18 ml). Add a solution of aminoacetic acid (3.75 g) and strong ammonia solution (1.5 ml) in water (18 ml) and dilute to 50 ml with *water*. Stir to complete solution.

Determination of the amount of Isoprenaline Sulphate retained in the oral adapter

Method Remove the adapter, rinse with methanol and dry. Gently shake the pressurised container, actuate the valve several times, wash the valve stem with methanol and allow to dry. Replace the oral adapter and fire 20 sprays carefully into the air, with gentle shaking between each spray. Remove the oral adapter and immerse it in a mixture of carbon tetrachloride (25 ml) and 0.005M H_2SO_4 (25 ml). Swirl the contents of the beaker, remove the adapter and continue as described above from the words 'wash it carefully with 0.005M H_2HO_4' but using a 4 cm cell for measurement of absorbance. Calculate the average amount of isoprenaline sulphate retained per spray. Hence, calculate the average content of isoprenaline sulphate per spray.

Uniformity of dose *Determination of the uniformity of dose from Isoprenaline Aerosol Inhalation*

The content of isoprenaline sulphate as determined above is an average value and for uniformity of dose each spray must be monitored in some way. Strictly, it is the metering valve that is being checked and as it may be required to deliver up to 400 sprays it is useful to apply a test for uniformity of dose at three stages, viz. when the container is full, when half-full and towards the end of its nominal number of sprays. The effect of sedimentation or caking of medicament or valve clogging may well be shown up in inaccurate dosage.

Method A Remove the adapter from the pressurised container and determine the loss in weight of the pressurised container after each spray for a total of 10 sprays. Calculate the coefficient of variation for the results which refer, in the main, to loss of propellant.

Method B Remove the adapter from the pressurised container, shake the container and invert it in a small beaker containing 0.005M H_2SO_4 (10 ml). Fire 1 spray below the surface of the liquid by pressing on the base of the container. Mix, filter if necessary through a sinter-glass filter (Porosity 3) and measure the absorbance of the filtrate in a 2 cm cell. Calculate the content of isoprenaline sulphate using an $A^{1\%}_{1cm}$ value of 100 for isoprenaline sulphate. Repeat the determination for a total of 10 results and calculate the coefficient of variation. Compare the result with that obtained under method A above as an indication of the relative precision of the two methods.

Identification of medicament

The direct firing of spray into an aqueous or ethanolic system is used to obtain a solution on which tests for identity can be carried out. They take the form of chemical tests for particular groups of characteristic reactions, e.g. the blue colour produced by ergot alkaloids with 4-dimethylaminobenzaldehyde solution. In addition, thin-layer or paper chromatographic tests are applied (Part 2).

Particle size

The medicament is usually intended to exert its action in the bronchioles and alveoli of the respiratory system. It is essential, therefore, that adequate penetration into the lungs takes place on inhalation. For this to occur the particle size of the emitted solution, or suspended particles in suspensions should be in the range 2–8 µm. The particle size must not be too small, however, as loss of medicament may occur on breathing out. Particle size is, therefore, an important part of the specification for aerosol inhalations, even though difficulties occur in its determination. These have been emphasised by Kirk (1972) who has designed a simple system to simulate the upper part of the respiratory tract (Fig. 11.1). The tubing is lined with 3% agar gel to reproduce the wet conditions of the lung.

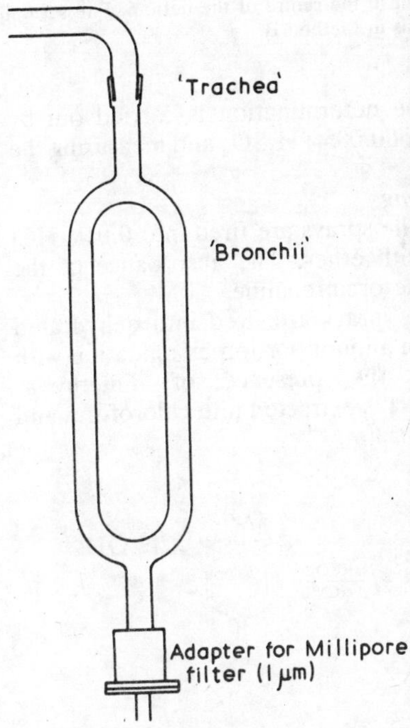

Fig. 11.1
Respiratory track simulator

With the vacuum pump set to give a flow of 16 l/min the aerosol adapter is fitted to the mouthpiece with a suitable gasket and a definite number of sprays fired into the apparatus. The fine particles collect on the Millipore filter which is removed for quantitative determination of medicament. The results are expressed as percentage of dose emitted from the valve to reach the filter or as percentage of dose leaving the mouthpiece to reach the filter. These particular doses are determined as described above. The results are related to particle size but the true value of the apparatus lies in the ability to compare aerosol inhalations for penetrative efficiency.

Isoprenaline Aerosol Inhalation *Determination of the particle size of the suspended isoprenaline sulphate*

The main difficulty associated with many methods is that of representative sampling and this is particularly so in the following.

Method A With the oral adapter in position fire 1 spray on to a clean dry microscope slide held about 5 cm from the end of the mouthpiece of the adapter and perpendicular to the direction of the spray. Rinse the slide carefully with carbon tetrachloride (2 ml), allow to dry and examine the residue under a microscope. Most of the individual particles have a diameter not greater than 5 µm; no individual particle has a length, measured along its longest axis, greater than 20 µm.

Method B Place several slides at the bottom of a tank of about 10 l capacity and fire 6 sprays into the top of the tank. Cover immediately and allow to stand overnight for the particles to settle. Measure the size of the particles deposited on the slides as in method above.
Method C Place several microscopic slides at the bottom of a tin of dimensions approximately 25 cm side, in which a hole is cut in the centre of the bottom. Fire several sprays upwards through the hole and proceed as in Method B.

COGNATE DETERMINATIONS

Ergotamine Aerosol Inhalation The determination is carried out by using 1% tartaric acid solution instead of 0.005M H_2SO_4 and measuring the absorbance of the solution at 317 nm.

Isoprenaline Aerosol Inhalation Strong

Orciprenaline Aerosol Inhalation The sprays are fired into 0.01M HCl and any fatty material is extracted with ether. The absorbance of the solution at 276 nm is used to determine orciprenaline.

Salbutamol Aerosol Inhalation The sprays are fired into dehydrated alcohol and a colour is developed on an aliquot portion by oxidation with alkaline potassium ferricyanide in the presence of dimethyl-*p*-phenylenediamine. The coloured product is extracted into chloroform and the absorbance determined at about 605 nm.

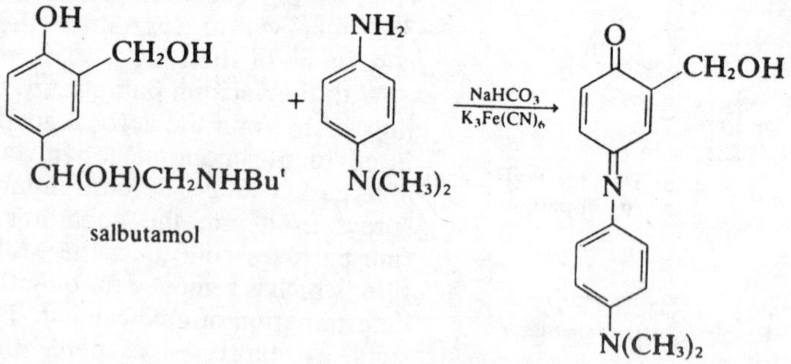

Capsules

Capsules consist of a medicament, in some cases mixed with a suitable diluent and stabiliser, enclosed in a shell which may be hard or soft. The shells are composed mainly of gelatin or other materials, the consistency of which may be adjusted using glycerol and/or sorbitol. In addition to controls on the content of active ingredient, capsules are subjected to uniformity of weight tests, disintegration tests and, where necessary, dissolution tests.

Disintegration test

The various national authorities specify the time within which capsules and tablets must disintegrate when tested in the apparatus and by the procedure described. Six capsules are used and the apparatus operated for

30 min. The capsules pass the test if no residue remains on the screen or if any remnant consists of fragments of shell or a soft mass without firm core. Where the active drug substance is readily absorbed from the stomach (e.g. *Cephradine* and *Flufenamic Acid*) the maximum time allowed for disintegration is reduced to 15 min. Should the capsules float and discs have been used, any residue remaining on their lower surface consists only of shell fragments.

Dissolution test

Where the method of production involves a procedure such as slugging of granules, a test to confirm that the drug is available in solution within a reasonable time (p.302) is necessary (e.g. *Tetracycline* and related compounds)

Uniformity of weight

Capsules vary in weight and a test for uniformity of weight of contents is applied. It is not always practicable to separate the contents quantitatively, since these may consist not only of active ingredient, but also fillers. The contents may also be present as a paste with a suitable liquid (usually an oil). In order to get a uniform test, which is generally applicable, it is simpler to remove the contents and weigh the shells. An intact capsule is weighed, opened without losing any part of the shell and the contents removed as completely as possible. With soft capsules containing a liquid or paste it may be necessary to remove material adhering to the shell by washing with ether or other suitable solvent, allowing the shell to stand until free from the odour of solvent. Shell material is weighed and the difference represents the weight of contents. The procedure is repeated on a further nineteen capsules selected at random and the average weight of contents is calculated. Not more than two of the individual weights should deviate from the average weight by more than the percentage deviation shown in Table 11.1 and none by more than twice that amount.

Table 11.1 Uniformity of weight of capsule contents

Average weight of capsule contents	Percentage deviation
Less than 300 mg	±10
300 mg or more	±7.5

Determination of active ingredients in capsules

Standard analytical procedures used include direct titration of acidic function; recovery of basic or acidic moiety followed by a weighing or non-aqueous titration; extraction of base, addition of internal standard and GLC determination of volatile derivatives; or simple dilution and measurement of absorbance. Such procedures are used to assay the active ingredients in the mixed contents of the capsules taken for uniformity of weight. In some cases, such as *Cycloserine*, an identity test for the drug substance can be modified to provide a quantitative colorimetric procedure. Biological procedures are still required for most antibiotics, such as *Tetracycline* and related products, though these are now being replaced wherever possible by HPLC or sophisticated chemical methods with a high degree of specificity, such as those used for penicillins.

Control of penicillins

The mercurimetric titration of penicillins serves two purposes: to control non-lactam impurities and to determine the undegraded penicillin. Several mercurimetric methods are available. Interaction with imidazole in the presence of mercury (II) chloride leads to rearrangement and formation of the corresponding penicillenic acid mercaptide, which can be determined spectrophotometrically (Bundgard and Ilner, 1972).

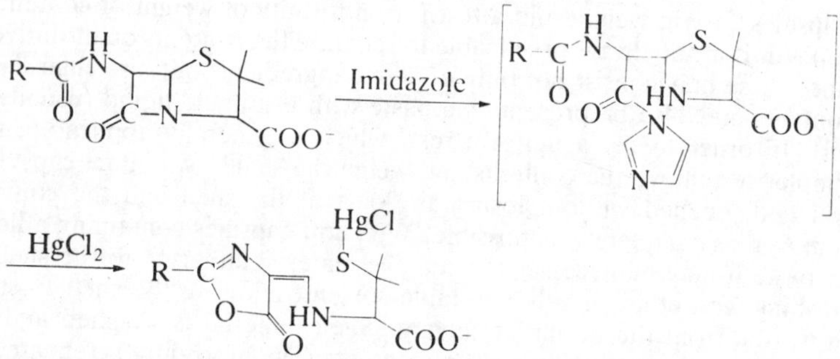

II

Only the intact β-lactam structure reacts, so that the titration gives a direct measure of undegraded penicillin. An additional test, such as that for the determination of iodine-absorbing substances, is necessary to control the presence of degradation products such as penicilloic acids. These react directly with iodine, in contrast with the intact β-lactam antibiotics, which only react after prior treatment with first alkali and then acid (see Benzylpenicillin assay, p.000).

An alternative mercurimetric method (Karlberg and Forsman, 1976) involves a double titration procedure with mercury (II) nitrate, firstly in buffer at pH 4.5 on the untreated penicillin, simply to determine degradation products, and secondly on a hydrolysed solution at the same pH to give the sum of degradation products and penicillin degraded by the hydrolysis. Two inflection points are observed in the potentiometric titration; the first at a molar ratio of penicillin and Hg(II) of 2:1 and the second when equimolar amounts of penicillin and Hg(II) have been reached. Ampicillin is determined by this method after preliminary acetylation of the side-chain amino group, which otherwise interferes with the reaction.

Determination of Ampicillin in Ampicillin Capsules

Method Weigh an amount of powder equivalent to 0.15 g of ampicillin into a 500 ml volumetric flask, dilute to volume with *water*, shake for 30 min and filter. To 10 ml in a 100 ml volumetric flask add 10 ml of 0.1M borate buffer, pH 9.0, then 1 ml of 0.2M acetic anhydride in dioxan, allow to stand for 5 min and dilute to volume with *water*. Transfer two 2 ml portions of this solution into separate stoppered tubes; to one tube add 10 ml of imidazole-mercury reagent (1.2M aqueous imidazole solution containing 10.3M mercury (II) chloride at pH 6.8), mix, stopper the tube and immerse in a water-bath at 60° for exactly 25 min, with occasional swirling. Cool the solution rapidly to room temperature (Note) and measure the absorbance at the maximum at about 325 nm (1 cm cell) using a solution of 2 ml of *water*, 10 ml of imidazole-mercury reagent as blank. Calculate the ampicillin concentration of the original preparation by reference to a suitable standard preparation.

Note The imidazole-catalysed rearrangement requires 3.5 h at 20–25° but is complete at 60° in about 25 min; thereafter, the penicillanic acid mercuric mercaptide is stable for more than 24 h.

COGNATE DETERMINATIONS

Amoxycillin Capsules
Cloxacillin Capsules
Phenethicillin Capsules

Sustained-release capsules

Fig. 11.2 Flow-through cell for determination of release rates

The method of testing sustained-release capsules involves the determination of the rate of release of the active ingredient. Although, ideally, *in vivo* methods should be employed, some measure of the reproducibility of the manufacturing process and release characteristics of batches of capsules can be obtained by *in vitro* methods. If the results of such tests show good correlation with those of *in vivo* methods the quality control analyst has some confidence in using the *in vitro* method routinely.

Many variations in the apparatus exist and, although that described for tablets (p.302) can be used, the nature of the formulation sometimes leads to difficulties. For example, the capsule contents may consist of small pellets coated with waxy or fatty materials which cause aggregation in the test. The rotating basket is better replaced by a flow-through system (Fig. 11.2) leading to a spectrophotometer if the active ingredient absorbs ultraviolet radiation.

The two sintered-glass filters are held together by a paraffin wax film (parafilm) and the capsule is placed on the porosity 1 sinter. A more uniform flow through the system is obtained than occurs when a single sinter is suspended in the dissolution medium, particularly if a small sinter is used as it soon becomes blocked. The system must be leak-proof and a rapid circulation of solvent is necessary to ensure satisfactory dissolution profiles.

Creams

Creams consist of water, fatty materials and emulsifying agents which yield either water-miscible (oil-in-water type) or oily (water-in-oil type) emulsions. The active ingredients may be in solution in the basis or present as a suspension, in which case, particle size of the medicament may be important, as, for example, in creams containing *Hydrocortisone Acetate* or *Clioquinol*. Physical characters such as appearance, uniformity, viscosity, pH and checks to detect cracking are important aspects of quality control.

Water can be determined using the Karl Fischer method (Chapter 6) and characterisation of the emulsifying agent used in proprietary preparations is possible by means of infrared spectrophotometry. For many creams where spectrophotometry is suitable as an end-method of analysis, the medicament can be determined without prior separation as the basis does not interfere with the assay process. The correct choice of solvent may also avoid a lengthy separation procedure, for example, toluene to extract dimethicone from a cream. Frequently, however, extraction and separation methods cannot be avoided, particularly where the active ingredient is present at low dosage.

Determination of Dimethicone in a cream

Dimethicone can be determined by an infrared absorption method, of which there are few examples in official compendia. It must, however, be separated from the water present in the cream. Silicones have strong infrared absorption bands but two, centred at about 1050 and 800 cm^{-1}, respectively, are so broad that they are not satisfactory for quantitative analysis. Of the remaining three absorption bands, only that at about 1270 cm^{-1} is suitable because other constituents of the cream, such as liquid paraffin, and the solvent interfere with absorption in the region of the other two bands at about 2900 and 1400 cm^{-1}. It would be difficult to compensate for these accurately and their choice would, moreover, be incorrect because they are not specific for silicones.

Method Mix the cream (about 5 g accurately weighed) with anhydrous sodium sulphate (10 g in a 50 ml beaker until a uniform mixture is obtained (Note 1). Extract the mixture with slightly warm toluene (6 × 5 ml), filtering each extract through a porosity 3 sintered glass filter into a receiver. Transfer the filtrate to a 50 ml volumetric flask, rinse the receiver with toluene, add to the volumetric flask and make up to volume with toluene.

Record the infrared absorption curve in 0.1 mm cells over the region 1350 to 1200 cm^{-1}, using as blank a 4% w/w solution of liquid paraffin in toluene (Note 2). Determine the absorbance due to dimethicone by the base-line technique (Part 2). Calculate the concentration of dimethicone by reference to a calibration curve prepared by examining solutions containing 0.2, 0.4, 0.6, 0.8 and 1.0% w/v of dimethicone in 4% w/w liquid paraffin in toluene.

Note 1: The anhydrous sodium sulphate abstracts the water as the decahydrate, leaving a readily filtered organic layer (compare Triamcinolone Cream, below).

Note 2. A 4% w/w solution of liquid paraffin in toluene is to allow for both liquid paraffin and cetostearyl alcohol (alkane absorption) in the sample solution. The actual interference at 1270 cm^{-1} is, however, quite small.

Triamcinolone Cream (0.1%) *Determination of Triamcinolone Acetonide*

If the cream is water-miscible, extraction procedures are necessary to obtain the acetonide in chloroform solution. The determination is based upon the reaction of α,β-unsaturated steroidal ketones with isoniazid to yield yellow products with an absorption maximum at about 415 nm. This is a particularly useful reaction in that it enables determinations, previously made difficult by the nature of the base or vehicle, to be carried out simply.

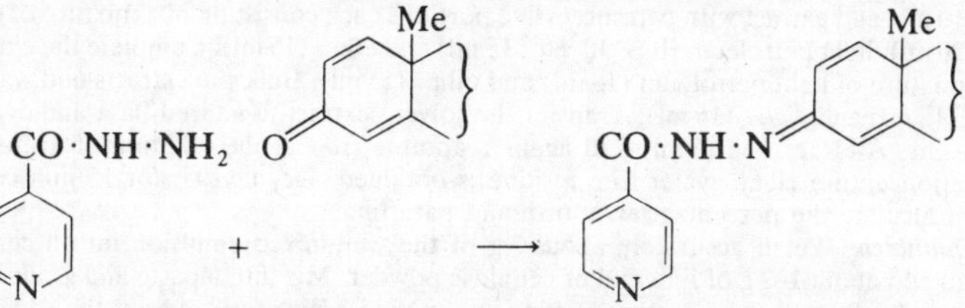

Reagent Isoniazid (0.5 g) in methanol (500 ml) containing hydrochloric acid (0.63 ml).

Method Transfer the cream (about 2 g, accurately weighed) to a 50 ml glass-stoppered flask, add chloroform (30 ml) and anhydrous sodium sulphate (0.5 g). Shake thoroughly (10 min) and filter the chloroform through a phase-separating paper (Note). Wash the flask and filter with chloroform (10 ml), transfer the filtrate and washings quantitatively to a 50 ml volumetric flask and adjust to volume with chloroform.

Transfer 5 ml of the solution to a 25 ml volumetric flask and add isoniazid solution (10 ml). Warm at 55° for 45 min, cool and adjust to volume with chloroform. Measure the absorbance at about 415 nm in a 1 cm cell using a blank 5 ml of chloroform treated in the same way as 5 ml of solution. The standard solution (5 ml) used for comparison is 0.005% triamcinolone acetonide in chloroform treated as described above.

Calculate the concentration of triamcinolone acetonide by reference to the two absorbances.

Note Where a mixture consists of organic solvent and solid matter, the use of a sintered glass crucible is satisfactory for filtration. If, however, droplets or a layer of water is present, phase-separating paper is the system of choice.

Whatman No.1 PS is a filter paper impregnated with a stable silicone to make it water-repellent. The method of use is as follows:

(1) Fold the paper in the normal way and place it in a conical filter funnel.

(2) Pour the mixed phases directly into the paper. It is not normally necessary to allow the phases to separate cleanly before pouring.

(3) Allow the solvent phase to filter completely through the paper.

(4) If required, wash the retained aqueous phase with a small volume of clean solvent. This is normally necessary when separating lighter-than-water solvents in order to clean the meniscus of the aqueous phase.

Do not allow the aqueous phase to remain a long time in the funnel after phase separation, or it will begin to seep through. This is caused by evaporation of the solvent from the pores of the paper, creating a local vacuum which, in turn, draws the water through.

Emulsions

Pharmaceutically, a distinction is drawn between emulsions which are oil-in-water preparations for internal use and creams and applications which are intended for external use.

Liquid Paraffin and Phenolphthalein Emulsion *Determination of Liquid Paraffin and Phenolphthalein*

The determination of liquid paraffin is based upon solvent extraction and that for phenolphthalein on the formation of a red colour with alkali. The conditions for development of the colour are critical and for a discussion of these the original paper by Allen, Gartside and Johnson (1962) should be consulted.

Method

Liquid Paraffin Weigh accurately about 5 g of the well shaken emulsion into a separator, add *water* (10 ml) and extract with two successive portions each consisting of a mixture of ethanol (96%; 10 ml), light petroleum (b.p. 40–60°; 15 ml) and ether (15 ml). Complete the extraction with a mixture of light petroleum (15 ml) and ether (15 ml). Bulk the extracts and wash with NaOH (15 ml) and *water* (15 ml). Transfer the solvent extract to a tared flask and evaporate the solvent. Add acetone (5 ml) and again evaporate; repeat the addition of acetone and evaporation until a clear, water-free residue is obtained. Dry at 105° for 15 min cool and weigh. Calculate the percentage w/w of liquid paraffin.

Phenolphthalein Weigh accurately about 3 g of the well-mixed emulsion into a centrifuge tube and add about 1–2 g of Filtercel or cellulose powder. Mix into a paste and gradually add ethanol (96%; 20 ml) with constant stirring and mixing. Centrifuge, decant the supernatant liquid into a 100 ml volumetric flask and complete the extraction by washing the residue with ethanol (96%; 2 × 20 ml). Add the washings to the flask and dilute to volume with ethanol (96%). Evaporate the solution (5 ml) to dryness in a small beaker (Note 1) and dissolve the residue in glycine buffer solution (Note 2). Transfer to a 100 ml volumetric flask and rinse the beaker with more buffer solution, adding the washings to the flask. Dilute to volume with buffer solution, mix and measure the absorbance at 555 nm in a 1 cm cell. Complete the measurement within 10 min of adding the buffer solution (Note 3) and calculate the percentage of phenolphthalein using an $A^{1\%}_{1\,cm}$ value of 1055 for phenolphthalein in glycine buffer solution.

Note 1 It is essential to remove all traces of ethanol, otherwise a reduction in the intensity of the pink colour occurs.

Note 2 Maximum colour develops over a narrow range of pH and the buffer solution should be adjusted, if necessary, to pH 11.1. It is made by adding 0.1M NaOH (100 ml) to a solution (100 ml) containing aminoacetic acid (0.75 g) and sodium chloride (0.58 g).

Note 3 The colour is stable up to about 15 min and decreases slowly thereafter.

Eye drops

Eye drops are sterile preparations and those that are aqueous and supplied in containers intended for use on more than one occasion contain a bactericide and fungicide. Of the preservatives, benzalkonium chloride and chlorhexidine acetate are likely to interfere in the determination of the active ingredient when the assay involves the use of sodium tetraphenylborate. As a result, three methods are used.

Method 1 is used where no interference is to be expected, for example, with strong solutions (1% or more) containing phenylmercuric acetate as preservative. The normal method outlined on page 281 is used.

Method 2 is used where benzalkonium chloride is present as preservative. A determination is carried out at pH 10 when only the preservative is titratable (compare p.282).

Method 3 is used for those preparations containing chlorhexidine acetate. The modification consists in precipitating the chlorhexidine as sulphate by addition of sodium sulphate. After allowing time for precipitation to occur, the assay is continued as described for Method 1.

Many of the eye-drop preparations can be assayed by direct measurement of ultraviolet absorption on a diluted aliquot portion of the sample. Fell (1978) described a procedure based on second derivative spectrophotometry (see Part 2) for the assay of pilocarpine and physostigmine in eye drops also containing benzalkonium chloride. Zinc sulphate preparations are readily titrated directly with 0.01M disodium edetate in alkaline buffer.

Sodium tetraphenylborate titrations

A 0.01M solution of sodium tetraphenylborate may be used to titrate a slightly alkaline solution of some quaternary ammonium compounds, such as benzalkonium chloride in the presence of chloroform and bromophenol blue. At the end point, the blue, chloroform-soluble base-indicator complex is decomposed and the chloroform layer becomes colourless.

The reagent is also very useful for precipitating, at pH 3.7, the tetraphenylborates of alkaloids and synthetic organic bases. The analytical application of the reagent has been studied by Johnson and King (1962) and the assay described below is based on that report.

The method may be applied to a large number of basic nitrogen compounds, either alone or in pharmaceutical preparations such as eye drops, injections, tablets and suppositories.

Physostigmine Eye Drops *Determination of the percentage of* $(C_{15}H_{21}N_3O_2)_2, H_2SO_4$

The following reagents are required.

Bromophenol blue solution Warm bromophenol blue (0.1 g) with 0.05M sodium hydroxide (3.0 ml) and ethanol (90%; 5 ml) until solution is complete and add ethanol (20%) to produce 250 ml.

Buffer solution pH 3.7 Dissolve anhydrous sodium acetate (10 g) in *water* (about 300 ml), add bromophenol blue solution (1 ml) and sufficient glacial acetic acid (35 to 40 ml) until the indicator changes from blue to a pure green. Dilute to 500 ml with *water*. Dilute with an equal volume of *water* before use.

0.005M *Cetylpyridinium chloride* Dissolve cetylpyridinium chloride (1.80 g) in ethanol (96%; 10 ml) and dilute to 1 l with *water*. Store in an amber bottle.

0.01M *Sodium tetraphenylborate* Dissolve sodium tetraphenylborate (3.42 g) in *water* (50 ml), add moist aluminium hydroxide gel (0.5 g) (Note 1) and shake for 20 min. Dilute to 300 ml with *water*, dissolve sodium chloride (Note 2) (16.6 g) in the solution and stand for 30 min. Filter through two thicknesses of No.42 Whatman filter paper under suction to yield a clear filtrate. Wash the filter with *water*, add the washings to the first filtrate, dilute to 1 l with *water* and adjust to pH 8.0 to 9.0 with 0.1M sodium hydroxide. Store the solution in an amber bottle.

0.01M *Potassium chloride* Dissolve potassium chloride (analytical reagent grade, 0.1491 g), previously dried at 150° for 1 h, in buffer solution (200 ml).

Standardisation of 0.01M sodium tetraphenylborate To 0.01M potassium chloride in acetate buffer pH 3.7 (10 ml by pipette) in a dry beaker add 15 ml of the sodium tetraphenylborate solution, allow to stand for 5 min and filter through a dry sintered-glass filter. To the filtrate (20 ml) add 0.5 ml bromophenol blue solution (Note 3) and titrate the excess of sodium tetraphenylborate with 0.005M cetylpyridinium chloride to the blue colour of the indicator.

Repeat the procedure omitting the potassium chloride. The molarity of the tetraphenylborate solution is then:

$$a \times w/[15\ (a - b) \times 0.07455]$$

where a is the volume of 0.005M cetylpyridinium chloride required when the potassium chloride is omitted, b is the volume required when the potassium chloride is present and w is the weight, in g, of potassium chloride present.

Method To a volume of preparation equivalent to about 20 mg of physostigmine sulphate, add pH 3.7 buffer (5 ml), 0.01M sodium tetraphenylborate (15 ml), mix and allow to stand for 10 min. Filter, wash the container and filter with water (5 ml), combine the filtrate and washings and titrate with 0.005M cetylpyridinium chloride using bromophenol blue solution (0.5 ml) as indicator. Repeat the operation without the preparation being examined; the difference between the titrations represents the amount of tetrabenylborate required. If a titration at pH 10 is required the difference between the titrations at pH 3.7 and pH 10 represents the amount of tetraphenylborate required.

Each ml of 0.01M sodium tetraphenylborate is equivalent to 0.003244 g of $(C_{15}H_{21}N_3O_2)_2$, H_2SO_4.

Note 1 The solution tends to be cloudy and addition of the gel assists in obtaining a clear filtrate.

Note 2 The precipitated tetraphenylborates are semi-colloidal, and the presence of sodium chloride allows ready filtration of the mixture.

Note 3 The volume of the indicator must be accurately measured as it introduces a blank of about 0.1 ml.

COGNATE DETERMINATION

Pilocarpine Eye Drops

Injections

In maintaining the quality of injections, consideration must be given not only to the amount of active ingredient present but also to factors affecting the stability and safety in use of the formulation. Specifications for containers, closures and particulate matter in solutions, therefore, become important in addition to factors affecting dosage and its determination, e.g. volume or weight in the containers, and the presence or absence of bactericides, which may influence choice of assay.

Limit test for particulate matter

Solutions for injection should not contain particles of foreign matter that can readily be observed on visual inspection using ordinary and plane-polarised light. In particular, containers of 100 ml capacity or more should be examined individually for the presence of undesirable particles. This inspection, however, is subjective and a more discerning test designed to limit the number of particles of particular size is applied solely to large-volume solutions (100 ml or more per container), such as intravenous fluids made from various sugars.

Apparatus. No particular apparatus is mandatory so long as it is capable of counting the numbers of particles having equivalent sphere diameters equal to or greater than 2 μm and equal to or greater than 5 μm. A Coulter counter is satisfactory (Part 2) and, as it is based upon electrical resistance, a filtered sodium chloride solution should be added to the injection if necessary, for example, for non-conducting dextrose injections.

Sampling. Contamination with particles on opening large-volume containers is not likely to be great but if ampoules containing 1 to 2 ml of injection are examined some precautions are necessary. Somerville and Gibson (1973), referring to a paper by Emerot and Dahlinder, adopted a procedure of immersing the ampoules in a water bath at 70°, followed by wiping off to reduce drastically misleading contamination before opening. Heating the ampoules to 70° produced a positive internal pressure, which reduced the possibility of airborne contamination. The use of a laminar flow cabinet supplied with filtered air for such manipulative techniques would be an additional precaution.

Standard. The generally accepted official standard for 100 ml or larger-volume injections is that the average count (as determined by the sampling procedure recommended for the particular instrument) determined on each container examined, for the undiluted product, does not exceed 1000 ml^{-1} greater than 2.0 μm and 100 ml^{-1} greater than 5.0 μm.

Extractable volume

Containers intended for use on one occasion must have a slight excess (overfill) of the injection solution. When the nominal volume does not exceed 5 ml, the average volume of five containers (usually inspected to confirm that volumes are approximately equal) should be not less than the nominal volume and not more than 115% of the nominal volume. The corresponding limits for the average of three containers with more than 5 ml are not less than the nominal value and not more than 110% of the nominal volume.

Method Select a syringe of volume not greater than twice the volume to be measured and fit it with a needle. Withdraw a small quantity of the injection to be examined, from a container not to be used in the measurement test, and discharge with the needle pointing upwards to remove air bubbles. Using this prepared syringe, transfer the injection from the containers under test to a measuring cylinder of capacity not greater than twice the total volume to be measured. Measure the total volume and calculate the average volume in each container. Reserve the solution for subsequent assay purposes.

Uniformity of weight

This requirement applies to those injections prepared by dissolving the contents of a sealed container before use. If the average weight of the contents is small (40 mg or less) or contains less than 2 mg or less than 2% w/v of active ingredient it is usual to omit this test and apply a uniformity of content test. Otherwise, the following method is used.

Method Remove any paper labels attached to the container, wash the outside with water and dry carefully. Open the container, weigh all parts and remove the contents. Wash with water and then with ethanol 96%, dry at 105° (or a suitable lower temperature for plastic containers), cool and weigh. The difference between the weights represents the weight of contents. Repeat with a further nineteen containers and determine the average weight. The usual requirements are that not more than two of the individual weights differ from the average weight by more than 10% and none deviates by more than 20%.

In this way, uniformity of dosage is ensured; accuracy of dosage is controlled by declared content test.

Uniformity of content

Powders for injection that have a total weight of less than 40 mg in each container, or contain less than 2 mg or less than 2% w/v of active ingredient are controlled, as for tablets, by a test for **uniformity of content** instead of a test for uniformity of weight. As in the case of tablets, (p.300) uniformity of content is determined by separately assaying the individual content of active ingredient in each of ten separate containers. As in the case of the uniformity of content test for tablets, typical limits are that the contents of each container is between 80 and 120% of the average, except that for one container the content may be between 75 and 125% of the average.

Mannitol Intravenous Infusion *Determination of the percentage of* $C_6H_{14}O_6$

The principle underlying the assay is oxidation of mannitol to formic acid and formaldehyde by sodium periodate followed by determination of excess periodate. The latter determination uses a different end-method of analysis from that described under Glycerol Suppositories (p. 297). The reactions taking place overall are as follows.

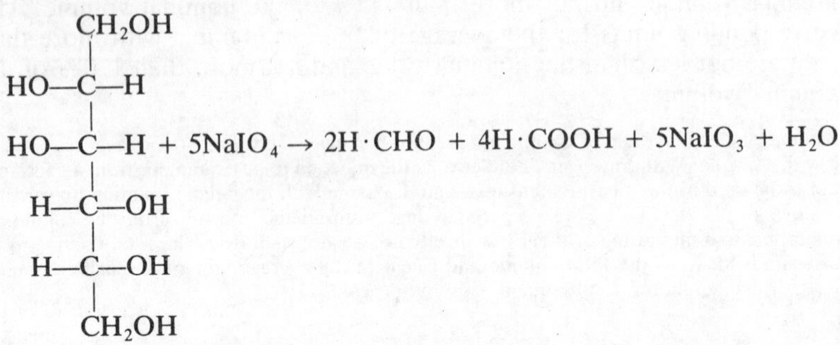

$$NaIO_4 + 2HI + H_2O \xrightarrow[\text{solution}]{NaHCO_3} NaIO_3 + I_2 + 2KOH$$

$$\underset{\text{(excess)}}{Na_3AsO_3} + I_2 + H_2O \longrightarrow Na_3AsO_4 + 2HI$$

$$2HI \xrightarrow{NaHCO_3} NaI + H_2O + CO_2$$

$$182.2\text{ g } C_6H_{14}O_6 \equiv 5NaIO_4 \equiv 5Na_3AsO_3 \equiv 5I_2$$

$$36.44\text{ g } C_6H_{14}O_6 \equiv 1000\text{ ml M iodine}$$

$$1.822\text{ g } C_6H_{14}O_6 \equiv 1000\text{ ml } 0.05\text{M iodine}$$

$$0.001822\text{ g } C_6H_{14}O_6 \equiv 1\text{ ml } 0.05\text{M iodine}$$

Method Dilute a volume of the injection containing about 0.4 g of mannitol to 100 ml with *water*. Transfer 10 ml of the dilution to a stoppered flask, add sodium periodate solution (2.14%; 20 ml) (Note 1) and M sulphuric acid (2 ml). Heat on a water-bath for 15 min (Note 2). Cool, add sodium bicarbonate (3 g) and 0.05M Na_3AsO_3 (25 ml) and mix. Add potassium iodide solution (20%; 5 ml), allow to stand for 15 min (Note 3) and titrate with 0.05M iodine to the first trace of yellow colour. Repeat the operation without the mannitol and calculate the percentage of mannitol from the difference in titrations.

Note 1 The volume of periodate solution must be accurately measured.

Note 2 Loosen the stopper of the flask to release any pressure.

Note 3 The liberated iodine immediately oxidises the arsenite present in the presence of bicarbonate but no reaction occurs between the iodide and iodate under neutral or slightly alkaline conditions. In the blank determination all the periodate is available to liberate iodine from potassium iodide and, hence, more of the arsenite is oxidised than occurs when sample is present. The iodine titre for the blank is, therefore, less than that for the sample.

COGNATE DETERMINATION

Sorbitol Injection For strong solutions Greenwood (1972) has shown that refractometry (Part 2) is a rapid and convenient method.

Clonidine Hydrochloride *Determination in an injection*

The principle underlying the method is the formation of an indicator-base complex which can be extracted into chloroform under mildly alkaline conditions. To increase sensitivity the colour of the complex is measured under mildly acidic conditions. For a fuller account of the factors involved in the formation and breakdown of the complexes, see page 247.

Many bases will give this reaction and the procedure is, therefore, no more specific than a measurement of ultraviolet absorption. In such cases non-specific assay procedures must be supported by adequate tests for identity of active ingredient and for limitation of the amount of related substances.

Method Dilute the injection to give a solution containing about 0.15 g of clonidine hydrochloride and add citrophosphate buffer (pH 7.6; 25 ml) and 1 ml of a solution containing 0.15% w/v of bromothymol blue and 0.15% w/v anhydrous sodium carbonate. Add chloroform (30 ml), shake for 1 min and centrifuge if necessary. To an aliquot (15 ml) of the chloroform extract add boric acid solution (Note; 10 ml) and measure the absorbance of the solution at the maximum at about 420 nm using a reagent blank. Repeat the procedure using a standard solution of clonidine hydrochloride (0.003% w/v) and calculate the percentage of Clonidine Hydrochloride in the injection.

Note. Boric acid solution is prepared by dissolving boric acid (5 g) in *water* (20 ml) and absolute ethanol (20 ml) and diluting to 250 ml with absolute ethanol.

Pyridoxine Hydrochloride *Determination in a vitamin preparation for intravenous administration*

Interference in the direct determination of pyridoxine by ultraviolet absorption is caused by thiamine hydrochloride, nicotinamide and riboflavine, depending upon the wavelength chosen. However, by a correct choice of buffer solutions a ΔA value (Part 2) may be obtained at 328 nm characteristic of pyridoxine alone. The spectral changes that occur when using solutions of pH 2–7 are shown in Fig 11.3.

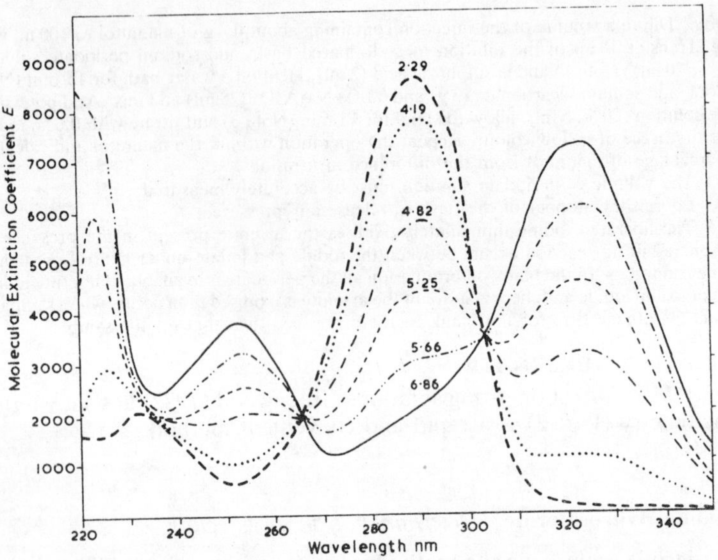

Fig. 11.3 Absorption spectra of pyridoxine at pH values less than 7

Method Dilute the injection with *water* to give a 0.01% solution of Pyridoxine Hydrochloride. Dilute the solution (10 ml) to 100 ml with glycerinated phosphate buffer pH 7.0 (Solution A) and glycerinated 0.1M hydrochloric acid (Solution B). Measure the absorbance 1 cm cells at the maximum at about 328 nm of Solution A using Solution B as blank.

Calculate the percentage Pyridoxine Hydrochloride using a value of 340 for the $A^{1\%}_{1\,cm}$ of Pyridoxine Hydrochloride under these conditions.

Lozenges

The active ingredients are usually incorporated in a flavoured basis consisting mainly of sucrose and the mass is then moulded or compressed into lozenges of the required shape and weight. Standards for uniformity of weight and content of active ingredient of compressed lozenges are those applicable to tablets.

Amphotericin Lozenges *Determination of Amphotericin A in the Amphotericin used in the lozenges*

The parent drug Amphotericin is required to contain not more than 15.0% of Amphotericin A. The method is of interest because it consists of a spectrophotometric determination of a mixture of two components as described in Part 2 (Chapter 7). The method is complicated, however, in that one of the two standards used, Amphotericin B reference, contains a small percentage, F, of Amphotericin A. Also, in order to calculate the appropriate $A^{1\%}_{1\,cm}$ values, S_1 and S_2, for the lozenges, the total content of Amphotericin must be known. Fortunately, a microbiological assay can be used for this purpose, and the formula for calculating the content of Amphotericin A in the Amphotericin used becomes:

$$F + \frac{100(BS_2 - bS_1)}{(aB - Ab)}$$

where

$A = A^{1\%}_{1\,cm}$ of Amphotericin A reference at 282 nm
$a = A^{1\%}_{1\,cm}$ of Amphotericin A reference at 304 nm
$B = A^{1\%}_{1\,cm}$ of Amphotericin B reference at 282 nm
$b = A^{1\%}_{1\,cm}$ of Amphotericin B reference at 304 nm
$S_1 = A^{1\%}_{1\,cm}$ of the sample at 282 nm*
$S_2 = A^{1\%}_{1\,cm}$ of the sample at 304 nm*
F = declared content of Amphotericin A in Amphotericin B

*with respect to the content of Amphotericin determined by microbiological assay.

The quality control of the lozenges consists of general appearance, uniformity of diameter, uniformity of weight, content of active ingredient and a limit on a likely impurity.

Method Powder the lozenges and accurately weigh a quantity equivalent to about 50 mg of amphotericin into a 100 ml volumetric flask. Add dimethyl sulphoxide (20 ml) (Note), mix well and dilute to 100 ml with dehydrated methanol. Filter and dilute the filtrate (10 ml) to 50 ml with dehydrated methanol. Measure the absorbance in 1 cm cells at 282 and 304 nm using a blank of 0.8% v/v dimethyl sulphoxide in dehydrated methanol. In the same way determine the absorbance of 0.008% Amphotericin B reference and 0.008% Amphotericin A in 0.8% v/v dimethyl sulphoxide in dehydrated methanol.

Calculate the appropriate $A^{1cm}_{1\%}$ values, and substitute in the equation to determine the percentage of Amphotericin A in the Amphotericin used to prepare the lozenges.

Note Dimethyl sulphoxide is a very useful solvent for the initial penetration and solution of constituents of even crude drugs.

Throat lozenges *Determination of* 2,4-*Dichlorobenzyl Alcohol and Amyl*-m-*cresol*

These moulded lozenges are about 2.5 g in weight and contain 1.2 mg of 2,4-dichlorobenzyl alcohol and 0.6 mg of amyl-*m*-cresol. In principle, the two compounds should be readily separated from the sugar basis by solvent extraction with ether under slightly acid conditions. Thereafter, the phenolic compound should be extractable with sodium hydroxide from ethereal solution in accordance with the principles described in Chapter 9. In practice, however, this latter separation does not occur and ultraviolet spectrophotometric method of assay for each component is not feasible.

A gas chromatographic procedure would seem ideal for the relatively small quantities involved. Versamide 900 is frequently used for phenols; polypropylene glycol adipate (Reoplex 400), 15% on Chromosorb W80-100 mesh is also very useful, both components being separated on this column (1 m) at 190°, with chloroxylenol as internal standard.

Typical retention times are:

2,4-dichlorobenzyl alcohol	10 min
amyl-*m*-cresol	12 min
chloroxylenol	14 min

It is common practice in many laboratories to restrict the number of packed columns in use to around six. OV 17 (3%) on Gas Chrom Q would be a likely candidate for inclusion in such a list. At a column temperature of 125° and an appreciably higher injector temperature it gives excellent results for this preparation.

Method Weigh and powder five lozenges. Dissolve a weight of powder equivalent to one lozenge in *water* (15 ml), add 2M hydrochloric acid (2 drops), sodium chloride (2 g) and internal standard solution (0.02%; 5 ml). Extract with ether (3 × 20 ml), bulk the ether extracts, wash with saturated sodium chloride solution (5 ml) and dry with anhydrous sodium sulphate (2 g). Decant most of the ether (Note) and evaporate to low volume (1–2 ml) using a little ethanol if necessary to obtain a clear solution free from traces of suspended water. Inject about 1 μl into the column and compare the result with that obtained for a mixture of the components (amyl-*m*-cresol, 0.012%, 4.5 ml; 2,4-dichlorobenzyl alcohol, 0.024%, 5 ml) treated in the same way.

Calculate the content of each component in the normal way (Part 2).

Note The presence of internal standard makes quantitative recovery at this point unnecessary.

Extension (*a*) Investigate by polarimetric and other means the nature of the sugar basis.
(*b*) An odour of menthol is detected on preparing the solution used for the assay. Confirm its presence by GLC.

Mixtures

The satisfactory presentation of unpalatable preparations requires the presence of flavourings, sweeteners and colouring agents in the formulation. Official Compendia draw a distinction between syrups and elixirs with the former based on preparations from natural sources with no added medicaments. On the other hand, the term syrup is widely applied in pharmaceutical and hospital situations to reconstituted antibiotic preparations which are truly suspensions. Syrup formulations that are free from cariogenic sugars are now also available in preparations for chronic use in children or diabetics.

Elixirs

The complexity of the formulation of many elixirs makes separation methods of paramount importance in the determination and identification of active ingredients. Ultraviolet spectrophotometry as the end-method of analysis lacks specificity and is subject to interference by irrelevant absorption. Methods of allowing for, or eliminating such interference have been extensively studied. The original method of Morton and Stubbs (1946) based on the mathematical treatment of data collected at three wavelengths around the absorption maximum of the analyte, has been extensively modified by many workers. In spite of being tedious and lengthy, these methods have found wide acceptability.

A simpler approach has been to employ difference spectrophotometry where the absorbance of analyte in the formulation matrix is measured at its λ_{max} against a solution of the matrix as reference. Derivative spectroscopy, where the first, second and sometimes higher mathematical derivative of the absorbance is automatically recorded as the spectrum is scanned, is now the subject of renewed attention as a means of resolving spectral overlap (Fell, 1980).

Promethazine Hydrochloride *Determination in a simple elixir*

Phenothiazine compounds form reddish-coloured complexes with palladium and this reaction is the basis of a procedure which detects intact drug.

Method Dilute a weighed quantity of elixir equivalent to about 20 mg of drug substance to 100 ml with *water*. To this solution (10 ml) add palladium chloride reagent (Note), dilute to 50 ml with *water* and measure the absorbance of a 2 cm layer at the maximum at about 472 nm, using a reagent blank in the reference cell. Calculate the content of $C_{17}H_{20}N_2S$, HCl, taking 110 as the $A^{1\%}_{1\,cm}$ value at that wavelength.

Note Palladium chloride reagent Dissolve, with gentle heating, 0.2 g palladium(II) chloride in 116 ml of M hydrochloric acid, cool, add 13.6 g sodium acetate and dilute with *water* to 1000 ml.

Promethazine Hydrochloride *Determination in a complex mixture containing Ephedrine Hydrochloride, Codeine Phosphate and Promethazine Hydrochloride*

There is such a vast difference between the volatility of ephedrine at one extreme and codeine at the other that gas-liquid chromatography offers no simple procedure. A separation method by thin-layer chromatography followed by solvent extraction of the bases from the plate and determination of each by a suitable method, such as ultraviolet absorption, is a possible, though not ideal, approach because ephedrine is weakly absorbing compared with codeine and promethazine. With no simple differential extraction procedure available, consideration of the individual components and the possibility of their determination by separate characteristic reactions forms an interesting exercise.

Davidson (1976) described a difference spectrophotometric technique for the determination of intact phenothiazine compounds in the presence of oxidative and photochemical decomposition products, colouring and flavouring agents. The method is based on the absorbance of the sulphoxide derivative of the parent drug relative to the absorbance of underivatized drug (Fig. 11.4).

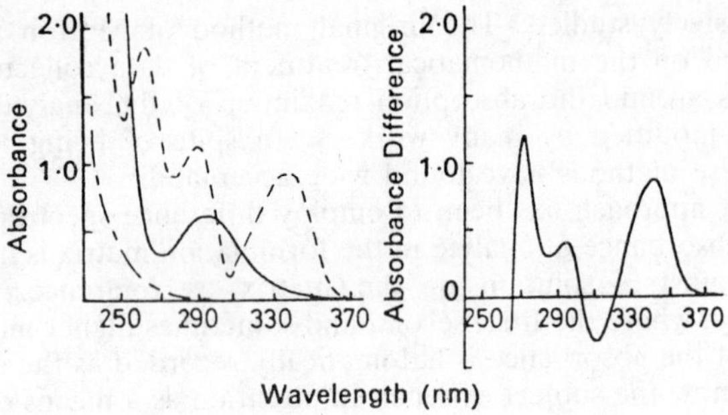

Fig. 11.4 (From Davidson, 1976). (*a*) The ultraviolet spectra of aqueous solutions (0.10 mmol l^{-1}) of promethazine hydrochloride (——), promethazine sulphoxide (- - -) and oxidizing reagent diluted (1 + 19) with water (— - —).
(*b*) The difference absorption spectrum of a solution of promethazine sulphoxide (0.10 mmol l^{-1}) relative to a solution of promethazine hydrochloride (0.10 mmol l^{-1}).

Codeine can be determined by spectrophotometry making due allowance for promethazine sulphoxide where it has been formed during storage of the mixture in a partly filled bottle, and ephedrine hydrochloride can be oxidized to benzaldehyde by sodium periodate.

Oxidising reagent Dilute hydrogen peroxide (100 volumes; 5 ml) to 500 ml with glacial acetic acid. Allow the solution to stand at room temperature overnight or heat the solution at 70° for 1 h. The reagent thereafter is stable for at least 2 months.
Standard reference solutions Dissolve promethazine hydrochloride reference substance (about 60 mg, accurately weighed) in *water* and dilute to 100 ml. Transfer an aliquot (5 ml) to two volumetric flasks (100 ml) and dilute the contents of one flask to volume with *water*. To the other flask add oxidising reagent (5 ml) and dilute the contents to 100 ml with *water*. Measure the absorbance of the oxidised solution using the unoxidised solution in the reference cell at the difference maximum in the wavelength range 335–355 nm using 1 cm cells (Note 1).
Method for the mixture Codeine Phosphate To a weight of mixture equivalent to about 10 mg of codeine phosphate add 0.1M hydrochloric acid (10 ml), extract with chloroform (4 × 25 ml) (Note 2) and wash the bulked chloroform layers with water (2.5 ml; Note 3). Add the aqueous washings to the acid layer and basify with ammonia. Extract immediately with chloroform (3 × 25 ml), bulk the chloroform extracts and wash once with saturated solution of sodium chloride (2.5 ml). Extract the bases from the chloroform layer with 0.05M sulphuric acid (3 × 25 ml); transfer the acid extracts to a 100 ml volumetric flask and wash the chloroform layer with *water* (20 ml). Add the washings to the flask, adjust to volume with *water* (Solution A) mix well and record the absorption curve over the region 260–400 nm in a 2 cm cell (Note 4). Prepare and examine an oxidised and an unoxidised solution as in the determination of promethazine hydrochloride in the preparation (25 ml aliquots), but using 4 cm cells. Calculate the concentration of codeine (Note 4) after allowing for the contribution of promethazine sulphoxide at the codeine absorption maximum at about 285 nm (Note 1).
Note 1 An absorption curve over the range 260–355 nm is required in the assay for codeine to calculate the correction at 285 nm based upon the absorbance at 340–345 nm.
Note 2 Shake gently to avoid the formation of stable emulsions. If there is a tendency for emulsion formation add solid sodium chloride sufficient to saturate the aqueous phase.
Note 3 Promethazine hydrochloride, which would contribute handsomely to the total (codeine + promethazine) absorbance at 285 nm, is readily soluble in chloroform; promethazine sulphoxide is not.
Note 4 The absorption curve of a standard solution of codeine phosphate should be run at the same time. If there is no interference due to promethazine sulphoxide this absorbance

value is valid. Strictly, it may be a three-component mixture but ephedrine does not absorb radiation at 285 nm.

Promethazine hydrochloride. Accurately weigh an amount of the preparation equivalent to about 12 mg of promethazine hydrochloride and dilute to volume with water in a volumetric flask (100 ml). Prepare an oxidised and an unoxidised solution using aliquots (25 ml) and dilute to 100 ml as described under standard reference solutions and measure the difference absorbance of the solutions at the same wavelength as was used for the standard reference solutions.

The concentration of promethazine hydrochloride in the mixture as mg per unit dose is given by:

$$\frac{\Delta A_T}{\Delta A_s} \times C_s \times D \times \frac{\text{weight of unit dose}}{\text{weight of sample taken}}$$

where ΔA_T and ΔA_s are the difference absorbances of the sample and the standard solutions, respectively, C_s is the concentration of the standard reference solutions in mg per 100 ml and D is the dilution factor.

Codeine Phosphate. Ephedrine hydrochloride. Dilute Solution A (codeine phosphate determination; 10 ml) to 100 ml with *water* and to the dilution (10 ml), in a separator, add saturated sodium bicarbonate solution (1.5 ml) and sodium periodate solution (5%; 3 ml). Allow to stand for 15 min, add 0.05M sulphuric acid (10 ml) and cyclohexane (spectroscopic grade; 15 ml). Shake well for 3 min, allow to separate and reject the lower aqueous phase. Wash the cyclohexane layer with a further 10 ml of acid. Filter the cyclohexane layer through a plug of cotton wool and record the absorption curve over the region 225–275 nm using as blank the cyclohexane layer obtained in the same procedure but using *water* in place of the solution to be examined. Calculate the amount of ephedrine hydrochloride per unit dose by treating known amounts of ephedrine hydrochloride in the same way.

COGNATE DETERMINATION

Determine the active ingredients in the following preparation:

Pseudoephedrine Hydrochloride	30 mg
Triprolidine Hydrochloride	1.25 mg
Codeine Phosphate	10 mg
Colour	q.s.
Syrup base to	5 ml

Hint Triprolidine contains a pyridine nucleus, a characteristic feature of which is that the ultraviolet absorption of an acid solution is greater than that of an alkaline solution at a particular wavelength.

Mouthwash *Determination of chloroxylenol*

Preparations containing chloroxylenol are readily assayed by a GLC procedure using either a Carbomax 20M column at 100°, or the conditions described under Lozenges (p.271), in both cases with 4-chloro-*m*-cresol as internal standard.

Method Dilute the sample (5 ml) with methanolic chlorocresol solution (1%; 5 ml) and sufficient methanol to produce 25 ml of dilution. Dilute the sample (5 ml) to 25 ml with methanol. Inject 1 µl volume into the gas-chromatography apparatus and calculate the percentage of chloroxylenol by the method of peak height ratios. Use a standard solution of chloroxylenol 1.0% (5 ml) diluted with chlorocresol solution (1%; 5 ml) and methanol to 25 ml.

Adopt the standard method of 3 injections of each solution and make sure that no interference from constituents of the preparation occurs at the position of the internal standard peak.

Suspensions

Chloramphenicol Palmitate Mixture *Determination of the percentage of Chloramphenicol and the content of Polymorph A*

Chloramphenicol Palmitate is known to exist in at least two principal forms. When administered as an oral suspension, one of the forms referred to as Polymorph A in the British Pharmacopoeia is virtually inactive. The two forms can be distinguished by their infrared absorption spectra and a limit on the amount of Polymorph A is imposed by means of this technique. The quality control, therefore, consists of a visual check for uniformity of the suspension, an assay for total chloramphenicol, a limit test and weight per ml.

Polymorph A Mix the suspension (about 20 ml) with *water* (20 ml) and centrifuge for 15 min (ideally at about 18 000 rpm). Remove the supernatant liquid, which contains flavouring and preservative agents, and resuspend the residue in *water* by mixing thoroughly with a small quantity (2 ml) first and diluting with more *water*. Centrifuge and repeat the washing and centrifuging twice more. Suspend the solid in *water* (10 ml), filter by suction and dry the residue first by pressing between filter paper and then in an oven at 20° for 24 h under vacuum.

Prepare a liquid paraffin mull (Part 2) of the dry residue so as to obtain an absorption band equivalent to about 20–30% transmittance at about 810 cm^{-1}. To achieve this intensity of absorption a concentrated mull is required.

Prepare in a similar manner the absorption curves for standards of Chloramphenicol Palmitate reference standard containing 20% and 10% of chloramphenicol palmitate Polymorph A reference standard. The curve obtained for the 20% mixture is used to determine the exact wavelengths of minimum and maximum absorptions at about 880, 790 and 858, 840 cm^{-1} respectively. Using a base line technique (Part 2), obtain the corrected extinctions at about 858 and 840 cm^{-1} and calculate the ratio A_{858}/A_{840} for sample and 10% mixture. The absorbance ratio for the sample should be less than that for the 10% mixture.

Chloramphenicol Mix the sample and weigh about 12 g accurately. Dilute to 1000 ml with *water* and mix thoroughly to disperse the suspension. Allow to stand (10 min) and mix thoroughly once more. Immediately pipette 5 ml of the suspension into a 100 ml volumetric flask and dilute to the mark with ethanol (96%). Measure the absorbance in a 1 cm cell at about 271 nm (maximum) and calculate the percentage w/w of chloramphenicol palmitate in the mixture, taking 178 as the $A^{1\%}_{1\,cm}$. Convert to terms of per cent w/v by means of the weight per ml of the preparation and to chloramphenicol with the factor 0.575.

Primidone Mixture *Determination of the percentage of* $C_{12}H_{14}N_2O_2$

The determination is difficult by normal extraction methods because of the poor solubility of primidone in water and organic solvents. It is, however, sufficiently soluble in methanol to allow simple dilution of the preparation and examination by gas-liquid chormatography, using an OV17 (3%) column at 260° and *N*-phenylcarbazole as internal standard.

Method Prepare solutions as follows:
Solution A Dissolve *N*-phenylcarbazole (about 0.25 g, accurately weighed) in methanol (100 ml).
Solution B To primidone reference substance (about 0.15 g, accurately weighed) add Solution A (25 ml) and methanol to volume (50 ml), warming to dissolve if necessary.
Solution C To a volume of the mixture equivalent to about 0.3 g of primidone, add methanol

(45 ml) and mix while warming on a water-bath (5 min). Cool, add Solution A (50 ml), dilute to volume (100 ml), shake well and allow precipitated solids to settle out.

Determination Inject suitable volumes (Note) of Solution B and Solution C onto the column. Calculate the ratio of appropriate peak heights in each. Determine the weight per ml of the mixture and calculate the content of $C_{12}H_{14}N_2O_2$ weight in volume.

Note As usual, it must be confirmed that the preparation contains nothing with a retention time similar to the internal standard.

COGNATE DETERMINATIONS

Fenfluramine Tablets
Primidone Tablets

Phenethicillin Elixir *Determination of Phenethicillin*

In addition to the official BP method, the syrupy suspension, freshly reconstituted by adding the specified quantity of water to the dry powder, may be assayed by the following procedure. The determination is based upon the formation of a hydroxamic acid on opening of the β-lactam ring in the presence of hydroxylamine and the formation of a characteristic red colour which such acids give with a ferric salt.

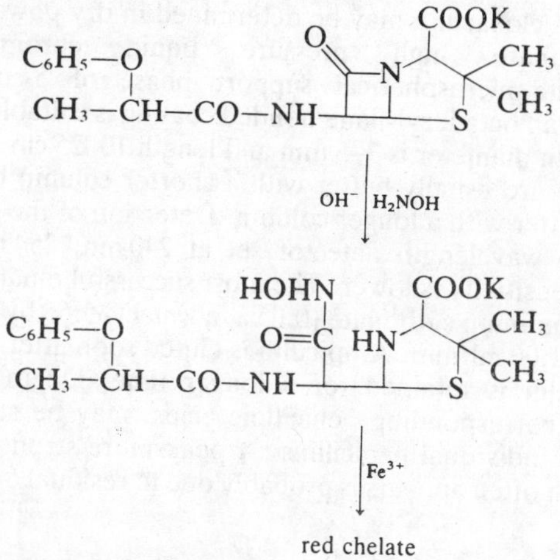

The reaction depending on the presence of the intact β-lactam ring imparts a degree of specificity and inactive hydrolysis products will not, therefore, be determined; the stability test is based upon this. As control of pH is important in maintaining stability a limit of 5.2–6.2 is essential in the regenerated elixir.

Method Weigh accurately about 3 ml of the freshly prepared elixir (Note 1), dilute to 50 ml with *water* and transfer 2 ml of the solution to each of two test-tubes.

To one tube add 2M NaOH (1 ml), allow to stand exactly 45 min and add 2M H_2SO_4 (1 ml). To the second tube add *water* (2 ml).

Add hydroxylamine hydrochloride solution (6 ml) (Note 2) to each tube and allow to stand for 40 min. Add ferric reagent (2 ml) (Note 3) to each tube and allow the colour to develop

(20 min). Measure the absorbance of the solution in the second tube in a 1 cm cell at 490 nm using the solution in the first tube as blank. Calculate the concentration w/w of phenethicillin by reference to a calibration curve prepared with phenethicillin potassium reference. Determine the weight per ml of the elixir and convert the per cent w/w to terms of w/v. The specification requires 95.0–120% of the stated amount as determined by the method above.

Stability Test Reserve a portion of the prepared elixir, store at 15° ± 1° for 7 days and repeat the determination. The result should not be less than 90% of that found for the freshly prepared elixir.

Note 1 3 ml contains the equivalent of 75 mg of phenethicillin if the preparation is stated to contain 125 mg in 5 ml. For stronger solutions the quantity should be correspondingly reduced.

Note 2 The reagent is freshly prepared by mixing equal volumes of hydroxylamine hydrochloride solution (35.76% in *water*) and a solution containing sodium hydroxide (17.3 g) and sodium acetate (3.18 g) in 100 ml. The mixture is adjusted to pH 7.0 by addition of the alkaline solution or hydrochloric acid, as appropriate, and diluted with 3 volumes of ethanol (90%).

Note 3 Ferric ammonium sulphate (30 g) dissolved in a mixture of *water* (70 ml) and sulphuric acid (9.3 ml) and diluted to 100 ml with *water*.

Ampicillin and Cloxacillin *Determination in powders for reconstitution with water*

Ampicillin and cloxacillin may be determined in dry powder preparations by reverse-phase high pressure liquid chromatography. A microparticulate-microspherical support phase of 5 or 10 µm mean diameter with an octadecylsilane bonded coating is suitable. The preferred internal column diameter is 3–5 mm and length 10–25 cm. Peak shape and retention time are usually better with a shorter column but separation is likely to be better with a longer column. Detection of the two penicillins is achieved with wavelength detector set at 240 nm; 254 nm readings are possible but sensitivity is lower. The most successful eluant composition is methanol:ammonium carbonate (0.05M in *water*) 35:65 by volume at a flow rate of 0.8 to 1.00 ml/min. Ampicillin is eluted soon after the solvent front while cloxacillin is retained for a longer time. Degradation products, probably the corresponding penicilloic acids, may be seen as shoulders preceding the individual penicillins. A peak more strongly retarded than cloxallin which often appears is probably due to residual 6-aminopenicilloic acid.

Method Dissolve an aliquot of mixed powder equivalent to about 60 mg of ampicillin in 100 ml of a mixture of methanol and *water* (35:65 by volume) (Note) and inject 20 µl volumes on to the column. Calculate the concentrations of ampicillin and cloxacillin in the preparation by comparison under the same conditions with solutions containing known amounts of ampicillin and cloxacillin reference substances. Alternatively, penicillin V, which elutes between ampicillin and cloxacillin, may be used as internal standard.

Note Methanol:ammonium carbonate solvent should not be used to dissolve the preparation. The slightly alkaline pH causes slow hydrolysis of the penicillins, resulting in anomalously low values for parent substances and high values for degradation products.

Compound powders and compound liquid mixtures *Determination of Bismuth, Calcium and Magnesium Carbonates and Sodium Bicarbonate*

Bismuth. This is determined directly at pH 1–2 using xylenol-methyl orange as indicator, since calcium and magnesium do not interfere under the conditions of the assay.

Magnesium. Bismuth, if present, is removed by precipitation as the oxychloride. Calcium is removed by precipitation as oxalate from an aliquot of the filtrate and the combined filtrate and washings titrated directly at pH 10.

Calcium. A further aliquot of the filtrate obtained in the magnesium titration is titrated, with prior removal of calcium. The titre is equivalent to both calcium and magnesium, and the calcium titre is obtained by difference.

Sodium Bicarbonate. The determination of sodium bicarbonate necessitates a correction for the appreciable amounts of calcium and magnesium bicarbonates which survive the attempt to decompose them by boiling, filtering twice and cooling. The filtrate is titrated with 0.5M hydrochloric acid using screened methyl orange (methyl orange-xylene cyanol FF) as indicator, strong ammonia–ammonium chloride added and the solution titrated with 0.05M disodium edetate using mordant black II solution as indicator. One-fifth of this titration is deducted from the volume of 0.5M hydrochloric acid required. The difference is equivalent to the sodium bicarbonate in the sample. The correction factor of one-fifth is derived as follows:

$$\mathrm{Ca/Mg(HCO_3)_2 + 2HCl \rightarrow Ca/MgCl_2 + H_2O + CO_2}$$

$$\therefore \quad \mathrm{Ca/Mg(HCO_3)_2 \equiv 2HCl \equiv 4000\ ml\ 0.5\text{M}\ HCl}$$

but

$$\mathrm{Ca/Mg(HCO_3)_2} \equiv 1 \text{ mole disodium edetate}$$
$$\equiv 20\,000 \text{ ml } 0.05\text{M disodium edetate}$$

hence, the equivalent of calcium and magnesium present, when expressed in terms of 0.5M is equal to 4000/20 000, i.e. one-fifth of the titre obtained in the disodium edetate titration.

See Chapter 8 for details of methods.

Ointments

Ointments are generally semisolid, greasy preparations and distinct from creams, in which water forms a significant proportion of the whole. The necessarily chemical inertness of an ointment basis and its solubility in organic solvents rarely lead to complications in the determination of active ingredients. Greenwood and Guppy (1974) have examined a number of preparations of benzoic acid and of salicylic acid by titration with 0.1M NaOH using phenolphthalein as indicator. The results were well within the range normally allowed for quality control.

Haemorrhoid ointment *Determination of Ephedrine Hydrochloride, Lignocaine Hydrochloride and Allantoin*

This example is chosen to illustrate how the solubility properties of the ingredients of an ointment can be used to full advantage to determine each, free from interference by the other. The preparation is usually made to the following formulation:

Ephedrine Hydrochloride	0.25%
Lignocaine Hydrochloride	0.5%
Allantoin	0.5%
Ointment basis (Paraffin-type) to	100%

The determination of ephedrine by normal procedures for synthetic organic bases (Chapter 9) is complicated by marked losses due to water-solubility of the base during the washing of the bulked ether extracts. This property, however, can be turned to advantage in separating ephedrine from lignocaine.

Allantoin is soluble in acid, neutral and alkaline aqueous media and, although showing end-absorption only in acid, peak absorption at about 225 nm appears in alkaline solution.

Isolation of ingredients Weigh the ointment (about 2 g accurately weighed) into a small beaker and disperse thoroughly in ether (25 ml). Transfer to a separator (Note 1) (100 ml) and complete the transfer with ether (20, 20 ml) and 0.05M H_2SO_4 (10 ml). Shake thoroughly, run off the acid layer into a separator (50 ml) containing ether (10 ml). Shake, allow to separate and run off the acid layer into a third separator (50 ml). Complete the extraction of allantoin and salts in the same manner with *water* (10, 10 ml).

Isolation of bases Make the bulked aqueous extracts alkaline with sodium hydroxide solution (20%; 2 ml) and extract the bases with ether (5 × 20 ml). Bulk the ether extracts and wash once with saturated sodium chloride solution (3 ml). Add the washing to the reserved alkaline aqueous liquid for the determination of allantoin.

Ephedrine Extract the ether layer with sodium bicarbonate solution (1%; 8 × 10 ml), washing each extract with ether (10 ml). Make up to 100 ml with more bicarbonate solution, mix well and record the absorption curve of the solution using 1% sodium bicarbonate as blank in 4 cm cells (Note 2). On the same chart record the absorption curve of a 0.01% solution of ephedrine hydrochloride in 1% sodium bicarbonate.

Calculate the percentage of ephedrine hydrochloride in the ointment.

Lignocaine Extract the ethereal layer with 0.05M H_2SO_4 (15, 10, 10 ml), washing each extract with the same ether (10 ml) as was used in the determination of ephedrine. Transfer to a 100 ml volumetric flask and make up to volume with 0.05M H_2SO_4. Record the absorption curve in 4 cm cells using a standard lignocaine hydrochloric solution (0.01% in 0.05M H_2SO_4) for comparison with 0.05M H_2SO_4 as blank.

Calculate the percentage of lignocaine hydrochloride in the ointment.

Allantoin Dilute the reserved alkaline aqueous solution to 100 ml in a volumetric flask with *water*. Dilute the solution (5 ml) to 50 ml with 0.1M NaOH and also 5 ml to 50 ml with 0.05M H_2SO_4. Record the absorption curve for each solution using respective blanks of 0.1M NaOH and 0.05M H_2SO_4 in cells over the region 220–300 nm.

Calculate the percentage of allantoin by means of the ΔA value of a standard allantoin solution (0.001% in 0.1M NaOH and 0.05M H_2SO_4, respectively).

Note 1 In all experiments involving precision ultraviolet absorption measurements it is advisable to use as little grease as possible on separator taps. Ideally, water alone should be used as there is only slight sticking of taps at worst.

Note 2 Ephedrine has zero absorbance at 285 nm and the pen should be set at zero at this wavelength to take account of any slight haziness that might occur in the solution over a 4 cm path length. The trace itself should show no evidence of irrelevant absorption except at about 230 nm and below. The advantage of recording the absorption of the pure sample on the same chart is the confirmation obtainable that this is so. A scale-expansion accessory is invaluable in this determination.

Suppositories

The bases for suppositories may be either fatty, for example, theobroma oil and hydrogenated vegetable oil, or water-soluble, magrogols. Standards imposed on suppositories include uniformity of both internal and external **appearance** as a control on the uniform distribution of medicament; **uniformity of weight** in which twenty suppositories are weighed singly and no suppository should deviate by more than 10% of the average weight except that two may deviate by not more than 5%; and a **disintegration test** under carefully controlled conditions whereby, unless otherwise specified, disintegration should be complete within 30 min.

Except for *Glycerol suppositories*, which are dissolved directly in water, representative samples of suppositories for assay are obtained either by cutting into small pieces and weighing the required quantity of these, or by melting together, allowing to cool, with stirring, until set and weighing the required quantity of the mass.

In spite of the variation in bases they rarely complicate the determination of active ingredients. Thus, non-aqueous titration is adopted for *Bisacodyl suppositories* (perchloric acid) and for *Phenylbutazone suppositories* (tetrabutylammonium hydroxide), whilst *Indomethecin suppositories* are examined by ultraviolet absorption.

Glycerol Suppositories *Determination of the percentage of* $C_3H_8O_3$

The determination depends upon the reaction expressed by the following equation:

$$\begin{array}{l} CH_2OH \\ | \\ CHOH + 2NaIO_4 \longrightarrow 2H\cdot COOH + 2NaIO_3 + H_2O \\ | \\ CH_2OH \end{array}$$

$$H\cdot COOH + NaOH \rightarrow H\cdot COONa + H_2O$$

$$\therefore \quad C_3H_8O_3 \equiv H\cdot COOH \equiv 1000 \text{ ml } \mathrm{M}$$

$$\therefore \quad 92.10 \text{ g } C_3H_8O_3 \equiv 1000 \text{ ml } \mathrm{M}$$

$$\therefore \quad 0.00921 \text{ g } C_3H_8O_3 \equiv 1 \text{ ml } 0.1\mathrm{M} \text{ NaOH}$$

The excess sodium metaperiodate is removed by the addition of propan-1,2-diol when the following reaction occurs:

$$\begin{array}{l} CH_2OH \\ | \\ CHOH + NaIO_4 \longrightarrow H\cdot CHO + CH_3CHO + NaIO_3 + H_2O \\ | \\ CH_3 \end{array}$$

It is essential to convert the periodate to iodate because is consumes sodium hydroxide with concomitant fading of the indicator and gradual appearance of a purple colour. No such effect is observed with iodate and a sharp colour change occurs at the end point at about pH 7.

Method Dissolve a sufficient number of suppositories to give about 8 g of glycerol and dissolve in *water*. Dilute the solution (5 ml) with *water* (about 150 ml), add 5 drops of bromocresol purple indicator solution and neutralise any acid by addition of 0.1M NaOH to the blue colour of the indicator. Add sodium metaperiodate (1.6 g) and allow to stand for 15 minutes. Add propan-1,2-diol (3 ml), allow to stand for 5 minutes and titrate the formic acid with 0.1M NaOH to the same blue colour.

Suppositories incorporating a number of ingredients offer good opportunities of applying separation techniques and illustrating difficulties introduced by the formulation.

Compound Bismuth Subgallate Suppositories *Determination of bismuth subgallate* and *Zinc Oxide*

Zinc and bismuth ions can be rapidly and conveniently determined without physical separation by edetate titration. The end point for bismuth, however, can be difficult to locate and the AOAC method for bismuth in drugs is sometimes useful. It is based on the reaction in which bismuth salts in acid solution give a yellow colour with thiourea. It has, however, been criticised by Plank (1972), who recommends a polarographic method as being more accurate and precise. This would appear to be the ideal method as zinc ions are also readily determined by polarography. However, macrogols from the suppository base interfere with the reduction of zinc ions and a reduction step for zinc at about −1.1 V does not appear. That for bismuth at −0.1 V remains unaffected and offers an alternative means of determining the ion.

Method Weigh five suppositories, melt together and allow to cool, stirring continuously until the mass is set. Weigh a quantity of the mass equivalent to about 0.6 g of bismuth subgallate and extract in a Sochlet apparatus with ether; discard the ether. Moisten the thimble with sulphuric acid (1 ml), ignite and ash (below 500°). Dissolve the residue, with the aid of heat, in a mixture of nitric acid (5 ml) and water (10 ml), filter, washing the filter with *water* and dilute the combined filtrate and washings to 50 ml (Solution A).

Bismuth Dilute an aliquot (5 ml) of Solution A to 250 ml with water. Pipette the solution (10 ml) into a 50 ml volumetric flask, add M hydrochloric acid (5 ml) and thiourea solution (10%; 5 ml), mix and dilute to volume with *water*. Measure the absorbance at 395 nm using a reagent blank. Calculate the amount of bismuth subgallate by reference to a calibration graph prepared from aliquots of a solution of bismuth subnitrate (70 mg) dissolved in hydrochloric acid (40 ml) and diluted to 500 ml with *water*.

Zinc oxide To 20 ml of Solution A add 2M hydrochloric acid (2 ml) and 200 ml of *water*. Add 5M ammonia until the solution is just acid to litmus paper, boil for 2 min, cool and filter off the undissolved bismuth salt through a porosity 3 sintered glass crucible using warm *water* to wash the residue. Cool, add ammonia buffer solution (p.000) and titrate the zinc solution with 0.05M disodium edetate using mordant black II as indicator.

$$0.004068 \text{ g ZnO} \equiv 1 \text{ ml } 0.05\text{M disodium edetate}$$

COGNATE DETERMINATIONS

Magnesium Carbonate in compound powders.

Magnesium Carbonate in compound powders and suspensions containing Kaolin. The filtrate from the determination of **acid-insoluble matter** is treated with ascorbic acid and potassium cyanide to mask iron impurity as hexacyanoferrate(II) and with triethanolamine to mask traces of aluminium derived from the kaolin. The solution is titrated with 0.05M disodium edetate in an ammonia buffer with mordant black II as indicator.

Tablets

A variety of controls are applied to tablets in addition to the assay for active ingredient content and appropriate tests for related substances. These include tests for Uniformity of Weight, Disintegration and, where appropriate, Uniformity of Content, and Dissolution.

Uniformity of weight

All uncoated tablets are subject to a test for uniformity of weight to control the extent deviation from the average. The permitted percentage deviation decreased with increasing size of tablet, as shown in Table 11.2. The weights of at least eighteen of the twenty tablets examined must fall within the permitted deviation; none must deviate by more than twice that limit.

Table 11.2 Uniformity of weight of uncoated tablets

Average weight of tablet	Permitted deviation (%) for ≯ two of twenty tablets weighed
80 mg or less	± 10.0
> 80 and < 250 mg	± 7.5
250 mg or more	± 5.0

Content of active ingredient

Since the individual tablets may vary considerably in weight, the assay is so adjusted that a reasonably average sample is taken for the determination. This is done by weighing a batch of twenty tablets, and calculating the *average weight* for each tablet. The final calculation of the content of active ingredient per tablet is based upon this average figure and, therefore, is itself only an average weight per tablet.

Where twenty tablets are not available, a smaller number, not less than five, may be used but to allow for sampling errors the assay tolerances are widened according to Table 11.3.

Table 11.3 Adjustment of assay tolerances for small samples of tablets

Weight of medicament in each tablet	Adjustment: Subtract from the lower limit			Adjustment: Add to the upper limit		
Sample size	15	10	5	15	10	5
0.12 g or less	0.2	0.7	1.6	0.3	0.8	1.8
> 0.12 g and ch014 0.3 g	0.2	0.5	1.2	0.3	0.6	1.5
0.3 g or more	0.1	0.2	0.8	0.2	0.4	1.0

In spite of the variation amongst individual tablets, the tablet granules from which the tablets are prepared should be of reasonably constant composition (but see below), so that determination of the active ingredient content of the powdered tablets is an accurate measure of the amount of active principle present. For this reason it is essential to express the result of any such determination in two ways:

(*a*) the percentage of active ingredient in the powdered tablet and
(*b*) the average weight of active ingredient per tablet.

Uniformity of content of active ingredient

A test for uniformity of content is based on the assay content in the individual dosage unit, to ensure uniformity of distribution of active ingredient when it is less than 2 mg or where the active ingredient accounts for less than 2% by weight of the total. If the dosage unit contains a mixture of several active ingredients the test applies only to those which comply with those conditions; the test is not applied to polyvitamins and trace-element preparations. The analytical method need not be the same as that used in the assay of average content so long as it has a coefficient of variation of 1 or less. It is usual to expect at least 95% of the units in a batch to have a content within ± 20% of the average content and all to be within ± 25%.

An acceptable sampling plan is to take 10 units and assay their content individually; the sample passes the test if the coefficient of variation of the contents found is 6 or less. If the value is greater than 6, the sample can be enlarged to check whether the limits for variability are satisfied.

Disintegration

The function of official disintegration tests is to ensure that tablets break up completely in water under the test conditions in the apparatus specified. *Uncoated Tablets* are normally required to disintegrate at 36–38° within 15 min. The requirement for *Sugar-coated* and *Film-coated Tablets* is not more than 1 h. The temperature (36–38°), but not the medium, is chosen to simulate physiological conditions. *Soluble Tablets* and *Dispersible Tablets* which are required to dissolve or disperse in water at ambient temperatures are required to disintegrate under the same test conditions in water but at 19–21°, within 3 min.

Apparatus. The Pharmacopoeia specifies the design and dimensions of the apparatus to be used for the test. It consists of a rigid basket-rack assembly supporting six cylindrical glass tubes, the lower ends of which are effectively sealed with stainless steel wire gauze of specified thickness and mesh size. As the basket is raised and lowered thirty times (complete cycle) per minute in a thermostatically controlled water bath the tablets are usually suspended in motion within the tube, touching neither the mesh nor the disc and thus avoiding any abrasive effect. Where tablets float, they are pressed constantly against the top disc and again mechanical damage is avoided.

Method Place *water* in the cylinder to a depth of not less than 15 cm and raise the temperature to 36–38°. Check that, when the basket moves up and down, the gauze just breaks the surface of the water at the highest position and that the rim of the basket remains just above the surface of the water at the lowest position.

Place six tablets in the basket and set the basket in motion. Note the time and proceed with the test until no particle, except fragments of coating, remains above the gauze which would not readily pass through it. Note the time at this point which, in the absence of special circumstances, should be not more than 15 min in the case of uncoated tablets.

The apparatus is also used in the disintegration test for *Enteric-coated tablets* but the water is replaced with 0.1M hydrochloric acid to simulate the acid conditions of the stomach. After 2 h (1 h for *Erythromycin tablets*), during which time particles of tablet coating only pass through the gauze, the tablets are washed rapidly with water, the acid solution is replaced with solution of pH 6.8 and discs are added to the tubes. The test is continued for 1 h by which time disintegration, as defined above, should be complete.

Dispersible tablets are also required to conform to a test for **Uniformity of Dispersion**, in which two tablets, when dispersed in water (100 ml), give a smooth dispersion which completely passes a sieve with a nominal mesh aperture of 710 μm. The disintegration of *effervescent tablets* is determined similarly on six tablets, testing one tablet at a time in 200 ml of water. Each tablet should either completely dissolve or disperse, with gas evolution.

Dissolution

The problem that branded and non-branded products of the same medicinal compounds may not be clinically equivalent is well recognised. Dissolution tests are more valid indicators of differences between brands and of bio-availability than are disintegration tests. It must be stressed, however, that strict comparisons of bio-availability can only be made in animal experiments and clinical trials. A suitable apparatus for the routine determination of dissolution rates on production batches of tablets is shown diagrammatically in Fig. 11.5a, but it must be emphasised that many modifications are possible to take account of variables that make meaningful results difficult to obtain with some products.

Excipients. Tablet excipients may have an adverse effect on the absorption of a drug as compared with that for a solution or dispersion of the drug.

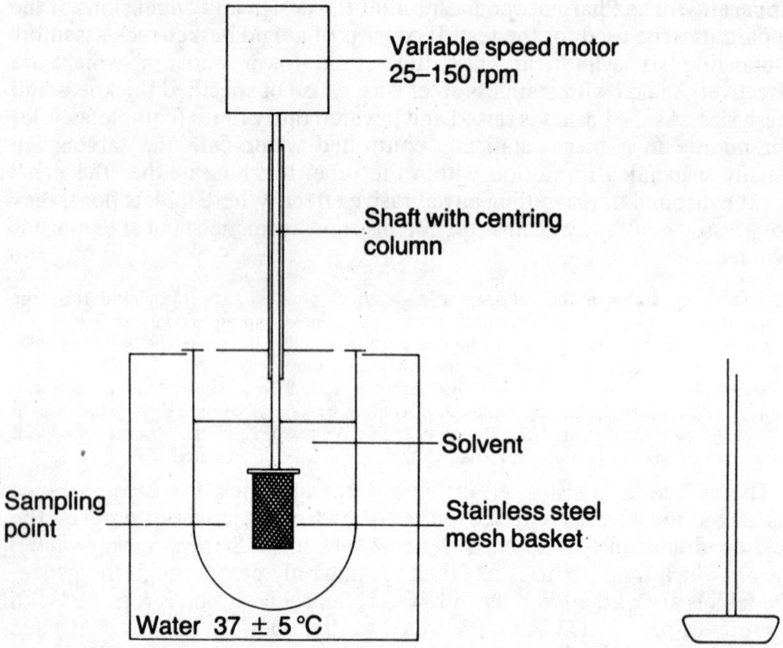

Fig. 11.5 (*a*) **Typical apparatus with rotating basket for tablet dissolution studies.** (*b*) **Paddle stirrer.**

Dissolution medium. The medium is generally chosen to suit the appropriate conditions, for example 0.6% hydrochloric acid for those drugs absorbed from the stomach and buffers of varying pH up to about pH 8 for those tablets designed to liberate the medicament over several hours or for enteric-coated tablets. Enzymes, such as pepsin for acid, or trypsin for alkaline media, are not normally added and the omission materially assists spectrophotometric end-methods of analysis.

The volume of dissolution medium is also important, perhaps more so for drugs of poor solubility in water or acid media. It is essential that sufficient medium is used such that the drug, even if completely dissolved, is present to no more than about 30–40% of its saturation solubility. Usually 600 or 900 ml of medium are used.

Shape of beaker. The results obtained with flat-bottomed beakers tend to differ from those obtained with round-bottomed beakers. An explanation may lie in solvent flow past the tablet in the rotating basket as the following variables are also known to be significant in dissolution studies.

Stirring. The *position* of the basket in the beaker and the *rate of rotation* should be fixed to obtain comparable results on different occasions. Normally, a speed of 100 rpm is used for any one laboratory, therefore conditions must be carefully standardised, even to the extent of deciding on whether or not the basket should be rotating as it is lowered into the dissolution medium. Clearly, solvent flow around the tablet in the basket is complex and modifications suggested to improve upon this position are continuous-flow dissolution apparatus, as described by Tingstad and Riegelman (1970), and a rotating filter-stationary basket system (Shah and Riegelman (1970), and Ochs (1974). In filtration systems used for continuous-flow studies it is essential to avoid clogging of the filter, even to the extent of providing a by-pass so that solution is passed through the filter and spectrophotometer cell at fixed intervals instead of continuously.

Temperature. To simulate body conditions a temperature of 37° is used.

Alternative methods

It seems likely that European Authorities will eventually adopt the USP XX paddle method shown diagrammatically in Fig. 11.5b.

Individual monographs in USP XX carry details of the volume of dissolution medium, its composition, rate of stirring and tolerance limits. A single tablet or capsule is placed in the apparatus, weighted, if necessary, to prevent flotation and samples (usually 1 ml) collected at specified intervals for assay, the volumes withdrawn at each stage being replaced by an equal volume of fresh dissolution medium.

The criteria for acceptance in this test are determined in stages, designated S_1, S_2 and S_3. Six tablets or capsules are tested at Stage S_1. If the quantity Q is the amount of dissolved active ingredient specified in the monograph required to be released in the stated time, expressed as a percentage of the labelled strength, then the batch of tablets or capsules is acceptable if each unit is not less than $Q + 5\%$.

If this criterion is not met a further six tablets or capsules must be tested (Stage S_2), when the criterion for acceptance is then that the average of 12 units ($S_1 + S_2$) is equal to or greater than Q and no unit is less than $Q - 15\%$.

If acceptance is still not reached, a further 12 units must be tested (Stage S_3). The final criterion for acceptance is then that the average of 24 units ($S_1 + S_2 + S_3$) is equal to or greater than Q and not more than 2 units are less than $Q - 15\%$.

A number of very insoluble compounds are not very amenable to test by the above methods because of the very large volumes of dissolution medium that would be required. Examples include such products as Griseofulvin tablets and Spironalactone tablets. Such products are, however, capable of being examined for their dissolution characteristics by means of specialised flow-through cells.

Tetracycline Hydrochloride Tablets *Comparisons of dissolution behaviour of different brands of tablets*

Method Place 0.1M HCl (1000 ml) in a beaker (Fig. 31b) and allow to stabilize at 37°. Place the tablet to be studied in the wire mesh basket, set the motor in motion at 100 rpm and lower into the dissolution medium, noting the time ($t = 0$). Attach a simple filtration system to a 10 ml pipette (Note 1) and take samples at t 5, 10, 20, 30, 60 and 90 min. Maintain the volume constant by adding 10 ml of 0.1M HCl after each removal of sample.

Determine the absorbance of each sample, making dilutions when necessary, over the region 250–400 nm (Note 2) and calculate the amount of tetracycline in solution, taking 320 as the $A^{1\%}_{1\,cm}$ value at the maximum, around 353 nm. Plot the amount of drug released as a percentage of stated tablet content. Carry out three determinations and repeat for different brands (Note 3). Construct a table of the results in the following form:

Brand name	Manufacturer	Lot No.	Per cent drug released at (min)			Time for 70% release
			10	30	60	
			(i) (ii) (iii)			(i) (ii) (iii)

Note 1 A small tube containing glass wool fitted to the end of the pipette should remove insoluble tablet basis or coating.
Note 2 The general shape of the absorption curves should be the same as that of a standard curve to show that tablet excipients are not interfering in any way with the tetracycline.
Note 3 As a comparison of different brands is being undertaken it is essential to obtain some idea of the reproducibility of the method and apparatus. The significance of between-brands variation could not otherwise be assessed.

Slow Lithium Carbonate tablets *Determination of solution rate*

In the treatment of some mental diseases it is essential that careful control of plasma concentrations of lithium in the range 0.6–1.6 mmol/l be maintained. Slow lithium tablets are designed to release lithium ions over a period of several hours, thus avoiding high concentrations which put the patient at risk. Even so, monitoring of lithium levels in plasma is carried out. The apparatus for solution rate is that described for the disintegration test for tablets except that an extraction thimble is used to hold the tablets.

Method Place three tablets in the thimble and using a suitable known volume of 0.6% v/v of hydrochloric acid instead of *water*, operate the apparatus for 2 h at 37°. Reserve the acid extract (A).

Replace the acid with a buffer pH 6.0 and continue to extract for 1 h. Reserve the extract (B).

Replace the buffer pH 6.0 with one of pH 6.8 and continue to extract for a further 5 h. Reserve the extract (C).

Examine each extract by atomic emission spectrophotometry (Part 2) for the lithium content.

The tablets comply if the following results are obtained.

Extract A	≯ 30%	
Extract A + B	≮ 30%	and ≯ 50%
Extract A + B + C	≮ 70%	and ≯ 95% of Li_2CO_3 available

COGNATE DETERMINATIONS

Slow Potassium Chloride Tablets The object of this preparation is to avoid high local concentrations of potassium chloride in contact with the lining of the intestine and so prevent irritation.

Slow Fe *Tablets* Use the apparatus described under Tetracycline Hydrochloride tablets and measure 1 ml samples for development of a colour with *o*-phenanthroline. As the experiment proceeds the concentration of ferrous ions increases so much that dilution of samples becomes necessary. This is one reason why *o*-phenanthroline is not added to the dissolution medium; the other reason is that with tablets containing the equivalent of 50 mg of Fe, more than 0.5 g of *o*-phenanthroline would be required for each determination, which becomes quite expensive.

Determination of medicaments by solvent extraction

The ingredients of tablet formulations, other than active ingredient, are largely insoluble in aqueous media and in organic solvents and, thus, cause physical interference in many assay procedures; recourse must, therefore, be had to extraction techniques. Solution of active ingredient and removal of insoluble debris may be regarded as extraction at its simplest. There are many instances where it is only necessary to add a suitable solvent to tablet powder containing a single active ingredient, filter, dilute to volume with solvent and measure the ultraviolet absorbance against solvent blank; aluminium hydroxide tablets and calcium gluconate tablets dissolved in suitable medium and filtered can be titrated directly with disodium edetate; acetazolamide tablets may be dispersed in water, filtered and determined polarographically.

In cases where it is necessary to use the property of partition between immiscible solvents as described in Chapter 9, the pH of the aqueous dispersion is adjusted to liberate the active moiety which is then extracted, usually in chloroform, and determined by titration or gravimetrically. For example, with tertiary base salts, such as codeine phosphate, the tablet powder is dispersed in water, ammonia is added to pH 10 and the free base is extracted into chloroform which is then evaporated. The residue can be determined by (a) addition of excess of standard acid and back-titration with alkali or (b) direct non-aqueous titration with perchloric acid. On the other hand, where the active moiety is acidic, as in barbiturates and related compounds, the pH is adjusted to around 3 and the residue from chloroform or ether is sufficiently pure to be weighed. Where sodium dodecyl sulphate is used as titrant the immiscible solvents are not separated and the mixture requires to be agitated mechanically.

In tablets with more than one ingredient the properties of each must be considered in devising a differential extraction procedure; for example, Tablets of Aspirin, Caffeine and Codeine.

Amylobarbitone Sodium Tablets *Determination of the weight of $C_{11}H_{17}N_2NaO_3$ per tablet of average weight*

Method Dissolve the sample of powdered tablets (equivalent to approximately 0.3 g of Amylobarbitone Sodium) as completely as possible in dilute sodium hydroxide solution. Saturate the solution with sodium chloride and acidify with hydrochloric acid. The latter liberates the amylobarbitone from its sodium salt and the sodium chloride ensures that it is displaced from solution in water. Extract with ether in the usual way, dry the residual amylobarbitone and determine gravimetrically.

Dicyclomine Tablets *Determination of the weight of $C_{19}H_{35}NO_2$,HCl per tablet of average weight*

Method Weigh and powder twenty tablets. To a weighed amount of powder equivalent to about 30 mg of parent substance, add *water* (20 ml) and shake. Add M sulphuric acid (10 ml), dimethyl yellow indicator and chloroform (40 ml), shake and titrate with 0.004M sodium dodecyl sulphate, shaking vigorously and allowing the layers to separate after each addition, until a permanent orange-pink colour appears in the chloroform layer. Each ml of 0.004M sodium dodecyl sulphate is equivalent to 0.001384 g of $C_{19}H_{35}NO_2$,HCl.

Aspirin, Caffeine and Codeine Tablets *Aspirin, Caffeine and Codeine Phosphate are all determined*

Method (*a*) *Acetylsalicylic acid* Weigh and powder twenty tablets and transfer an accurately weighed sample (1.2 g) to a separator. Add M NaOH (20 ml) and shake for 2 min to dissolve the aspirin as its sodium salt. Extract the caffeine and codeine by shaking gently (to prevent emulsification) with successive portions of chloroform. Extraction is complete when 2 ml of the chloroform fails to yield a residue on evaporation. Wash each chloroform extract with the same 10 ml of *water* and reserve the combined chloroform extracts for determination of caffeine and codeine.

Combine the alkaline solution and the aqueous wash liquors in a flask, add sodium hydroxide solution and warm (water-bath) to remove dissolved chloroform. Reflux (beware frothing) for 10 min to hydrolyse the aspirin to sodium salicylate. Cool, neutralize with dilute hydrochloric acid and adjust to 500 ml. Transfer 50 ml of the solution to a glass-stoppered bottle, add 50 ml of 0.05M bromine and concentrated hydrochloric acid (5 ml). Shake repeatedly during 15 min and set aside for 15 min. Add aqueous potassium iodide solution, shake thoroughly to replace excess of bromine by iodine. Wash the stopper and back-titrate the excess of iodine with 0.1M $Na_2S_2O_3$.

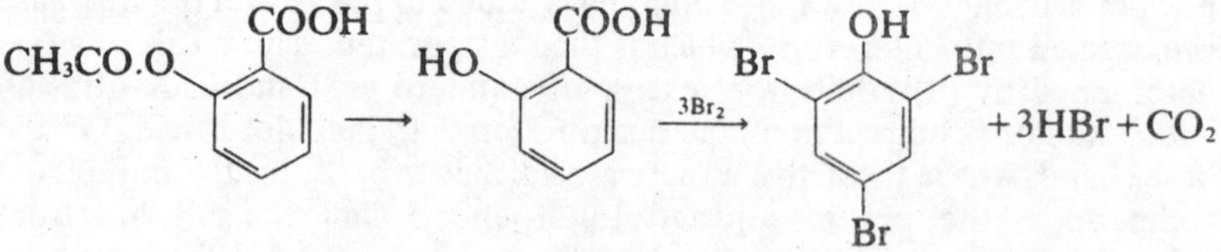

$$\therefore \quad 180.2\text{ g } C_9H_8O_4 \equiv 3Br_2 \equiv 60\,000 \text{ ml } 0.05\text{M}$$
$$\therefore \quad 0.003003\text{ g } C_9H_8O_4 \equiv 1 \text{ ml } 0.05\text{M}$$

(*b*) *Caffeine* The first chloroform solutions, obtained as above by extraction of the alkaline solution, contain both caffeine and codeine. Shake out with three successive portions of 0.05M H_2SO_4 to remove the codeine. Wash the mixed acid solutions with two 10 ml portions of chloroform and wash each chloroform solution with the same 10 ml of *water*. Combine the chloroform solutions, filter into a tared flask and evaporate to dryness. Weigh the dry residue of caffeine and calculate as caffeine citrate.

(*c*) *Codeine Phosphate* Mix the acid extracts and aqueous wash liquors, make alkaline with ammonia to precipitate codeine and extract with successive portions of chloroform (20 ml). Wash each chloroform solution with *water*, filter and evaporate to dryness. Dissolve the residue in 1 ml of neutralised ethanol, add excess of 0.02M HCl and back-titrate the excess with 0.02M NaOH to methyl red.

Further control is exercised over the tablets by a limit test for salicylic acid which arises from possible hydrolysis of the aspirin on storage.

Determination of medicaments by derivatisation

Primary and secondary amines, alcohols and alkanolamines of low volatility can be derivatised quantitatively with trimethylchlorosilane to form trimethylsilylamides or trimethylsilyl ethers, which are sufficiently volatile for determination by GLC. The internal standard is usually a related compound which forms a similar derivative under the conditions used.

Atropine Sulphate Tablets *Determination of* $(C_{17}H_{23}NO_3)_2H_2SO_4,H_2O$

Atropine sulphate is converted to its trimethylsilyl ether, using homatropine hydrobromide as internal standard, and determined by GLC at relatively high temperature (220°) on a 1.5 m column of 4 mm internal diameter, packed with OV17 (3%) coupled to a flame ionisation detector. A modification of the procedure described can be used to check on the uniformity of content of a sample of *Atropine Sulphate tablets*.

Method Prepare reagent Solutions A, B and C as follows, such that

(a) Solution A contains atropine sulphate reference + homatropine hydrobromine reference

(b) Solution B contains atropine sulphate sample + atropine sulphate reference + homatropine sulphate reference

(c) Solution C contains atropine sulphate sample.

Solution A To a 0.05% w/v solution (10 ml) of atropine sulphate reference substance, add 0.5% w/v solution (1 ml) of homatropine hydrobromide reference substance (internal standard) in methanol and 5M ammonia (1 ml). Extract with two quantities of chloroform (10 ml), dry the combined extracts with anhydrous sodium sulphate and evaporate the filtrate to dryness. Dissolve the residue in dichloromethane (2 ml).

Solution B Shake a quantity of the powdered tablets equivalent to about 5 mg of atropine sulphate with 0.1M HCl (10 ml), add Solution A (1 ml) and extract with two volumes of chloroform (10 ml) and discard the chloroform extracts. Add 5M ammonia (1 ml) and proceed as for solution A.

Solution C Proceed as for Solution B, but omit Solution A.

Derivatisation To separate volumes (1 ml) of Solutions A, B and C, add 0.2 ml of a mixture of *NO*-bis(trimethylsilyl)acetamide (4 volumes) and trimethylchlorosilane (1 volume), mix and allow to stand for 30 min to yield Solutions 1, 2 and 3, respectively.

Determination Inject suitable volumes of Solutions 1, 2 and 3 on to a column maintained at 220°. Calculate by peak height or integration measurements the ratio of atropine sulphate to internal standard and so determine the amount of atropine sulphate per tablet of average weight.

COGNATE DETERMINATIONS

Atropine Eye Drops and *Eye Ointment*
Homatropine Eye Drops
Hyoscine Eye Drops and *Eye Ointment*

Radiopharmaceuticals

The standardisation of radiopharmaceuticals poses a number of special problems arising from the radioactivity of the materials and the natural instability which this confers on the products. The approach to standardisation is essentially the same as for the corresponding non-radioactive products with the normal requirements for identification and control of the chemical purity of the labelled compound by assay and appropriate limit tests, and such other controls as tests for sterility and pyrogens as the finished product would normally require. The additional requirements arise in the identification of the radionuclide, the measurement of radionuclidic and radiochemical purity, the determination of specific activity and stability and labelling.

Radionuclidic identification

Radiopharmaceuticals are prepared from radionuclides, i.e. atomic species (nuclides) which are radioactive and, hence, capable of emitting radiation in the process of spontaneous nuclear fission to other nuclides. Radionuclides are characterised by the nature (α,β,γ) and energy of their radiations and by the rate of radioactive decay expressed in terms of half-life, the time in which the amount of radionuclide is reduced to half its initial value. Measurement of half-life and determination of the nature and energy of the radiation (Part 2), therefore, form the basis for the identification of radionuclides.

The nature and energy of the radiation is usually determined either by construction of an absorption curve and calculation of the **coefficient of mass absorption**, μ, for β-emitters or by **gamma spectrometry** for γ-emitters. Briefly, absorption curves are constructed by measuring the counting rate from the source with a Geiger-Müller or proportional counter, without and with a succession of aluminium screens of increasing mass per unit area. The absorption curve is obtained by plotting the logarithm of the number of impulses per unit of time on the ordinate against the mass per unit area of screen on the abscissa. The coefficient of mass absorption, μ, is calculated from the most nearly straight line portion of the 'curve' according to the expression:

$$\mu = \frac{2.303(\log A_1 - \log A_2)}{(m_2 - m_1)}$$

where m_1 = mass in mg/cm of the lightest screen

m_2 = mass in mg/cm of the heaviest screen

A_1 = the number of impulses/unit of time for m_1

A_2 = the number of impulses/unit of time for m_2

Gamma-ray spectra are characterised by the number and energy of gamma rays in the spectrum. Resolution of detectors in common use is such that peaks 5 keV apart in the spectrum are capable of detection, the energy of each peak being determined by proportion to the electrical output of the detector, which is measured after amplification.

Radionuclidic purity

Standards for most radiopharmaceuticals incorporate a test for **radionuclidic purity**. This, in effect, is the proportion of the total radioactivity present in the form of the named radionuclide, including that due to daughter radionuclides, which by convention are not considered as impurities. Ideally, the activity of every radionuclide present should be measured. This, however, is seldom practicable and the usual method is to examine for certain specific radionuclides by a defined method. This is usually gamma scintillation spectrometry, either directly or, if long-life impurities are present in short-life radionuclides, by observation of time-dependent spectrum changes.

The usual level of radionuclidic purity demanded is a general one of not less than 99% within the life of the product specified by the expiry date. Monographs may, however, also set limits for specific radionuclidic impurities, such as ^{199}Au($\not>$ 10%) in *Colloidal Gold* (^{198}*Au*) *Injection*. Likewise, *Sodium Pertechnetate* (^{99m}*Tc*) *Injection* (*Fission*) places special limits on ^{99}Mo, ^{131}I, ^{103}Ru, ^{89}Sr, ^{90}Sr, α-emitters and other radionuclidic impurities.

Some short-lived radionuclides also present special problems. *Sodium Pertechnetate* (^{99m}*Tc*) *Injection* is usually prepared in a generator from molybdenum-99. The resulting technetium-99m has a half-life of only 6 h and the standard for the Injection is such that, up to the date and hour at which the product may be used, not more than 0.01% of the total radionuclides must be due to radionuclides other than technetium-99m or its decay product technetium-99. The detection of precursor molybdenum-99 can only be properly achieved by waiting until the product technetium-99m has decayed to a sufficiently low level. Such a procedure, however, is precluded by the short shelf-life of the Injection. A much more approximate test is, therefore, used in which the gamma-ray spectrum of the injection solution is examined through a lead shield (6 mm). The latter absorbs the lower energy (0.140 MeV) technetium-99m radiation, so that any response recorded by the detector is due to the more energetic (0.780 MeV) molybdenum-99.

A further problem arises with certain radiopharmaceuticals, such as *Sodium Pertechnetate* (^{99m}*Tc*) *Injections* (*Fission* and *Non-fission*) and similar short-life injections, which are sterilised by filtration. The half-life of the radionuclides is such that the *test for sterility* required of an injection prepared by filtration, cannot be completed until after the injection has been used. The Pharmacopoeia, however, nonetheless requires a sterility test to be completed, so that the general procedures used in preparation are subject to examination, albeit in retrospect. Thus, whilst the control of each preparation is nil, there is a positive on-going supervision of the general preparation process.

Radiochemical purity

The **radiochemical purity** of a radiopharmaceutical product is a statement of the proportion of radionuclide in the designated chemical form. Thus, not less than 95% of the iodine-131 in *Sodium Iodohippurate* (^{131}I) *Injection* is in the form of sodium 2-iodohippurate. The determination of radiochemical purity usually involves some form of chromatography which is specific for the particular drug substance. Thus, a dilution of the *Sodium Iodohippurate* (^{131}I) *Injection* is diluted to contain about 1 mCi/ml. Paper chromatography of this dilution under standardised conditions should yield a chromatogram with the main spot characteristic of 2-iodohippuric acid, visible in ultraviolet light at 366 nm, in which not less than 95% of the radioactivity is concentrated.

In some cases, non-radioactive parent compound is used as a carrier to aid detection of the spots on the chromatogram. In the special case of L-*Selenomethionine* (^{75}Se) *Injection*, L-methionine, rather than L-selenomethionine, is used as the carrier in order to limit the extensive oxidation of sample selenomethionine which would otherwise occur.

Radiochemical methods are described in more detail in Part 2.

References

J. Allen, B. Gartside, and C.A. Johnson (1962), *J. Pharm. Pharmac.*, **14,** 73T.
H. Bundgard and K. Ilner (1972), *J. Pharm. Pharmac.*, **24,** 790.
A.G. Davidson (1976), *J. Pharm. Pharmac.*, **28,** 795.
A.F. Fell (1978), *Proc. Analyt. Div. Chem. Soc.*, **15,** 260.
A.F. Fell (1980), *Br. J. Pharm. Pract.*, **February,** 7.
N.D. Greenwood (1972), *Pharm. J.*, **208,** 290.
N.D. Greenwood and I.W. Guppy (1974), *Pharm. J.*, **212,** 30.
C.A. Johnson and R.E. King (1962), *J. Pharm. Pharmac.*, **14,** 77T.
K. Karlberg and U. Forsman (1976), *Analytica Chimica Acta*, **83,** 309
W.F. Kirk (1972), *J. Pharm. Sci.*, **61,** 262.
R.A. Morton and A.L. Stubbs (1946), *Analyst*, **71,** 348.
W.M. Plank (1972), *J. Assoc. Official Anal. Chem.*, **55,** 155.
A.C. Shah and J.F. Ochs (1974), *J. Pharm. Sci.*, **63,** 110.
T.G. Somerville and M. Gibson (1973), *Pharm. J.*, **211,** 128.
J.E. Tingstad and S. Riegelman (1970), *J. Pharm. Sci.*, **59,** 692.

Index